THE NEUROPHYSICS OF HUMAN BEHAVIOR

Explorations at the Interface of Brain, Mind, Behavior, and Information

THE NEUROPHYSICS OF HUMAN BEHAVIOR

Explorations at the Interface of Brain, Mind, Behavior, and Information

Mark Evan Furman
Fred P. Gallo

CRC

CRC Press
Boca Raton London New York Washington, D.C.

Library of Congress Cataloging-in-Publication Data

Furman, Mark Evan
The neurophysics of human behavior : explorations at the interface of brain, mind, behavior, and information / Mark Evan Furman, Fred P. Gallo
p. ; cm.
Includes bibliographical references and index.
ISBN 0-8493-1308-2 (alk. paper)
1. Cognitive neuroscience. 2. Neuropsychiatry. 3. Biophysics. I. Gallo, Fred P. II. Title.
[DNLM: 1. Biological Psychiatry. 2. Biophysics. 3. Mind-Body Relations (Metaphysics). WM 102 F 986n 2000
612.8′2—dc21 00-023640
CIP

Direct all inquiries to CRC Press LLC, 2000 N.W. Corporate Blvd., Boca Raton, Florida 33431.

International Standard Book Number 0-8493-1308-2
Library of Congress Card Number 00-023640
Printed in the United States of America 1 2 3 4 5 6 7 8 9 0
Printed on acid-free paper

Preface

Even though life can be the greatest joy, unfortunately several millennia of human interaction have unearthed an ever-increasing morass of human unhappiness, discontent, and suffering that seems in retrospect to be an inescapable condition of being *human*. While nature appears to be in such harmony, we, a part of nature, are not. Why is this so? And how can we change it?

This book was written for the purpose of providing a progress report of a lifelong endeavor to answer several mind-twisting questions that could potentially influence the course of human development. What is life? What does it mean to be human? What is our place in nature? Is it our fate to endure an existence of relentless unhappiness, discontent, mental suffering, and disease? If not, how can we change? What is mind? Where does it come from? How are brain, mind, matter, and energy related? How do they interact? Why does this interaction seem to be the source of our suffering? What could we learn about being human if we were to weave the psychological sciences, neurosciences, biological sciences, and the physical sciences into a single integrated picture? Can we create a comprehensive model of mind and brain so that we may be able to perceive and influence the network of interactions that we are embedded within and influenced by? What is the most fundamental way in which we can describe their interaction so that we may understand who we are and ultimately improve the quality of human life? The answers to these and an even longer list of questions have developed into an interdisciplinary branch of science we refer to as *cognitive neurophysics*.

The psychological and psychotherapeutic sciences, since their inception, have been developing in isolation, all but ignoring the fact that we, and all that we call *self*, are a transient result of a physical process — a property of the interaction of matter and energy in the physical world. We have thus far neglected to see ourselves as *process* and not *thing*, and that we are governed by the same physical laws as all of nature. The processes of nature have illimitable dominion over the development of all forms and their interaction. The last 70 years of research and development in the physical sciences have taught us that it is pure folly to conceive of brain, mind, behavior, thoughts, emotions, or man as existing separately from each other or nature itself. The idea that any *thing* can exist apart from events has been demolished by the recent discoveries in high-energy particle physics and quantum mechanics. Yet the human sciences continue to branch off and develop in isolation, rarely, if ever, attempting to integrate their disparate worldviews into a single, unified whole that we can embrace. Cognitive neurophysics and the present work intend to synthesize such a perspective.

Thus far, the expansive perspective afforded by cognitive neurophysics has permitted the development of a theory and a model, which we believe will significantly alter our current worldview and the course of human development. We refer to the theory as the *Standard Theory of Pattern-Entropy Dynamics* and the application model as *NeuroPrint*.

The Standard Theory of Pattern-Entropy Dynamics constructs a systemic perspective from which we can view the relationship between humans and nature. We use this theory to answer the many questions posed above by exploring the ramifications of two fundamental conclusions. First, information *is* pattern in space and time; that is, a piece of *information* is equivalent to a particular *state of motion* or movement pattern found in nature. *Pattern* — states of motion — is the fundamental process of nature permitting the development of certain forms and governing their interaction while constraining the development of others. Second, brain, mind, behavior, thoughts, and emotions are *properties of interaction* between numerous information fields — both internal and external patterns or states of motion in time and space, arising in nature. It is from this way of seeing that we may dissolve many human paradoxes.

NeuroPrint was developed in order to provide a way of perceiving the effects of this network of interactions between information fields on our dynamic bioarchitecture and our quality of life. It brings into focus the network of interactions that permits the development of brain, mind, behavior, thoughts, and emotions, and, by the same methods, it redefines the very meaning and precision of psychotherapeutic intervention. From the perspective of NeuroPrint and cognitive neurophysics, intervention is simply precise microscopic and macroscopic changes in the state of motion of a neurocognitive system. In physics, this state of motion is referred to as a *phase path*. NeuroPrint was designed to afford the scientist, practitioner, and student of human behavior and cognition the ability to predict and influence the transition probabilities between any two or more behaviors, thoughts, emotions, or physiological states available to a particular human being, and thus predict and alter the course of human thought and behavior.

We believe that the questions posed earlier are answerable, albeit obscured by a limited perspective. Our intrinsic tendency to consume and produce order and pattern in efforts to counterbalance the destructive, disorganizing force of entropy causes us to artificially abstract and divide our experience — an indivisible, interdependent whole — resulting in a debilitating misalignment of our expectations with the ubiquitous, relentless laws of nature. For as long as we unwittingly continue to set our expectations by this limited perspective in direct opposition to the natural inclinations of nature, we will till the soil of mental suffering.

This misalignment between our expectations and nature's immutable laws is further perpetuated by the failure of our formal educational systems to teach us to *see* the patterns of nature to which we owe our very existence and with which we must align our expectations and understandings of human behavior, our environmental relationships, and life itself. Our failure to *see* that we ourselves are products of, and governed by, the illimitable dominion of nature's processes over all things deprives us of the deep pleasure that comes from experiencing our own life as an intrinsic part *of* nature.

The lifelong practice of science engenders within its most avid students an uncommon equanimity — inspiring understanding, affinity, awe, and wisdom, which can only come from a unified perspective. Such insight allows us to more appropriately realign our expectations of human beings, and of life itself. These expectations are aligned, not with the multitudes of fabricated myths we are so often force-fed, but instead with the ubiquitous inclinations of nature itself.

To profoundly understand the paradox of human mind and behavior and the seemingly inescapable suffering and discontent it so reliably engenders, we must deeply examine the nature of pattern — and the patterns of nature — and thus gain a clearer view of the weaving of the tapestry we call our lives.

Even as Newton admitted that he arrived at his wider perspective by standing on the shoulders of those discoverers who came before him, we humbly acknowledge some of the giants upon whose shoulders we stand. Taking the chance of neglecting to acknowledge so many who have had an influence on our thinking, we nonetheless would like to thank Socrates, Plato, Aristotle, Popper, Kuhn, Einstein, Minkowski, Schrodinger, Dyson, Heisenberg, Bohr, Bohm, Poincare, Feynman, Penrose, Darwin, Gould, Loewenstein, Margulis, Cairns-Smith, Maynard Smith, Lovelock, Dawkins, Haken, Kelso, Prigogine, Mandelbrot, Kauffman, Smolin, Pribrim, Hameroff, von Newman, Hofstadter, Minsky, Ashby, Powers, Weiner, Pavlov, Skinner, Festinger, Korzybski, Chomsky, Whorf, Hebb, Edelman, Kandel, Damasio, Gazzaniga, Posner, Roland, Kosslyn, Bandler, Grinder, Erickson, Csikszentmihalyi, Bateson, Buckminster Fuller, and Whitehead.

Mark Evan Furman
Fred P. Gallo
June 2000

The Authors

Mark Evan Furman is a scientist, author, and international lecturer. He is the founder of Cognitive Neurophysics, a branch of science that studies the effects of information and information processing on human brain function, structure, and development. His scientific research and development have contributed to the fields of psychotherapy, education, communication, medicine, learning disabilities, and human relations. His work has branched into three well-known and widely practiced fields of development, referred to as Intelligent Learning Systems (ILS), NeuroPrint, and Human Performance Modeling and Engineering, respectively.

Mr. Furman graduated from the College of Staten Island (CSI), New York, in 1984 with a B.A. degree in Psychology. He is a Certified Practitioner of Neuro-Linguistic Programming (NLP). In the last 5 years, he has authored 18 seminal papers of international significance, which were subsequently published in 42 countries and registered with the Library of Congress in Washington, D.C. Selected papers have recently been translated into Russian, German, French, Spanish, and Portuguese. By the age of 37 a record of his prolific contributions was placed within the pages of the *Marquis' Who's Who in the World*, a comprehensive chronicle of the contributions of living world leaders from 215 countries and territories.

Mr. Furman is the founder of the International Society for Education Neuroscience, and the Society for Cognitive Neurophysics. He is the director of education and research at the Keys to Success, Inc. in Coral Springs, FL, where he continues his pioneering research and development of intelligent learning systems.

He is a member of the American Association for the Advancement of Science, the New York Academy of Sciences, and the International Society for the Study of Peace, Conflict, and Violence, division 48 of the APA in Washington, D.C. He can be reached via e-mail at neurosci@gate.net.

Fred P. Gallo is a psychologist, researcher, author, and international lecturer. He is the founder of Energy Diagnostic and Treatment Methods (ED×TM™), an advanced psychoenergetic therapy rooted in causal diagnostic procedures. In addition to articles on PTSD, energy psychology, and brief therapies, he has authored several books, including *Energy Psychology: Explorations at the Interface of Energy, Cognition, Behavior, and Health; Energy Diagnostic and Treatment Methods;* and *Energy Tapping: How to Rapidly Eliminate Anxiety, Depression, Cravings, and More Using Energy Psychology.* Since 1980, he has been training professionals in approaches such as Neuro-Linguistic Programming (NLP), Ericksonian Hypnosis, Thought Field Therapy (TFT), and his own ED×TM.

He began his professional career as a teacher and counselor after undergraduate and graduate study in philosophy at Duquesne University. He later attended training in clinical psychology and child development, receiving an M.A. from the University of Dayton and a Ph.D. from the University of Pittsburgh. He has lectured at Pennsylvania State University and has also worked in the fields of corrections, mental retardation, child welfare, chemical dependency, and hospital psychology. In addition to private practice, he is associated with the University of Pittsburgh Medical Center (UPMC) at Horizon.

Dr. Gallo is a member of the American Psychological Association and the Pennsylvania Psychological Association, and is on the advisory board of the Association for Comprehensive Energy Psychology. He can be reached via e-mail at fgallo@energypsych.com. His Energy Psychology and Psychotherapy Web site is at www.energypsych.com.

This book is dedicated to Beth, for her love, support, and unbounded belief in me, and to our children, Lauren and Jonathan. MEF

To my family and friends. FPG

Contents

Introduction

CONTENTS

THEORY, PARADIGM, AND THE ADVANCEMENT OF SCIENCE

Science is a method of perceiving and describing. In this regard it also determines how we see the world and our relationship to it. In that the perceiver in science is a human being, by virtue of his or her sensory organs, there are limitations constraining what can be and is seen at any given time. As such, science changes as our ability to perceive changes.

What exactly is being perceived by a human brain? Our current limit of perception into the human brain itself informs us that the brain is designed to perceive difference. As Gregory Bateson (1979) observed, "All receipt of information is news of difference and all perception of difference is thus limited by threshold" (p. 29). Let us explore this further.

This threshold that is imposed by our sensory organs, which are coarsely tuned to filter incoming stimuli/information (low resolution with respect to atomic level), also allows us to adapt effectively to our environment. However, this factor simultaneously imposes profound limitations on our ability to understand the workings of our world and universe. Differences that are too slight or too slow to activate our sensory circuitry are not perceived and therefore, for purposes of everyday understanding, do not exist. This has profound implications as to what can be expected of a scientific theory or model and the accuracy by which it corresponds to that being described.

The invention of the microscope extended our ability to perceive difference within the realm of the very small or microuniverse, just as the invention of the telescope increased our ability to perceive difference in that which is far away. Both of these inventions had profound ramifications for science. As our ability to measure time to within a fraction of a nanosecond became a reality, again science took a leap forward as we came to perceive difference with greater precision. As technology improves our ability to detect difference, theories developed during periods of less refined perception are sometimes proved absurd. Such theories beg immediate revision, extension, and sometimes replacement.

Sometimes a change in our perception does not come from a technological extension of our sensory acuity, but instead from the development of a new theory or model itself. Such was the case with the paradigm shift created by Newton's laws of motion and later by Einstein's theories of relativity. In these cases new ways of seeing, as suggested by a new theory, refocused our attention to areas of difference that were previously outside of our awareness, and thus left unmeasured. New theory suggests new avenues for experiment and, in turn, new avenues of experiment beg the development of new technology capable of extending our sensory acuity. Thus science is driven to continually extend its limits of description by both the continued development of theory as well as our means of perceiving and measuring difference. Buried deep within each such cycle, a new paradigm lies sleeping, waiting to be aroused. And, as it awakens, a new way of seeing our world and our relationship to it reverberates rapidly through the scientific community, as it stimulates new research and raises new questions.

Nature does not voluntarily reveal its deep structure. Mature practitioners of science realize that what is observed is not reality itself, but rather reality exposed to and altered by our nature and methods of questioning. As a new paradigm is advanced, new questions are asked; and as new questions are asked, the nature of things reveals itself in new ways. In addition to providing a practical means of diagramming neurocognitive functioning so as to facilitate intervention, this book offers a theory that suggests a new paradigm.

BRAIN, MIND, BEHAVIOR, INFORMATION, AND A UNIFIED THEORY OF INTERVENTION

A scientific theory or model strives to craft a new perceptual lens so that we may look at the same facts in a new way, thus extending our ability to describe and predict that which we formerly could not. At the same time, science continually strives to develop unified laws capable of assisting predictions across contexts. Thus laws that would predict behavior equally well in biological systems as they would in physical systems are preferred over laws that are context specific. The major objective of this book is to describe a set of relationships between brain, mind, behavior, and information so as to extend the scope and precision by which the process of change can be directed. The description of these relationships is universally adaptable to a wide range of therapeutic tools.

Theory has a structure. Its survival and utility depend upon how well its structure adheres to certain basic laws and rules of assembly. One of the most fundamental rules governing the stability of a theory is that there must be a need to both explain the newly observed phenomenon, which does not match currently available theories or models, as well as to extend the precision and scope with which predictions can be made about the aggregates modeled.

In his seminal book, *The Structure of Scientific Revolutions*, Thomas Kuhn (1996) refers to such a need as the recognition of an anomaly. This recognition by mature practitioners of science has a three-part structure. First, discovery begins with awareness of the existence of an anomaly and the recognition that nature has somehow violated the paradigm-induced expectations that govern the practice of normal science. Second, this violation of expectation causes the commencement of an extended exploration into the area of anomaly. Third, this exploratory behavior ceases when the paradigm theory has been adjusted so that the anomalous has become expected. Such a need has, in fact, existed since the birth of the science of psychology and psychological intervention.

All scientific disciplines during their infancy attempt to unify apparently disconnected behavioral observations. At one time electricity, magnetism, and light were considered to be completely unrelated phenomena in nature. Through the practice of unification, all three apparently disconnected phenomena have come to be seen as different properties of a single aggregate, namely, an electromagnetic field. Unification such as this is the primary objective of science today, the field of psychology being no exception. Therefore it is our objective to advance a unified theory, aggregating brain, mind, behavior, and information, to significantly increase the scope and precision of psychological research, development, and intervention. The scientist and practitioner both must be able to perceive these elements as different properties of a single aggregate, in order that the science of psychology may advance significantly.

This brings us to our second most important rule in the development of a unified theory. The theoretical scientist must search for the most fundamental level of description possible, a description of behavior that is ubiquitous to all of the elements that make up the aggregate described. That is, we must describe brain, mind, behavior, and information in terms of a fundamental process that is ever present, that can be found in all of the elements at all times: the common denominator.

In *What is Life?* (1944), theoretical physicist Erwin Schrodinger articulated such a fundamental level of description. This ubiquitous property we shall refer to as *pattern* (or order). Schrodinger's fundamental insight was that all things, living and nonliving, attempt to counterbalance *entropy* by the ever-present behavior of the production of order, structure, and pattern. Entropy, as defined by the second law of thermodynamics, is the degree to which relations between components of any aggregate are mixed up, unsorted, undifferentiated, unpredictable, and random. Entropy is the tendency of all things to lose pattern, structure, and thus information — to eventually regress to a state of *thermodynamic equilibrium*, the complete absence of pattern (order). Schrodinger coined the term *negentropy* to describe the degree of ordering, sorting, or predictability in an aggregate. The most extreme example of this property of counterbalancing entropy by spontaneous pattern formation can be found in the dynamic balance between life and death itself. In the language of thermodynamics, death is the state of maximum entropy between aggregates of a biological system. Maximum entropy is reached when a system of aggregates is in a state of thermal, chemical, and mechanical equilibrium or thermodynamic equilibrium. When this state is reached, the incorporation of pattern, order, structure, and information is impossible.

One of Schrodinger's deepest insights was his distinction between living and nonliving things, observing that all life counterbalances the tendency toward entropy or thermodynamic equilibrium by consuming or incorporating order, pattern, and information available within the environment. As we come to view a human biological system through this paradigm, it becomes clear that we are not only carnivores and herbivores, but in essence we are *informivores*. Just as we consume food in order to maintain our biological stability, we consume information (pattern) in order to maintain our neurocognitive stability.

Section I develops a solid foundation from which a unified theory of psychological intervention can be constructed, by expounding on this ubiquitous property intrinsic to brain, mind, behavior, and information. We explore the tendency of these aggregates to counterbalance entropy by the incorporation and production of pattern. In this regard, we describe brain, mind, behavior, and information in terms of *pattern-entropy dynamics*. We draw from the fields of physics, chemistry, molecular biology, neuroscience, and memetics, which have been highly successful in describing and predicting the dynamics of pattern-making within self-organizing systems. In this section, the reader will come to appreciate that memory, emotions, perception, behavior, protein production, phobias, trauma, understanding, confusion, ambiguity, meaning, stress, relaxation, boredom, waking, dreaming, rigid mind sets, creative intuition, neural networks, DNA, atoms, and life and death itself reflect and are born from the dynamic counterbalancing of pattern and entropy.

We refer to this property of matter–energy interaction and its emergent phenomena as *pattern-entropy dynamics*, and the theory as the *Standard Theory of Pattern-Entropy Dynamics.* The purpose is to unify the emergent phenomena born out of the interaction between brain, mind, behavior, and information. With such a foundation in place, the anomalies of psychological science can be adequately explained and incorporated, such that the anomalous become the expected.

GEOMETRIZING THAT WHICH DEFIES VISUALIZATION: THE EINSTEIN/MINKOWSKI SOLUTION

In Section II, we delineate a visual method of modeling neurocognitive patterns, the relationship between those patterns, and the neurocognitive topology that results from the dynamic counterbalancing of entropy. The necessity of such a visual representation is highlighted by an important historical development in the history of physics. In 1905, Albert Einstein published four original papers on what seemed to be very diverse branches of physics. These papers were so revolutionary in their scope that they became paradigm theories that shifted the way scientists saw the universe. One of these came to be known as the special theory of relativity. However, Einstein's theories did not begin to gain wide acceptance by physicists until about the middle of 1907. When an emerging scientific theory necessitates a paradigm change, there is always a delay in its wide acceptance, as it takes time for our perceptions of observation and experiment to accommodate to a new organizational structure. But this was not the only problem inherent in Einstein's special theory of relativity.

The primary reason for resistance among practitioners in the scientific community was first noticed by Hermann Minkowski, one of Einstein's former mathematics lecturers (White and Gribbin, 1993). The special theory of relativity defied visualization. Obviously for a paradigm theory to gain widespread acceptance, it must first be understood. Once Einstein translated his concepts and observations into complex mathematical equations, this made them understandable to a small number of theoretical physicists who were able to translate the equations back into a visualizable model of the relationships among its elements. To overcome this problem, Minkowski developed a mathematical interpretation for the special theory of relativity in terms of geometry. This highly pictorial way of representing the implications of Einstein's theory dramatically accelerated its spread and acceptance among the scientific community and the nonspecialist. Minkowski's geometrizing of Einstein's theory made the insights of a scientific genius readily accessible, resulting in worldwide acceptance of the new paradigm. We refer to this methodology of theory articulation and representation as the Einstein/Minkowski solution.

The Einstein/Minkowski solution gives us the third major criterion that must be satisfied for the proper development of a paradigm theory. The complex interaction among aggregates must be represented or representable in visual form. This translation process not only assists understanding, but also allows an increase in the scope and precision by which the theory can be applied. In Section II, we present *NeuroPrint*, a visual geometric model of the complex relationships among interacting patterns of brain, mind, behavior, and information.

The critical step of geometrizing a unified theory of neurocognitive intervention can only be accomplished through the selection of the most appropriate descriptive tools available. At certain points along the scientific timeline, new theories required the development of entirely new systems of mathematics, as was necessary when Newton described his laws of motion. Science has significantly advanced since that time and provides many valuable descriptive tools, which require only slight modification as we develop this theory.

The most valuable tools currently available for describing the behavior of patterns formed by any set of aggregates can be found in the field of statistical physics. There are four branches of statistical physics from which we draw tools and vocabulary in order to advance a pictorial method of modeling the neurocognitive topology of a human being and the relationships among brain, mind, behavior, and information. In this section, we integrate principles from thermodynamics, quantum mechanics, nonequilibrium dynamic systems theory, and synergetics, each of which has been highly successful in predicting the behavior of interacting systems of aggregates. These branches of physics also possess a well-developed conceptual vocabulary, useful in perceiving essential connections between the microscopic and the macroscopic. With these tools we are able to address a plethora of anomalies, including the structure of various psychiatric and other disorders and the range of emotional, behavioral, and cognitive phenomena.

MODELING NEUROCOGNITIVE TOPOLOGY IN PREPARATION FOR SKILL TRANSFER

In Section I, we develop a fundamental level of description necessary to unify the emergent phenomena from the interaction of brain, mind, behavior, and information.

In Section II, this description is developed into a pictorial representation of neurocognitive topology detailing the implications of the Standard Theory of Pattern-Entropy Dynamics, to extend the scope and precision of research, development, understanding, and intervention.

In Section III, we introduce an extended methodology for modeling the aggregates of brain, mind, behavior, and information, to prepare a human being for accelerated skill or knowledge transfer or education. We further integrate principles of neuroscience and its many branches, as well as cybernetics, control theory, and behavioral engineering. In greater depth, we also cover several conceptual tools discussed in the previous sections. Additionally, there is detailed discussion about the proper use of, and intricate cognitive neuromechanics behind, numerous tools for influencing and changing cognitive neurodynamics. Deeper exploration increases understanding and, by implication, the precision with which new therapeutic applications are developed.

EXPLORATIONS AT THE INTERFACE OF THEORY AND APPLICATION

Practical application is an essential element when advancing a unified theory that is intended to change a paradigm. Theory must always be articulated together with applications to some concrete range of natural phenomena. Sections II and III develop practical applications by isolating the fundamental ingredients of some of the most successful cognitive interventions of the last decade. The Standard Theory of Pattern-Entropy Dynamics draws attention to how each intervention successfully influences the ubiquitous properties of pattern and entropy in order to modify neurocognitive topology, resulting in rapid therapeutic change. Many of these interventions are formerly recognized by the psychological community under the classification of brief therapy.

In addition to discussing some of the pattern dynamics that are common to hypnotherapy, Neurolinguistic Programming (NLP), Eye Movement Desensitization and Reprocessing (EMDR), and various approaches to energy psychology, such as Thought Field Therapy (TFT), we also suggest some specific guidelines for the effective design and development of future neurocognitive interventions that will be of interest to the scientist, practitioner, and educator.

In Section I, we trace some of the 25-year history of NLP, which helped to advance the field of psychological therapeutic intervention in four important ways:

First, the initial developers of NLP — Richard Bandler, John Grinder, and Robert Dilts — observed that internal subjective experience has a structure that varies from person to person. Through prediction and experiment, they found that the structure determined how someone experiences an event currently being incorporated (perception) or previously incorporated (memory) from the external environment. This was a discovery of monumental importance in that the early practitioners were able to produce compelling evidence that the meaning of an event is not created by the external event of a stimulus field alone imposing its pattern on our neurological receptors. Rather, the meaning of an external event is created by and dependent upon the internal structure and organization of representations made by the nervous system in response to the stimulus pattern. Thus NLP helped to remove the lid of the great black box which had previously hidden the interface between stimulus and response. In so doing, many anomalies of human thought, emotion, and behavior became the expected. Our scope and precision in predicting human behavior took a leap forward.

Second, these developers were among the first to recognize the importance of pattern in describing the structure of subjective experience. At that time, the ubiquitous and pervasive nature of pattern in the human organism could not be fully realized or appreciated for its great beauty and complexity. Yet the fact that pattern, organization, and structure were somehow important in maintaining mental health was strongly suspected. During the evolution of NLP, however, the notion of pattern was only applied to the specific organization of internal subjective experience and how that pattern of organization might affect thoughts, emotion, behavior, and meaning. The idea of pattern was still trapped within the very limiting conceptual framework of programs. The notion or metaphor offered at that time was that the human brain is a computer, consisting of a labyrinth of programs waiting to be activated, and that certain stimuli would cause the activation of certain programs.

Third, in its later stages of development, NLP was employed to study the process of change. Although other schools of psychological thought had previously explored change methods, NLP was the first used to observe the intricate structure of change in terms of linguistics, sensory modalities, etc.

Last, NLP made a significant step toward a unified theory of psychological intervention by aggregating numerous models, which could be used to solve mental health problems, which were previously resistant to change, by other approaches. During its early evolution, NLP became a powerful amalgamating force, as it attracted disenchanted proponents of various schools of thought, assimilating parts of their models. Some of these models worked well together and formed greater synergies, while others did not. Hence, NLP became an impressive aggregation of models without a unified or guiding theory. Largely, this feature of NLP owes itself to the original "mission statement" of its founders, which was to elaborate effective models without reference to theory and truth.

Unfortunately, as a result of this elaborately loosely connected mosaic, NLP had many fits and starts in its formal development, and has fallen short of successfully articulating a unified theory of intervention. Yet there is much to be learned from the pieces from which the discipline owes its existence today. By providing a brief history of NLP, we hope to assist the reader in apprehending the essential missing pieces necessary for a complete articulation of a unified theory of intervention. This is the very mechanism of science alluded to earlier. NLP is like the microscope, which extended our ability to see difference in the very small, and in doing so, uncovered many new anomalies that begged the development of a new theory that would make the anomalous predictable.

REFERENCES

Bateson, G. (1979). *Mind and Nature: A Necessary Unity.* New York: E. P. Dutton.

Kuhn, T. (1996). *The Structure of Scientific Revolutions*, 3rd ed. Chicago: University of Chicago Press.

Schrodinger, E. (1944). *What is Life? With Mind and Matter and Autobiographical Sketches.* New York: Cambridge University Press.

White, M. and Gribbin, J. (1993). *Einstein: A Life in Science.* New York: Penguin Books.

Section I

The Standard Theory: Pattern-Entropy Dynamics of Matter and Energy Interaction

A Unified Theory of Matter, Energy, and Information as Applied to: Brain, Mind, Behavior, and Information

1 The Five Missing Pieces: Advances toward a Unified Science of Neurocognitive Intervention

CONTENTS

It is partly from developments in neurolinguistic programming that we have been able to make further advancements toward a science of neurocognitive intervention and a unified theory of human change. NLP has provided us with the outer scaffolding of a magnificent skyscraper. By detailing its contributions and its errors, we will develop an understanding of the essential missing pieces necessary for the advancement of a unified theory.

NEUROLINGUISTIC PROGRAMMING

NLP is the study of Dilts et al. (1980) and Dilts (1983):

- The *structure* of internal subjective human experience
- *Correlative patterns* arising between internal cognitive representations of experience and macroscopic, observable behavior
- The human therapeutic *change* process

Earlier schools of psychological thought failed in their attempts to accurately describe mechanisms of cognitive activity that intercede between an event in a stimulus field and resulting behavior. This failure resulted in the surfacing of innumerable anomalies of human behavior, forcing psychology into a state of crisis. When such a crisis occurs, numerous schools of competing thought spontaneously sprout, voraciously arguing their explanatory models to no end. Each school attempts to explain the anomalies within the context or trap of their own paradigm. One school of thought completely sidestepped the task of explaining internal experience by conducting most of its experiments on animals with less complex neural structures. That approach was behaviorism (Watson, 1970; Skinner, 1953), which refused to peer into the black box of internal structure, holding that we cannot know scientifically what is going on in there. Behaviorism suggested that knowing about internal experience was not necessary for understanding an organism's behavior.

NLP approached the task of understanding behavior differently. All predictions were tested by constructing experiments with human beings, while building from advancements made by schools of thought that did not. The origins of NLP are important and we will get to that shortly. While NLP did incorporate many of the tools of behaviorism, the earlier developers noticed that all internal representations were not created equal and that slight differences in the internal representation of an external event (initial conditions) could potentially result in dramatically different classes of thought, emotion, and resulting adaptive behavior. NLP had turned on the light in the great black box. A central operating presupposition of NLP is that by changing the internal representation of an event, there is also a change in how that event is thought about, the emotional reactions to the event, the resulting adaptive behavior, and thus the entire meaning of the event as understood by the individual. That is, a change in meaning and resultant behavior can be achieved by rearranging the structure, organization, or pattern of related internal representations. Thus, a change in the meaning of an event is equivalent to a reorganization of the structure of related internal representation.

However, this was not a new notion. Victor E. Frankl, the psychiatrist who developed *logotherapy* or meaning therapy (1959), believed that if we wanted to change someone's behavior, it is only necessary to change the meaning of the events related to that behavior. The main difference between NLP and logotherapy is that NLP provided the conceptual tools with which meaning could be discretely altered. Those tools were found in the very structure, organization, and pattern of internal representation.

By internal representation, we are referring to the fact that events in the external environment are experienced not as they are, but as we are, as our nervous systems are organized to process and respond to those events. Each time we experience a pattern of stimuli available in a stimulus field, we re-present that pattern of stimuli within each of the appropriate sensory areas of our cortex and our association cortices. Every pattern in a stimulus field, when processed by different nervous systems, will yield significantly different internal representations, and in turn will alter the way future representations are constructed from external stimuli. In this way, we build an internal model of the external world such that we correspond in a significant way with our external environment in order to adapt to it and operate effectively within it. How and why we develop our model of the world is the subject of later discussion. In the meantime, here is an example of how internal representation affects thoughts, emotions, meaning, and behavior.

Imagine three women, each at home, waiting for their spouses to return from work at their expected times. In each case, on this day, after a great many months of consistency which led to their expectation of time of arrival, each of these three men were now 45 minutes late, and there has been no contact between them and their wives.

Wife no. 1 had a very trying day and while juggling the normal frustrations of that day, spent 2 hours with her friend helping her to cope with a recent separation as a result of her friend becoming fed up with her husband's promiscuous sexual behavior. After listening to 2 hours of infidelity stories, she settled down that evening to watch her favorite weekly sitcom only to again be tortured emotionally when she finds the heroine of the show wrestling with the same problem. As she now sits and becomes concerned that her own husband is 45 minutes later than usual, different internal representations of possible causes of this tardiness flicker back and forth in her mind's eye, competing for cortical space and control of a single behavioral response pattern to the dilemma. Within a short while, she begins to feel, seemingly unwarranted, bursts of anger and jealousy while she fights to suppress visual images of her husband with another woman. (How and why certain information patterns and internal representations emerge victorious under competitive conditions is the subject of memetics, which is discussed more thoroughly in a later chapter. In the meantime, we are just noting the effect of an internal representation on a behavioral response given similar stimulus conditions.) As her husband enters the threshold of the house, he is not greeted amorously and with "How was your day, Honey," but rather by a clench-fisted, tight-jawed likeness of his wife, demanding to know where he was, what he was doing all this time, and what witnesses she could call, as she thought to herself, "All men are the same." She has constructed a behavioral adaptation, not to the actual stimulus of 45 minutes of lateness, but rather to her internal representation or model of the world. What appears to be bizarre, unwarranted behavior to her unsuspecting husband is completely justifiable when viewed within the context of the newly revealed contents of the black box of subjective experience.

Next let us turn our attention to wife no. 2. She is currently in her second marriage. Being previously widowed as a young mother of two, it was difficult for her to fall deeply in love a second time. Death took her first husband in the shadow of a tragic car accident while driving home from work. While she may have watched the same sitcom this evening, when the big hand of the kitchen clock hit the 45th minute, a single internal representation emerged victoriously: unsuppressible images of a tragic car accident rushed through her mind's eye. In the next few minutes, she internally lived the death of her current husband, the funeral, the search for a job to feed her children, and the emotional pain so familiar to those experienced in the loss of a loved one. Minutes later her husband crosses the threshold only to find one of the most emotionally intense and loving receptions in the history of his marriage. He is hardly able to escape his wife's intense grip long enough to change his clothes and eat dinner.

Wife no. 3 had a relatively uneventful day. She had some coffee with friends earlier in the day, went food shopping, helped her children with their homework, and ended their day with a few special bedtime stories. Grateful for the 45-minute breather between her children's bedtime and making dinner for her husband, she relaxed with a good book and her favorite classical music. The few times that her husband was late in the past, he was stuck late at work trying to close a deal. Unsuppressible visual images of new shoes, dinners out, and a nice vacation danced joyously within her mind's eye. Minutes later when her husband arrived, they kissed and hugged, sat down for dinner over candlelight and classical music, business as usual.

These examples illustrate how internal representations — our individual models of the world — can radically alter the effect of stimuli on our thoughts, emotions, behavior, and the very meaning we place on stimulus patterns — events in the stimulus field. In NLP parlance, internal representations are constructed from *4-tuples*, which are the product of all sensory systems sampling an event at any given time — visual, auditory, kinesthetic, olfactory/gustatory — represented as (V, A, K, O/G). While a subject may pay conscious attention to only one sensory system representation of an event at any given time, the entire 4-tuple is available to consciousness when attention is shifted. Seamlessly connected 4-tuples, called *strategies*, are among the cognitive building blocks that create our ongoing perception of continuous experience. Each 4-tuple represents simultaneously accessed sensory information, punctuated by time.

STUDYING THE STRUCTURE OF INTERNAL SUBJECTIVE HUMAN EXPERIENCE

NLP's study of the structure of internal subjective experience had its modest beginnings with the development of *reframing*. The concept of *frame* was proposed by American philosopher Marvin Minsky and deployed by some psycholinguists in the early 1970s. According to Minsky, a frame is an important type of schema, a cluster of common ideas about some domain of experience activated in processing texts and utterances to form textual words. Minsky referred to this as a framework for representing knowledge (Bothamley, 1993).

Early on, the NLP founders began to realize that the meaning of any event was dependent upon the frame in which it was re-presented internally, and that altering the frame or cluster of ideas currently related would change the meaning of that event and the resulting adaptive behavior. Although two types of reframing were demarcated by NLP — context and content reframing — the distinctions separating them tend to blur. Restating the theory in other terms, the meaning of a stimulus is dependent upon the stimulus field in which it is presented, whether external or internal. Thus by changing the context in which something is represented, its meaning and our response to it also change. This was the first explicit conceptual tool developed from the work of Frankl. The following example illustrates how this may work.

> There was a midwestern farmer who purchased a horse for his 17-year-old grandson. His grandson loved the horse and rode it daily until one night when it escaped from its stable. The next-door neighbors came by to console the heart-broken boy. They said

> to the grandfather, "How tragic." The grandfather simply responded, "Maybe." A few days later, the boy's horse returned with another wild stallion. The boy was overjoyed. When the neighbors came over to play cards with the grandfather that night they said, "How fortunate." The grandfather simply responded, "Maybe." The following morning, the youth got up extra early, saddled the new stallion, and attempted to ride him. However, the stallion threw the boy and the harsh landing broke his leg. When the neighbors came by with some freshly cooked soup, they exclaimed, "How tragic." The grandfather simply responded, "Maybe." Less than 1 week later, a draft notice appeared in the morning mail, requesting the grandson's immediate presence to serve in the armed forces. When the neighbors found out that he was rejected due to his broken leg, they said to the grandfather, "How fortunate." As expected, the grandfather simply responded, "Maybe."

What can we learn from this example? The example is really a story about the shifting of contexts in which the events are perceived, also referred to as framing. The cluster of events considered in the frame continued to grow as the story developed. Each addition of a new event or stimulus had the potential of changing the meaning, and effectively did so, for the neighbors. As the frame expanded, the meaning of the events contained within oscillated back and forth between tragic and fortunate. The early developers of NLP realized that many of the clients that they saw therapeutically had this in common. As in the case of the neighbors, the stimulus field by itself seemed to control the meaning of the event, the person having no active part. The grandfather, however, took an active role in the meaning-making process. By suspending immediate judgment, he could decide the meaning he placed on the cluster of events by opening or closing the size of the frame.

Drawing additionally on the work of Bateson (1979), reframing became an explicit tool for emotional and behavioral change by the early 1970s (Bandler and Grinder, 1982). While reframing evolved extensively over time, the basic mechanics involved influencing one to re-present a "tragic or troubling" event internally, within a new context. Context can be thought of as pattern in space and through time: spatial context and temporal context. For example, let us say that in a family therapy session, the parents vehemently complain about the stubbornness of their 13-year-old daughter. They emphasize, "She seldom does anything that we tell her to do. She has too much of a mind of her own." Perhaps the therapist can choose to place this troubling behavior in another context. Knowing from previous sessions that the parents themselves tend to be rather tenacious and that they are concerned with their daughter's coming of age — along with the potential for promiscuity, drug experimentation, and peer pressure — the therapist might casually state, "Well, at least we can feel confident that she will easily be able to withstand the dangerous effects of peer pressure." Assuming that this communication impacts the parents, this context reframe of the behavior can effectively change the meaning of stubborn and the parents' response to it, leading into productive conversations about how this "troublesome" behavior can be the greatest asset to everyone involved.

On the surface this technique may seem to be linguistically inane, yet it has been found to have profound effects on the structure and organization of cognitive elements and resulting behavior. As noted, our behavior is not a response to reality itself, but rather to the internal model we have constructed, which is a mere representation of the external world. Each time we alter our internal map, we also alter our emotional and behavioral responses.

THE EVOLUTION OF STRATEGIES

By the early 1970s, NLP had aggregated a number of conceptual tools whose origins could be found in numerous, loosely connected, conceptual models attempting to describe the human experience. One of these models was cybernetics (control theory). In his 1948 landmark book, *Cybernetics: Or Control and Communication in the Animal and the Machine*, Norbert Wiener shook the very foundations of our collective model of the world by introducing a paradigm theory capable of predicting certain classes of human behavior which formerly appeared random and structureless. Wiener was one of the first pioneers of self-organizing systems theory. He made the first extensive study of the effect of feedback and information on self-organizing systems, applying cybernetics to psychopathology and the study of brain wave patterns. In his chapter entitled "Information, Language, and Society," he laid the foundation for one of the operating presuppositions of NLP, which states that "the meaning of any communication is the response elicited." And the response elicited is dependent upon the listener's or the receiver's model of the world. Contrary to much of the early work of Russian physiologist Ivan Pavlov, who spent his lifetime investigating stimulus–response relationships, Wiener introduced the idea that the internal organization of the black box could have dramatic effects on the stimulus–response relationship. Wiener stated, "The value of a simple stimulus, such as an odor, for conveying information depends not only on the information conveyed by the stimulus itself but on the whole nervous constitution of the sender and the receiver of the stimulus as well" (1948, p. 157). Wiener had advanced a theory, which later gave birth to a conceptual tool that extended the scope and precision of the investigation of human behavior.

The next development in cybernetics, later put to use by NLP, occurred when Ross Ashby (1952) published *Design for a Brain: The Origin of Adaptive Behavior.* This was the first serious attempt at describing human behavior as a dynamic system, a predecessor to dynamical systems theory. Ashby added to the collection of conceptual tools applied to human behavior the idea of "adaptation as stability" and the explanatory tools of *phase-space*, *field theory*, and *transition between states*. These tools of description were formerly known only to the field of statistical physics. Most importantly, he defined a system's field as "the phase-space containing all the lines of behavior found by releasing the system from all possible initial states in a particular set of surrounding conditions" (p. 23). Within this field theory of the origin of adaptation, Ashby also introduced the idea of a *state-determined system*, a system whose behaviors are *state-dependent*. Later Ernest Rossi (1993) expanded this idea and led to the notion that all human learning, memory, and behavior are state-dependent or state-determined, formally notated as SDLMB. Thus a system capable of producing multiple states or fields could potentially entrap memories, behaviors, and other resources that would not be available to the system in another collective state of organization. This was a monumental hypothesis, which explained many anomalies of human memory and behavior and formed the guiding theory for the development and application of the NLP tool known as *anchoring*, the application of various stimuli to elicit and utilize internal states for intervention.

With the publication of *An Introduction to Cybernetics*, Ashby (1956) contributed further to the development of cybernetics and eventually NLP. In this work he articulated the concepts of transformation and change in biological systems, which later formed the basis for NLP's study of change in human systems. Here Ashby outlined the hypothesis that change in any system of aggregates, human or otherwise, must be preceded by the disturbance of a stable state in which the system exists, and that this disturbance will, if sufficiently profound, initiate a transformation from the existing state to a new one defined by the phase-space. This hypothesis led to one of the most important and least understood conceptual tools incorporated by NLP, referred to as *pattern interruption* (Dilts et al., 1980; Dilts, 1983). Like Schrodinger, Ashby also noticed the ubiquity and pervasive nature of pattern, but instead of the semantic environment of order and disorder he spoke of the "ubiquity of coding," "variety," and "gene-pattern" in biological organisms (1956, p. 140). Although they used different semantic environments to articulate the importance of pattern, both Ashby and Schrodinger elucidated the inseparable relationship between pattern and information, that the loss of pattern *is* the loss of information with a resultant increase in noise, disorder, and entropy. It is difficult to separate the study of structure, pattern, and change. This "difficulty of separation" is always a good place to start looking for a fundamental level of description capable of supporting the articulation of a unified theory.

Although NLP included many of its early operating principles and presupposition from cybernetics and control theory, the first formal tools for modeling the structure of internal subjective human experience came from later contributors to cybernetics, who further investigated adaptive behavior, through the paradigm theory of cybernetics. In *Plans and the Structure of Behavior,* Miller et al. (1960) advanced a conceptual tool for modeling and representing behavior, referred to as the TOTE. The TOTE was to replace Pavlov's S-R model as the smallest unit of behavior to account for "activities" within the black box. With this tool in hand, NLP made great advances in our understanding and modeling of subjective human experience. Once incorporated by NLP, *plans* were referred to as *strategies*. A strategy can be thought of as a series of states a system moves through in order to affect an outcome — a behavioral trajectory.

TOTE is an acronym for Test → Operate → Test → Exit. It stands for the notion that a behavior is only initiated by an organism when a *test* performed in one of its control systems sends a feedback signal called an *error condition*. An error signal is said to arise from a sensed difference between an *internal reference condition* held for some controlled quantity and the current *perceptual condition* as transmitted by one or more sense organs or collections of receptors. Let us suggest an example in order to clarify this. The model proposes that if you wanted to pick up a pencil, each *operator* (piece of behavior) in the sequence, called a plan, would be punctuated by a test between what that part of the behavior will look, feel, or sound like when it is finished (reference condition), and what it looks, feels, or sounds like now (perceptual condition).* As long as the reference condition and the perceptual condition tested for some controlled quantity do not match, an error signal is fed back and the behavior continues. Once the two conditions match, and the second test is complete, the system exits, that part of the behavior stops, and the system tests for a new operator (behavior). In other words, when you feel your fingers touch the pencil, reaching behavior stops and gripping behavior starts.

According to NLP, all behaviors, no matter how complex, have a structure that can be broken down, modeled, and understood in terms of totes which, when strung together, form strategies or behavior trajectories. It should be noted that 4-tuples are smaller conceptual units than totes. Their action can be found within each test phase. When the behaviors being modeled are not easily observed — because they are primarily cognitive — the developers of NLP widened the scope of their investigations from structure to pattern, the pattern of observable behavior that correlated with the cognitive activity being modeled. For this task, both eye movement patterns and language patterns proved indispensable.

The last major contribution to cybernetics and control theory, incorporated by NLP, occurred when William T. Powers (1973) published *Behavior: The Control of Perception*, which elaborated a control theory of far greater scope than that of Miller et al. (1960). Although this text was never formally cited, its influence on the later development of NLP tools, presuppositions, and methods is apparent. Carrying forward the same premise as Ashby, Powers believed that behavior was initiated by disturbance in a controlled quantity.

> The brain scans behavior for its results. The behavior pattern that reduces intrinsic error to zero stops the process of spontaneous reorganization, and that behavior pattern will persist. Punishment is anything that causes intrinsic error. The purpose of any given behavior is to prevent controlled quantities perceived, from changing away from the reference condition. (Powers, 1973, p. 187)

This premise reinforced the value of the TOTE model, guiding investigations by NLP developers to look for solutions to human problems, not only in the external stimulus field or perceptual condition, as suggested by the behaviorists, but also in the internal stimulus field or reference condition.

> A stimulus is not, except by chance, the same thing that the organism is controlling. Far more likely to be identified as the stimulus is the event tending to disturb the controlled quantity; disturbance always calls for a response .... The behavior of organisms is not organized around the control of actions or their effects; it is organized around the control of perception. (Powers, 1973, p. 187)

This important paradigm shift led to the development of many of the therapeutic methods known collectively as NLP today, such as *Change History, New Behavior*

---

* NLP developers found a tentative relationship between the performance of these tests and simultaneous patterns of eye movements easily visible to the trained observer. This apparent correspondence led to the development of NLP's *Eye Accessing Cues* model, which assisted in the tracking and recording of *strategies*.

Generator, and the *Swish Pattern*. Thus NLP developers realized that they could influence a specific behavioral outcome by altering a person's internal reference condition for a specific controlled quantity, by altering one's model of the world.

> Once one has identified what the organism is controlling and the reference condition, the relationship of a whole family of seemingly unrelated responses to a whole family of seemingly unrelated stimuli becomes completely predictable. There is no longer any reason to investigate such stimulus–response relationships after they have revealed what is controlled. (Powers, 1973, p. 187)

FROM STRATEGIES TO SUBMODALITIES, THE ATOMS OF COGNITION

Although framing, reframing, and strategies allowed an unprecedented depth of insight into the hidden structure of subjective experience, NLP was far from complete in its search for new and more precise tools. Turning up the power on the cognitive microscope, NLP borrowed from more recent neuroscience studies of perception, memory, sensory abstraction, and *submodalities*. NLP had found its next tool for studying subjective experience and, again, successfully revealed a new level of hidden structure that was unprecedented in the field of psychological therapeutic intervention. Like the atomic building blocks of molecules, submodalities proved to be responsible for the assembly of its larger counterparts of 4-tuples and strategies. Deep within the seamless continuum of sensory motor states, called strategies, resides a hidden domain of structure where the tests performed by the TOTE compare much finer degrees of difference. Submodalities appeared to be the atomic building blocks of our model of the world, each submodality in itself being an analogical control parameter capable of influencing thoughts, emotions, and behavior from the inside (reference condition), just as effectively as it did from the outside (perceptual condition).

Neuroscience evidence strongly suggests that submodalities are incorporated initially from our external environment via mechanisms of sensory perception. Learning to direct attention to submodalities dramatically expanded the scope of conscious awareness and the range of control conscious awareness acquired, as a tool for custom designing experience.

What are submodalities? If each sensory system is considered to be a system in itself, submodalities are the collection of all possible states of that sensory system as defined by its phase-space. For example, submodalities for the visual system include location of the image, brightness, color, contrast, movement, velocity, field size, image angle, image resolution, clarity, and all other analogical differences detectable by that system. Auditory submodalities are distinct in that they include volume, pitch, rhythm, location, duration, cadence, and so forth.

It became clear that neurocognitive states have a structure as unique as the individual, which can be described not only in terms of strategies, but also far more richly in terms of submodalities. Emotions like fear, confusion, confidence, frustration, anger, happiness, ecstasy, and compulsiveness all have a unique submodality pattern or structure capable of influencing the meaning of an event and one's response to it. It was found that meaning could now be altered independent of context or content. Meaning itself could be altered by the submodality building blocks, the very structure of an experience. Submodalities did for the domain of cognition what the microscope had done for cellular biology. Here is one example of the utility of submodalities.

A brief NLP therapeutic technique, Visual/Kinesthetic Dissociation (V/KD), utilizes submodality-level intervention to effect a rapid alleviation of long-standing phobias. It was predicted that the key to a phobic response could be found within the structure of the internal representations, and subsequent tests supported this hypothesis.

Take the case of a person with a phobia of snakes. A series of questions about the phobia would reveal that it has an anticipated structure when represented internally. When thinking about snakes, they are most probably seen as being larger and having greater velocity than their contextual background. We can also anticipate that the field size of the image would take up most, if not all, of the visual field space available in working memory. Often this would be visually represented as having the experience as if it were being seen through one's own eyes (visually associated), rather than looking at oneself in the picture (visually dissociated). Most commonly, the internal representation would be in the form of a movie rather than a still photograph and color rather than black and white. The sounds corresponding to the image would generally tend to be very loud, have a great deal of bass, and seem to emanate from a location behind the person. Another very important distinction is the phase-velocity of the visual and auditory representations, that is, the speed at which they move from minimum to maximum range analogically in each critical submodality (i.e., an image appearing to move toward you quickly would elicit a more intense physiological response than one with a lower phase-velocity). NLP developers, having thus modeled the structure of phobia, were able to shift critical submodalities individually in order to effect a rapid change in physiological response.

While there are a number of ways to accomplish this, one way would be to direct the client to imagine seeing himself from the back, sitting in a movie theater. Next the client would be asked to maintain this image, but "float up" into the projection room. From the projection room, he would then view a representation of the snake by turning on the projector and seeing only a freeze-frame, in black and white, on the movie screen. Next, our subject would be asked to begin to run the film slowly forward, all the way to the end, and then rapidly backward, repeating this several times. The next step would be to imagine himself now sitting in the theater watching the movie through his own eyes. When he could do this with no physiologic phobic response, he might be asked to actually hold a picture of a snake, to view a movie of a snake, or to even handle a live snake. This would be done to test the effectiveness of the structural changes made at the submodality level.

Since V/KD proved to be enormously successful in eliminating long-standing phobic responses, it became the lead demonstration used by early developers to illustrate the value of NLP to clinicians. The effectiveness of this method was initially explained in terms of interrupting the visual-kinesthetic stimulus–response bond or synesthesia pattern via introduction of dissociation. The dissociation was in effect an interruption of the visual-associated position that produced the negative kinesthetic response. Later this method was elaborated in terms of submodalities, in that the introduction of a dissociated position is in actuality a submodality distinction. V/KD's effectiveness was later theoretically articulated in neuroscientific terms by Furman (1995, 1996a). Hence, a new way to understand human experience emerged.

FROM STRUCTURE TO PATTERN

The next major movement in the early development of NLP led to an increase in investigative scope. The frame of investigation changed from structure to pattern. In this regard NLP began to search for the patterns that connected discrete internal experience with observable external behavior. NLP developers noticed correlation between eye movements and internal sensory representation called *eye-accessing cues*. They made and tested predictions about observable eye movements and eye positions that correlated with the sensory system being immediately favored by the subject when constructing an internal representation. While these patterns provided some valuable insight into internal representations and the subject's strategies, they prove to be only a loosely connected correlation in need of significant revision in order to be useful.

For the purpose of revealing the structure of internal representations, language and language patterns prove to have much greater utility. Two major contributors to the field of linguistics, Alfred Korzybski and Noam Chomsky, made invaluable contributions to NLP.

In the early 1900s, Korzybski pioneered the movement known as *general semantics*. While semantics is the study of how language creates meaning, Korzybski significantly advanced the field by introducing the importance of the abstraction process. His first contribution to general semantics was made in 1921, with the publication of *Manhood of Humanity: The Science and Art of Human Engineering*, in which he formulated a basic theory encompassing the biological and psychosymbolic nature of human experience. Korzybski viewed language as a time-binding mechanism, which uniquely allows humans to advance their state of knowledge from generation to generation. In view of the fact that he coined the term "neurolinguistic," his impact on NLP is apparent. He emphasized that both neurolinguistic and neurosemantic environments create an inescapable internal environment for representation of events, which conditions the reactions of the human organism as a whole. Additionally he observed that different cultures, having different language patterns, created uniquely inescapable representation environments.

In 1933, Korzybski published a thorough articulation of his theory in *Science and Sanity: An Introduction to Non-Aristotelian Systems and General Semantics*. He drew extensively from the fields of physics, mathematics, behavioral science, rhetoric, and neurophysiology, attempting to integrate them under the auspices of general semantics. General semantics was the largest aggregation of disparate scientific fields of its time ever to be integrated within a single field. Korzybski's intention was to create a unified picture of the human organism, and his work provided a ready-made template for the NLP aggregation 50 years later.

Korzybski advanced the *structural differential*, the first comprehensive model of the human abstracting process, which made it possible to differentiate the structures of abstraction. His life's work strongly suggested that languages could be used as maps, accurately describing internal representation, and that changes in language patterns would cause corresponding changes in internal representation. He provided compelling evidence that language accurately reflects one's model of the world, as represented internally by sensory motor systems, and, in turn, influences the construction and evolution of that model. He also advanced many key presuppositions that were later incorporated into the foundation framework of NLP. One of these important presuppositions is the notion that *the map is not the territory.* No map is capable of representing all of its presumed territory and, in essence, the map is a map of the mapmaker's assumptions, skills, and worldview. This presupposition was made explicit in the early stages of NLP and led to the development of the *Meta Model*, a simplified model of the abstracting process that presupposes that all human communication carried through the medium of language contains deletions, distortions, and generalizations of the territory described. The NLP founders learned to use this tool in order to test for, represent, and understand the limits of a communicator's model of the world. By the early 1980s, it was observed that language reveals the structure of a speaker's model of the world, even at the level of submodalities. Correlative patterns between words and submodalities seemed inseparable. For example, it was found that a phrase such as "bright future" had a corresponding internal visual representation of the future as brighter than that held of the present and/or past internal representation. It was also found that a listener's submodalities could be influenced by the application of submodality language. This process was assumed to be the means by which humans transfer meaning linguistically.

Incorporated into NLP were other less explicit presuppositions advanced by Korzybski, such as the notion that structure is the only content of knowledge and that meaning is strictly a function of the order or level of abstraction at which the term is used. Additionally, a term's meaning is so context-driven that it does not mean anything definite until the context is specified or understood. Korzybski recognized that consciousness of abstracting was essential for "fully functioning" humans. He made this a primary goal of general semantics training, as did NLP 50 years later. Korzbyski also wrote extensively on the use of neurolinguistics and neurosemantics as they could be applied to therapeutic procedures and to the prevention of psychological problems. He understood humans as being neurolinguistic systems organisms.

Recognizing the ability of language patterns to represent hidden cognitive structure, NLP needed a system of symbolically recording what was heard. To accomplish this end, NLP developers drew extensively from Noam Chomsky's (1957) *Syntactic Structures*. The text itself, a mere 102 pages, transformed American linguistics from a branch of anthropology to a mathematical science. Chomsky had drawn from the work of Leonard Bloomfield (1935), who had previously suggested a scientific approach to linguistics. As a proponent of logical positivism, Bloomfield attempted to reduce all meaningful statements to a combination of propositional logic and sense data, referred to in NLP as *sensory specific language*. In tracking such patterns of correlation the NLP developers believed that they could record the sequences of internal sensory system accesses (strategies) simply by listening to casual conversation. Chomsky's mathematical methods, as applied to the study of language, made such recording child's play. He believed that unique patterns were hidden within our words that could identify us and our model of the world just as accurately as fingerprints can be used for identification. He used algebra to capture some of the patterns of language that we all share, and one of his methods proved useful in being able to identify people from the words they wrote, given a sufficiently long passage. Chomsky accomplished this by tracking the relative frequencies of different words people typically used, and found that they formed a definite numerical profile, capable of accurately identifying the speaker (Devlin, 1998). His later work captured the structure of human grammar by analyzing its nature mathematically. Chomsky's work also proved to be invaluable in advancing NLP's modeling methods.

FROM STRUCTURE TO PATTERN TO CHANGE

With Chomsky's mathematical tools of linguistic modeling and Korzybski's suggested application to therapy and foundational presuppositions in hand, Bandler and Grinder wrote their first book in 1975, *The Structure of Magic*, directly followed by *The Structure of Magic II* in 1976. In these books they modeled the process of human therapeutic change by applying their tools to the study of several leading pioneers: Fritz Perls, Virginia Satir, Paul Watzlawick, Gregory Bateson, and Jay Haley. In Volume II, they also incorporated the modeling techniques of Miller et al. (1960), Ashby (1952, 1956), and Bateson (1972).

Drawing from the work of Watzlawick (1967, 1974), who studied problem formation, Bandler and Grinder began to study and consider change as structure through time, and they looked at the way in which leading therapeutic "wizards" influenced this process with their clients. They meticulously observed and recorded linguistic patterns used by these effective therapists, which appeared to affect the structure of problems, and then segmented these linguistic techniques into precise models of therapy. The result of their efforts was the meta model for language, which proved to be not just a model of how therapists utilize language to produce change, but also a set of linguistic tools that allowed for the cognitive exploration of a problem's structure, the first step toward change. Expanding on the notion that human beings construct an internal model of the world only after deleting, distorting, and generalizing the information available in a given stimulus field or event, Bandler and Grinder parsed the meta model in such a way that therapists would have precise linguistic tools capable of uncovering each of the three permutations.

As the developers began to trust linguistic structure to reveal internal activation and processing of sensory data, they noticed blatant mismatches or incongruity between observable behaviors. While some instances of incongruity in subsystems of a larger system were as obvious as the language of sadness paired with expressions of happiness, other distinctions were as subtle as a visual eye-accessing cue simultaneously paired with kinesthetic language patterns. Although the accuracy of eye-accessing cues was dubious at best, compelling evidence suggested that incongruity was the doorway to the structure of human problems and the key to change. Incongruity to Bandler and Grinder meant that the neurological system was in conflict. Although never explicitly mentioned, the guiding light to this discovery came from cybernetics and the best articulation of the theory came from Powers (1973), who believed that conflict was synonymous with malfunction.

> I have become more and more convinced that conflict *itself*, not any particular kind of conflict, represents the most serious kind of malfunction of the brain short of physical damage, and the most common even among 'normal' people. The reasons for the extraordinarily bad consequences of conflict are not to be found in specific behavioral effects, although disruptions of overt behavior certainly can make life difficult. Our model, however, tells us that mere practical consequences of specific conflicts are secondary to their major consequences, which is to remove parts of the brain's organizations from action as effectively as if they had been cut out with a knife, yet without getting rid of their undesirable influences on the whole hierarchy. The worst aspect of conflict between control systems is that the higher the quality of the control systems, the more violent and disabling is the result of conflict .... The basic mechanism behind conflict is *response to disturbance*.
>
> Conflict is an encounter between two control systems, an encounter of a specific kind. In effect, the two control systems attempt to control the same quantity, but with respect to two different reference levels. For one system to correct its error, the other system must experience error. There is no way for both systems to experience zero error at the same time. Therefore the outputs of the systems must act on the shared controlled quantity in opposite directions. (Powers, 1973, p. 253).

Given Power's articulation of conflict, Bandler and Grinder had sufficient reason to believe that unresolved conflict was the cause of many psychological problems and that behavioral incongruity was the best observable clue to conflict. The meta model became an indispensable tool for investigating internal conflict. Rather than referring to control systems, as did Powers and other proponents of cybernetics, Bandler and Grinder referred to *parts*. The use of the word *part* in a client's language was itself indicative of a conflict between control systems. For example, "Part of me wants to go to work and make more money and the other part of me wants to stay home and spend more quality time with my family." It is important to note that cybernetics was not the only field to make this distinction about the devastating effects of conflicting parts. This can also be found in the work of Pavlov (1927), that utilized conflicting stimulus field patterns in order to induce confusion and severe neurotic behavior in dogs. Bandler and Grinder found that the most effective way to deal with a conflict between parts (control systems) was to construct an intervention at a completely different hierarchical level of control, either above or below the level at which the conflict presented itself. Tools such as submodalities provided this opportunity at the micro level, whereas meaning reframing provided the same utility with control systems at the macro level. Bandler and Grinder created a formal design for negotiating between parts which incorporated hypnosis, *Six-Step Reframing* (1982).

Transitioning from the study of pattern to change, NLP developers adopted another concept from cybernetics and control theory, originally applied only at the level of strategies, with the intention of extending its range of application. The concept of *perceptual condition/reference condition* was renamed *present state/desired state*, forming the theoretical basis for the notion of "well formed conditions for outcomes" (Densky and Reese, 1986, p. 16). Specifically a well-formed outcome must be stated in the positive in sensory-based language, maintained by and within the control of the individual. The outcome must also be ecological, meaning that its establishment must not produce any other error condition. Last, the outcome must be testable and measurable through the individual's sensory experience, preserving the criteria of TOTE.

In 1975, Bandler and Grinder also published another important work on the language patterns of a leading figure in medical hypnosis at the time, *Patterns of the Hypnotic Techniques of Milton H. Erickson, M.D. Volume I*, which provided structural insight into the language of change. Erickson's language patterns were quite complex, however, and Chomsky's linguistic codifying system was insufficient to capture both the structure of his language and how that structure produced therapeutic change. To reveal these discrete patterns, they drew extensively from the field of hypnosis (Weitzenhoffer, 1957; Haley, 1967) and numerous branches of linguistics, including Watzlawick's work in pragmatics and change (1967; Watzlawick et al., 1974), and that of Benjamin Whorf (1956) and Edward Sapir (1963), who studied the relationship between the structure of language and human behavior as it varied across cultures. In order to complete their task, they also had to familiarize themselves with research from the field of neuroscience available at the time. They did so by primarily considering the work of three researchers: Eccles (1966), Pribram (1971) and Gazzainga (1974). However, this was still insufficient. Erickson's language patterns seemed to be infinitely complex and multileveled in structure. Implicit in the structure of his language were the experimental findings of behaviorists such as Pavlov, Watson, and Skinner. Erickson was so ingenious that he was able to incorporate into his hypnotic techniques language patterns that utilized the motor disabilities, from which he had sustained two early bouts of poliomyelitis. It was not uncommon for Bandler and Grinder themselves to be drawn into the hypnotic influence of Erickson when he would intentionally slur a few key words in a sentence, intended to induce ambiguity, confusion, and a hypnotically receptive mental state. Needless to say, the modeling of Erickson was a long and tedious, yet mesmerizing and fruitful, experience.

In 1977, they released *Patterns of the Hypnotic Techniques of Milton H. Erickson, M.D., Volume II*, including the input of co-author and co-developer, Judith DeLozier. The purpose of this volume was to distinguish Erickson's nonverbal from explicit linguistic change patterns. The concept of the 4-tuple was also introduced and employed extensively in this volume. It was found that Erickson's nonverbal behavior proved to be just as complex as his linguistic behavior. Here one of the most important guiding presuppositions of NLP came from Erickson: there is no such thing as a resistant client, only inflexible communicators. The concept of client/therapist rapport was central to all of Erickson's techniques. He had a unique way of gaining rapport with his clients; he would become them. Matching every observable behavior that he could detect, from eye movements to breathing patterns, he would systematically and completely diminish the differences between himself and the client. In NLP this came to be referred to as *pacing*. Once Erickson had achieved a deep level of rapport through this method, he would begin to slowly change his own behaviors, leading the client in an intended direction, but only as fast as the client would follow. In NLP this technique was referred to as *leading*. Pacing and leading, as a method of rapport building, later became a core technique taught in NLP practitioner training.

While NLP had its inception within academic walls, by 1979 research and development moved outside the academic arena and incorporated the general public within seminar/workshop formats. Steve Andreas, a first-generation NLP practitioner and subsequent co-developer, transcribed and edited one of these early workshops entitled *Frogs into Princes* (1979). It was here that the developers explicitly demonstrated the use of reframing, representational systems, accessing cues, and anchoring. Anchoring, which is equivalent to the notion of conditioned stimulus (Pavlov, 1927, 1928, 1941), became a multidimensional tool with a wide range of application.

By 1980, another first-generation NLP practitioner and early co-developer, Robert Dilts, assisted in the development of the first comprehensive text in the field, *Neurolinguistic Programming: Volume I: The Study of the Structure of Subjective Experience*. The intention of this volume was to organize a myriad of previously demonstrated tools into a three-part structure for change: elicitation, design, and installation. This volume arranged the NLP amalgamation of conceptual tools under these three categorical headings, and the change process primarily targeted the utilization of strategies. The developers found that strategies for thinking, feeling, and doing anything could be elicited, interrupted, and functionally cut and spliced with other existing strategies or with ones that were artificially designed. A somewhat hidden concept implicit in the design of this model, and ubiquitous to all types of strategies, was the notion of pattern interruption. Simply stated, experience suggested that the installation of a new strategy or pattern was impossible without prior interruption or disruption of the existing one. Considering the great importance of pattern interruption to the successful installation of a new strategy, only four and one half pages of the text were devoted to it. The developers were implicitly utilizing this fundamental key to change, while explicitly teaching only the technique itself and not the critical theoretical foundation. This omission had profound implications as each generation of practitioners taught NLP to the following generation, producing anomalies for future generations of practitioners. Without explicitly understanding the importance of pattern interruption and disruption, therapeutic changes produced by later generations of practitioners were not lasting changes, as were those that were demonstrated by the developers.

In 1981, Bandler and Grinder published *Trance-Formations: Neurolinguistic Programming and the Structure of Hypnosis*. Carrying the theme of structure and pattern, now into their seventh published work, they created a simplified reiteration and amalgamation of accumulated concepts of NLP and hypnosis. By this time, anchoring had been incorporated as a core technique to capture and hold constant neurocognitive states to be studied or changed, and the developers recognized the need to package change processes that could be used as templates for practitioners who would not invest the time necessary to study the guiding paradigms that made NLP's results possible. Hence, the *New Behavior Generator* was born, one of the first in the onslaught of instant, ready-to-use therapy templates, with no detailed knowledge required.

With the release of *Using Your Brain for a Change* (1985), Bandler unlocked the door to a new domain that forever shifted NLP research and development. This publication officially marked the transition from utilizing strategies to utilizing submodalities for change. He showed how the manipulation of even the smallest units of cognition could produce profound systemic effect on thought, emotion, and behavior. Uncovering the submodal structure of motivation, understanding, confusion, belief, fear, and other common human experiences, Bandler showed how the scaling of minute analogical control parameters available to conscious awareness could make the difference between mental health and disturbance.

Following closely behind Bandler's rendition of submodalities, Steve and Connirae Andreas published *Change Your Mind and Keep the Change* (1987). This marked another important step in NLP development. In addition to many other submodality-based techniques, there were two significant contributions contained in this text. The first was the *threshold pattern*. This technique illustrated that pattern interruption and disruption operated at the level of submodalities just as effectively, if not more so, than at the level of strategies, with regard to precipitating change. The second was the discovery of *time lines*, which illustrated that one's way of internally processing time has predictable impact on the re-experiencing of an event stored in memory. The discovery of time lines came from studying the close connection between patterns of language relating to space and time, as well as submodalities such as the size, angle, location, and clarity of visual images made by a subject in working memory. It was discovered that representations of time were organized in different ways from person to person, and that language reflects and affects that organization. This discovery had profound impacts on the future directions of therapeutic change and led to the publishing of *Time Line Therapy and the Basis of Personality* (1988) by Tad James and Wyatt Woodsmall, two first-generation practitioners and co-developers.

Time Line Therapy and the Basis of Personality succinctly developed numerous ready-to-use applications of the conceptual tool of time lines to thoughts, emotions, behaviors, and a host of human cognitive dysfunctions. Furman (May 1996; December 1996) suggested that a neuroanatomical model could be used to explain why certain organizations of time lines and submodalities were made possible or rendered impossible as a result of differentiation of cellular morphology throughout the brain. He also showed evidence that submodality distinctions would involuntarily be altered as a result of positional displacement of an image in working memory due, in part, to this differentiation of cellular morphology.

The last major work on submodalities, *An Insider's Guide to Sub-Modalities*, was published by Bandler and MacDonald (1988). This work proved useful in integrating many NLP tools for use at the submodality level. Here it was shown explicitly how anchoring and submodalities could be used together and how many of the older NLP techniques could be fine-tuned for greater precision.

While investigating the structure of beliefs, Dilts tied together existing knowledge in a new way with the publication of *Changing Belief Systems with NLP* (1990). This work dealt with the structure, pattern, and change of belief systems via submodalities. NLP had finally come full circle from studying the macroscopic to the microscopic and now revisiting the macroscopic. Bandler followed this trend with the publication of *Time for a Change* (1993). Both books carried forward the submodality units of cognition to intensively study the structure of beliefs and how they could be changed at this microscopic level. Again they implicitly make use of the ubiquitous tools of pattern interruption and disruption to effect change, now at the level of belief systems.

Let us take a moment to attend to one very important commonality between NLP and the aggregates from which it was assembled. Implicit in NLP is the understanding that the precursor to rapid lasting change is the interruption of an existing pattern at some level of cognitive organization. Erickson demonstrated his understanding of this by extensive use of confusion and ambiguity as a technique for inducing a transition from a normal waking state to a trance state, and for the installation of new thoughts, emotions, and behaviors. Watzlawick illustrated his understanding of this principle with his extensive use of paradoxical language to produce change. Pavlov employed symbolic confusion in his research and development to change animal behavior and to induce neurotic states. Korzybski illustrated a deep insight into the effect that language abstractions and the process of abstracting can have on sanity, understanding that words had the ability to alter the nervous system's functional and structural organization and, subsequently, its cognitive maps of reality. He saw, as did Pavlov, that improper use of language yielded symbolic confusion, leading to neurosis.

Although there were numerous books published during the development of NLP, only the ones mentioned demarcated significant advancements of the field. The others were mainly attempts to apply NLP to specific domains, such as persuasion, business, and education, essentially repackaging the same conceptual tools into ready-made templates, to be used in specific situations. Latecomers to the field, who hoped to continue its development, were somewhat lost. The plethora of books available to them sufficiently obscured the theoretical foundations that guided early development. These later attempts at development were made by assembling new templates from old templates. In other words, later generations of NLP practitioners developed applications from applications, leading the field into chaos. During this time, NLP fell from the status of a scientific discipline to that of a pop psychology. It is not uncommon to find numerous books of the NLP flavor sitting on self-help shelves of the local bookstore. Unable to control this regressive trend, Bandler abandoned the field of NLP and has since attempted to aggregate more serious practitioners under the guise of *Design Human Engineering* (DHE).

WHAT WE LEARN FROM THIS

Like all previous schools of psychological thought, the wake of NLP has left the field of psychology in a state of crisis. Its development traces essentially the same cycle as those who came before it, its tools and techniques randomly strewn across the floor, devoid of connective tissue or signs of life. The reader will recall that crisis is born from mounting anomalies for which a school of thought or theory is unable to account. While NLP was one of the most serious attempts at the development of a unified theory of therapeutic intervention, it was missing essential pieces necessary for its continued development and its ability to extend its range of application. The result of studying such cycles of theory development provides lucid revelations of what is essential to a unified theory.

MISSING PIECES NECESSARY FOR THE DEVELOPMENT OF A UNIFIED THEORY

1. A theory that describes and predicts the behavior of a complex system must utilize the most fundamental property available, one that is ubiquitous to all elements of the system. Cybernetics used the idea of control — comparing a reference condition to a perceptual condition of a controlled quantity to determine initiation and cessation of behavior. This fundamental approach significantly aided the advancement of NLP. However, control theory is quite limited in its range of application, and it leaves much anomaly intact when describing the complexities of human behavior. While control theory proved to be highly useful at the level of representational systems, its effectiveness broke down when attempting to address emotions, beliefs, and other human complexities of similar nature. In order to advance a sound theory of therapeutic neurocognitive intervention, we must seek to describe human complexity from a more fundamental level. This task should be approached from the level of pattern (structure, order). While NLP and its amalgamated disciplines only alluded to the importance of pattern, the remainder of the first section of this book is devoted to the illumination of the ubiquitous and pervasive nature of pattern in humans and nature. When approached from such a fundamental level, the scope of description and prediction, as well as the range of theoretical application, is significantly extended.
2. A system whose behavior is being described, predicted, and influenced must be represented in such a way that its state can be visible both before and after being acted upon. NLP, as well as the myriad of other psychological disciplines of this century, have not proffered a suitable method of representation for the human neurocognitive system, whereby the elements of the system, their interconnectivity, and changes made to the system could be precisely evaluated by the practitioner/therapist. Such a blind approach to therapeutic change of a neurocognitive system can never be accepted as the practice of science. It is for that reason that Section II is devoted to the development of a pictorial, topological model of representation for the human neurocognitive system, which is referred to as NeuroPrint.
3. A unified theory of therapeutic neurocognitive intervention must delineate the critical steps necessary for a complete and stable change process. Other than the incomplete and incompatible NLP models for change represented earlier, there is no existing model to be found in the field of psychology or cognitive science capable of accomplishing this task. A practitioner/therapist must know the following:
   a. *Where* in the neurocognitive system to intervene
   b. *What* the probable effect of that intervention will be on the rest of the system

c. *When* the neurocognitive system is ready for change and how to affect system readiness
d. *How* to locate leverage points and make that change most effectively with the least number of steps (i.e., elegance)
e. *Which* of their collection of intervention tools would be most applicable to the situation
f. *How* to *measure* the *stability* of the change made and design changes to last

A unified theory of intervention must make significant steps toward satisfying these criteria. The specifics of this process are detailed with conceptual tools from statistical physics and NeuroPrint in Section II.

4. The human brain is a highly complex system of biological aggregates, which obeys the laws of physics. Therefore, a unified theory of intervention must suggest tools of intervention whose effects on the neurocognitive system can be described and predicted by utilizing knowledge from the governing paradigms of physics, systems science, and neuroscience. The system itself must also be describable in these terms.
5. A unified theory of intervention cannot be complete without describing and predicting the behavior of information. Since all therapeutic psychological intervention can be seen as the introduction of new patterns of information to the neurocognitive system, it is essential to have an understanding of how patterns of information are translated between subsystems within a human being, and how information patterns compete to gain control of the neurocognitive system being acted upon. Such an understanding is vital to the proper construction of intervention tools, so as to eliminate unwanted interference from other competing information patterns. The study of competing, self-replicating information patterns is the domain of memetics and memetic engineering, discussed in detail in this section.

REFERENCES

Andreas, S. (Ed.). (1979). *Frogs into Princes*. Moab, UT: Real People Press.
Andreas, S. and Andreas, C. (1987). *Change Your Mind and Keep the Change*. Moab, UT: Real People Press.
Ashby, W. R. (1952). *Design for a Brain*. New York: John Wiley & Sons.
Ashby, W. R. (1956). *An Introduction to Cybernetics*. London: Metheun & Co. Ltd.
Bandler, R. (1985). *Using Your Brain for a Change*. Moab, UT: Real People Press.
Bandler, R. (1993). *Time for a Change*. Cupertino, CA: Meta Publications.
Bandler, R. and Grinder, J. (1975). *Patterns of the Hypnotic Techniques of Milton H. Erickson, M.D.* Volume I. Cupertino, CA: Meta Publications.
Bandler, R. and Grinder, J. (1979). *Frogs into Princes: Neuro Linguistic Programming*. Moab, UT: Real People Press.

Bandler, R. and Grinder, J. (1981). *Trance-Formations: Neuro-Linguistic Programming and the Structure of Hypnosis*. Moab, UT: Real People Press.

Bandler, R. and Grinder, J. (1982). *Reframing: Neuro-Linguistic Programming and the Transformation of Meaning*. Moab, UT: Real People Press.

Bandler, R. and MacDonald, W. (1988). *An Insider's Guide to Sub-Modalities*. Cupertino, CA: Meta Publications.

Bateson, G. (1972). *Steps to an Ecology of Mind*. New York: Ballantine Books.

Bateson, G. (1979). *Mind and Nature: A Necessary Unity*. New York: E. P. Dutton.

Bloomfield, L. (1935). *Language (Revised Edition)*. New York: Henry Holt Publishing.

Bothamley, J. (1993). *Dictionary of Theories*. London: Gale Research International Ltd.

Chomsky, N. (1957). *Syntactic Structures*. The Hague: Mouton.

Densky, A. and Reese, M. (1986). *Programmer's Pocket Summary* (Revised 1993). Indian Rocks Beach, FL: Southern Institute Press.

Devlin, K. (1998). *The Language of Mathematics: Making the Invisible Visible*. New York: W.H. Freeman and Company.

Dilts, R. (1983). *Roots of Neuro-Linguistic Programming*. (Part I, 1976) Cupertino, CA: Meta Publications.

Dilts, R. (1990). *Changing Belief Systems with NLP*. Capitola, CA: Meta Publications.

Dilts, R., Grinder, J., Bandler, R., and DeLozier, J. (1980). *Neuro-Linguistic Programming: Volume I: The Study of the Structure of Subjective Experience*. Cupertino, CA: Meta Publications.

Eccles, J. (1966). *Brain and Conscious Experience*. New York: Springer-Verlag.

Furman, M. E. (1995). Pattern Interruption: Redesigning the Pathways of Thought in the Brain. October. Unpublished paper, submitted to Florida State University during "Active Ingredient" study.

Furman, M. E. (1996a). Submodalities through the eyes of a neuroscientist. *Anchor Point*, 10(5).

Furman, M. E. (1996b). Foundation of neurocognitive modeling: eye movement — a window to the brain. *Anchor Point,* 10(12).

Gazzainga, M. (1974). *The Bisected Brain*. New York: Appleton-Century-Crofts.

Grinder, J. and Bandler, R. (1975). *The Structure of Magic*. Palo Alto, CA: Science and Behavior Books.

Grinder, J. and Bandler, R. (1976). *The Structure of Magic II*. Palo Alto, CA: Science and Behavior Books.

Grinder, J., DeLozier, J., and Bandler, R. (1977). *Patterns of the Hypnotic Technique of Milton H. Erickson, M.D.* Capitola, CA: Meta Publications.

Grinder, J. and Bandler, R. (1981). *Trance-Formations: Neuro-Linguistic Programming and the Structure of Hypnosis*. Moab, UT: Real People Press.

Haley, J. (1967). *Advanced Techniques of Hypnosis and Therapy: Selected Papers of Milton H. Erickson, M.D.* New York: Grune & Stratton.

James, T. and Woodsmall, W. (1988). *Time Line Therapy and the Basis of Personality*. Cupertino, CA: Meta Publications.

Korzybski, A. (1921). *Manhood of Humanity: The Science and Art of Human Engineering*. New York: E.P. Dutton & Company.

Korzybski, A. (1933). *Science and Sanity: An Introduction to Non-Aristotelian Systems and General Semantics, 5th Edition*. Englewood, NJ: Institute of General Semantics.

Luria, A. R. (1932). *The Nature of Human Conflicts: An Objective Study of Disorganization and Control of Human Behavior*. New York: Grove Press.

Miller, G., Galanter, E., and Pribram, K. (1960). *Plans and the Structure of Behavior.* New York: Adams–Bannister–Cox.

Pavlov, I. (1927). *Conditioned Reflexes: An Investigation of the Physiological Activity of the Cerebral Cortex.* New York: Oxford University Press.

Pavlov, I. (1928). *Lectures on Conditioned Reflexes, Volume I.* New York: International.

Pavlov, I. (1941). *Lectures on Conditioned Reflexes, Volume II, Conditioned Reflexes and Psychiatry.* New York: International.

Powers, W. T. (1973). *Behavior: The Control of Perception.* Hawthorne, NY: Aldine de Gruyter.

Pribram, K. H. (1971). *Languages of the Brain: Experimental Paradoxes and Principles in Neuropsychology.* New York: Brandon House.

Rossi, E. (1993). *The Psychobiology of Mind-Body Healing: New Concepts of Therapeutic Hypnosis*, Revised Edition. New York: W.W. Norton & Company.

Sapir, E. (1921). *Language: An Introduction to the Study of Speech.* Orlando, FL: Harcourt Brace & Company.

Sapir, E. (1963). *The Selected Writing of Edward Sapir.* Mandelbaum, D. (Ed.), Berkeley: University of California Press.

Skinner, B. F. (1938). *The Behavior of Organisms: An Experimental Analysis.* New York: Appleton-Century-Crofts.

Skinner, B. F. (1953). *Science and Human Behavior.* New York: Free Press.

Watson, J. B. (1970). *Behaviorism.* New York: W. W. Norton & Company.

Watzlawick, P., Bavelas, J. B., and Jackson, D. D. (1967). *Pragmatics of Human Communication: A Study of Interactional Patterns, Pathologies, and Paradoxes.* New York: W.W. Norton & Company.

Watzlawick, P., Weakland, J., and Fisch, R. (1974). *Change: Principles of Problem Formation and Problem Resolution.* New York: W.W. Norton & Company.

Weitzenhoffer, A. (1957). *General Techniques of Hypnotism.* New York: Grune & Stratton.

Wiener, N. (1948). *Cybernetics: Or Control and Communication in the Animal and the Machine.* Cambridge, MA: MIT Press.

Whorf, B. (1956). *Language Thought and Reality.* Cambridge, MA: Massachusetts Institute of Technology.

2 The Ubiquity of Pattern in Man and Nature: Exploring the Fundamental Properties of Brain, Mind, Behavior, Information, and Change

Pattern is nature's means of communicating and translating information. Pattern-making is ubiquitous in humans and essential to life, part of nature's deep design. The human brain and body are an amalgamation of nature's most ingenious and complex pattern-making biomechanisms, operating on multiple fractal scales of four-dimensional space–time at both the quantum and classical levels. In this chapter, we articulate and connect more thoroughly profound insights made by various scientists into an anomaly of nature's deep design, the ubiquity of pattern.

Bandler and Grinder observed and explored the relationship between pattern, structure, and change in human subjective experience. They provided considerable evidence that suggests the importance of pattern, order, and structure in influencing the quality of human experience. In this respect they detailed the inner subjective structure in terms of parts, linguistic structures, 4-tuples, strategies, and submodalities. Their findings are in concert with that of thinkers in a variety of fields.

Korzybski (1933), the central figure in general semantics, devoted the better part of his life to studying patterns of abstraction, the abstracting process, and its relationship to sanity. He believed that we incorporate within our nervous system a model of the world — the pattern, structure, and order of our external environment — with the purpose of creating a *correspondence* between that which is inside and that which is outside of us. We have already detailed the influence of Korzybski on the development of NLP.

Schrodinger (1944) believed that life is the process of incorporating within itself order found in the external environment. He believed that life is dependent upon the counterbalancing of entropy (i.e., loss of pattern) with order and structure. Thus, as living beings, we also incorporate within ourselves the order found in the external environment so as to counterbalance the pull toward entropy.

Bateson (1979) taught his students to look for the "pattern that connects" in order to reveal nature's deep secrets. He believed that the loss of pattern is synonymous with the loss of information. Thus pattern and information are integral to each other.

Turning to the field of organic chemistry, Cairns-Smith (1996) convincingly argues from an atomic-molecular paradigm that living organisms must constantly keep energy and entropy in dynamic balance in order to maintain existence.

Penrose (1994), in *Shadows of the Mind*, further developed the notion that human beings must continually decrease entropy in order to stay alive.

In *The Cerebral Code: Thinking a Thought in the Mosaics of the Mind*, neuroscientist Calvin (1996) lucidly develops the theory that the *representation* of an object or idea in the human mind is dependent upon the cloning or self-replicating of an electrochemical pattern at the synapto-dendridic level of the brain.

In *Biomimicry: Innovation Inspired by Nature*, Benyus (1997) discusses how life uses shape — patterns in three-dimensional space — to exchange and transfer information. She discusses in detail how pattern-making is an act of information processing.

Hameroff (1987), in *Ultimate Computing: Biomolecular Consciousness and Nanotechnology*, offers compelling evidence that the human brain translates information patterns from system to system by a type of *shape-based communication* at the subneural level. Through such mechanisms, neurons do not sum or average their thousands of inputs, but are rather capable of *representing* each of them through *shape* — pattern in three-dimensional space.

Finally, Csikszentmihalyi (1990) conducted extensive investigations into the psychology of optimal human experience and found that such a mental state can be achieved and maintained by engaging in activities that strike a delicate balance between order and disorder. He coined the term cognitive entropy to describe the mental state responsible for compelling a human being to seek a structured or patterned activity.

What mechanisms have led so many scientists, separated by space, time, and field of research, to come up with such similar insights? The answer can be found in the mechanisms of human perception: the dynamic balance between the ability to perceive fine degrees of difference in an aggregation of elements and the ability to perceive statistical, collective behavior of those elements. Information can be thought of as news of difference, and our perception of information is dependent upon the ability of our sense organs to detect highly refined degrees of difference (change). The degree of difference that we can detect, therefore, controls the richness and quality of information that we can perceive. Thus, part of our ability to perceive pattern comes from our ability to detect difference at ever finer degrees. Yet Schrodinger (1944) points out that we could most certainly hit a point of diminishing returns if our sensory systems were not tuned coarsely enough with respect to atomic and molecular levels.

> [W]hat a funny and disorderly experience we should have if our senses were susceptible to the impact of a few molecules only … heat motion tosses them like a small boat in a rough sea … the smaller their number, the larger the quite haphazard deviations we must expect …. For our organs of sense, after all, are a kind of instrument. We can see how useless they would be if they became too sensitive …. [A]n organism must have a comparatively gross structure in order to enjoy the benefit of fairly accurate laws, both for its internal life and for its interplay with the external world. For otherwise the number of co-operating particles would be too small, the 'law' too inaccurate. (Schrodinger, 1944, p. 13)

Bateson and Schrodinger clarify the fact that our sense organs must achieve and maintain a delicate and dynamic balance in order to support our ability to perceive pattern.

Thus, it seems that we are both pattern-producing and pattern-detecting bio-aggregates. And this process is so fundamental and essential to our proper functioning, our quality of life, and our very existence that the absence or excess of pattern would have serious and profound consequences to us. As it turns out, professional literature abounds with studies on the absence and excess of pattern, and their effects on neurocognitive function. These studies can be readily listed under any of four categories:

- Brainwashing and thought reform
- Cult conversion
- Isolation and sensory deprivation
- Information disease

Conway and Siegelman (1978) give a thorough and chilling account of their extensive research into the phenomenon of sudden, drastic personality alteration in *Snapping: America's Epidemic of Sudden Personality Change*. Conducting detailed investigations into the areas of cult conversion, mind control, brainwashing, post-traumatic stress, new-age therapies, and mass suicides, they searched for common ingredients that could explain the profound, lasting changes in the physical organization of the human brain and behavior, so common to these environmental conditions. In essence, they had made the first serious attempt at the development of a unified theory of sudden personality change, holding constant the hypothesis that this change in thought, emotions, and behavior was due to spontaneous, radical reorganization of the brain's information processing pathways. They searched each of the domains in question for common precipitating effects capable of initiating such spontaneous reorganization. Integrating knowledge from communication and information theory, mathematics, biology, neuroscience, and physics, they concluded that both significant overstimulation and understimulation of our sensory organs for prolonged periods of time were capable of destabilizing and completely destroying existing patterns of information flow through the information processing networks of the brain, resulting in massive functional and structural breakdown of predicted emotional and behavioral response. After such prolonged periods of overstimulation or understimulation, they noted changes in subjects' physical appearance, such as eye catalepsy, posture reorganization, voice tone, and patterns of emotional response to familiar stimuli, making subjects appear to be completely different people.

> [T]he latest research suggests, new and intense experiences may physically sever long-standing synaptic connections in the brain. Like a sudden trauma or electric shock, the new information-stresses people were being subjected to in the intense physical, mental and emotional experiences of group rituals and therapeutic techniques, were often powerful enough to destroy and replace lifelong patterns of mind and personality. They also appeared to alter and, in many cases, physically destroy long-standing information-processing pathways in people's brains and nervous systems. [T]he brain's living

> information networks are perpetually being shaped, changed, organized and reorganized by both the kind and quality of each individual's day-to-day experiences. (Conway and Siegelman, 1978, p. 123)

Much of their research was guided by Ashby (1952), who demonstrated scientifically that human survival and adaptation require a steady flow of information that is rich and varied in both kind and quality. Crucial to the organizing and adapting process of a human organism is the *law of experience,* which states that new information entering a communication system will tend to result in the destruction and replacement of earlier information of a similar nature.

During their investigations into brainwashing and thought reform, Conway and Siegelman accumulated compelling evidence that both overstimulation and understimulation result in a significantly altered state of suggestibility, not unlike the effects of Erickson's confusion technique. Such patterns of sensory overstimulation included the result of prolonged exposure to rhythmic drumming, dancing, singing, prayer, chanting, and breathing. In all, what mattered most in the production of an altered state of heightened suggestibility was not the activity itself, but the regularity of the rhythmic pattern of that activity. An extreme example of this is the notorious Japanese water torture technique used for interrogation and punishment. Sleep deprivation was also used effectively to precipitate a more rapid response to these rhythmic patterns. Similar altered states of heightened suggestibility were achieved by understimulation of the subject's sensory organs. Some of the mediums used to reduce or eliminate sensory stimulation were sensory deprivation chambers, fasting, meditation, breath-holding exercises, darkness, and social and perceptual isolation.

Drawing extensively from the work of Pavlov, Skinner, Ashby, Weiner, Sargant, Lifton, Hebb, and Pribram, their studies concluded that both intense repetition of pattern and the absence of pattern produced profound, spontaneous reorganization of information networks in the human brain, resulting in a radical alteration of thoughts, beliefs, emotions, behavior, and personality. How can sensory patterns affect so many different functions of mind and brain? Our biological tissues are so thoroughly integrated and complex that a change anywhere in the system can be expected to be experienced everywhere in that system, due to changes in boundary conditions between morphologically disparate tissues. We shall discuss this in greater depth shortly.

In *The Brain Benders: A Study of the Effects of Isolation*, Brownfield (1965) compiled the work of over 300 researchers who investigated the effects of sensory deprivation, perceptual and social isolation, phantom limb phenomenon, POW interrogation techniques, and brainwashing. Together these studies provide incontrovertible evidence that severe reduction or absence of patterned stimulation of sensory organs over a prolonged period can result in profound damage to the human neurocognitive system.

Early in this book, Brownfield discusses Hebb's (1949) hypothesis that monotonous, unchanging (no news of difference) stimulation results in a disorganization of the ability and capacity to think. Hebb attributed this disorganization to an interference with *phase sequence functioning* in the brain, reasoning that both the type and patterning of sensory stimulation are significant to the quality of neurocognitive function a human being can experience. Hebb was convinced that the absence of varied stimulation would result in impaired brain functioning. In other words,

Hebb believed that not only the complete absence of stimulation, but also simply a lack of changing or varied stimulation, would be enough to result in impaired brain functioning. Brownfield cites numerous studies confirming Hebb's predictions. Some of the controlled studies done simply reduced, rather than eliminated, sensory stimulation and movement, by employing such things as sound-attenuated cubicles, low levels of illumination, and translucent goggles. Repeated tests under these conditions predictably produced a profound disorganization of the thought process.

> Success in meeting most of life's basic needs involves becoming responsive to appropriate cues from the physical and social milieu in which we live. Sometimes inappropriate perceptions taking the form of misinterpretations or distortions of reality lead to poor judgment; if persistent and inflexible enough, this may lead to a state of emotional imbalance. People have even committed suicide because they believed in or acted upon false, delusionary, misleading perceptual cues. (Brownfield, 1965, p. 9)

A particularly interesting set of studies compiled by Brownfield indicates that the reticular formation of the human brain — a substantial netlike region of the lower brain stem — is highly vulnerable to the absence of varied, changing sensory stimulation. Studies reveal that as sensory stimulation becomes monotonous or severely reduced, there is an extinguishing of the arousal and attention-directing functions of the brain until such time as the reticular formation detects a change in stimulus pattern such as intensity, frequency, etc. These studies indicate that changes in sensory pattern are necessary for focusing attention, and that attention can only be maintained as long as a significant variation in sensory pattern is detected. If there is little or no variation, attention will lapse, resulting in a trance-like state.

> While a stimulus evokes or guides a specific bit of behavior, it also serves the nonspecific purpose of maintaining a normal state of arousal through the RAS [reticular activating system].* … Modification or alteration of sensory input appears to produce concurrent modification and alteration of response or output, so that changes in subjective experience are frequently reported. (Brownfield, 1965, p. 73)

Professional literature abounds with reports of experimental studies in the area of early development, suggesting that conditions of social isolation and maternal deprivation were causative, contributing to the deaths of some infants and to the subsequent intellectual and emotional deficits among those few who did survive such conditions. The mortality rate among the babies studied was exceptionally high, nearly 100%. Brownfield concludes:

> At this point it should be apparent that human beings are individually, socially, and physiologically dependent not only upon stimulation *per se*, but upon a continually varied and changing sensory stimulation in order to maintain normal, intelligent, coordinated, adaptive behavior and mental functioning. (p. 74)

---

* The reticular activating system (RAS) is an area of the brain stem which includes part of the reticular formation in addition to other communicating anatomical structures. It is responsible for the regulation of the level of consciousness and cortical alertness.

Many of the controlled isolation studies compiled by Brownfield indicated that whenever there was a rapid change from a condition of stimulation to one of relatively no stimulation, there was also an increase in hallucinatory activity, which was defined as internal representations made in the absence of actual stimuli which were uncontrollable by the subject and appeared to be external reality. It was also noted in many of these studies that the controlled reduction of patterned stimulation resulted in the subject expending an increasing amount of energy to structure perception. Note the similarity to the cognitive entropy work of Csikszentmihalyi and the importance of these findings in relation to insights and predictions made by Schrodinger, who concluded that life is the dynamic and consistent counterbalancing of entropy, and that the tendency of biological aggregates to lose structure, pattern, and information results ultimately in death.

Before proceeding, let us take a moment to summarize a few important points.

- Studies suggest that human beings seek a delicate balance of perceptual pattern (sensory stimulation), attempting to keep frequency, intensity, variation, and duration, as well as other factors, within an optimum range. If the balance is tipped too far in one direction or another, the human being will attempt to return to the optimum range for each of these parameters. Those overstimulated will seek sensory reduction. Those understimulated will seek ways of assimilating structured or patterned stimuli in order to structure and organize their own internal neurocognitive activity. This may help to explain why some stressed workers seek vacations in quiet, secluded places and why bored children turn to television and active-paced video games.
- Prolonged reduction or absence of changing, varied pattern, detectable by sensory organs, can produce neurocognitive effects such as boredom, hallucinations, phantom limb phenomenon, heightened suggestibility, confusion, disorientation of thought, inability to concentrate, bizarre behavior, psychosis, and, in some cases, even death. Experimental studies seem to provide compelling evidence of Schrodinger's deep insight that living organisms must consume order in order to counterbalance entropy and maintain life.
- Drastic increases in pattern intensity, variability, frequency, and duration can also produce similar effects to those mentioned above.
- Pattern, order, organization, function, and structure seem to be inextricably linked elements of human, neurocognitive life.
- The human being, as an organized biological system of aggregates, requires not only stimulation, but also a continually varied (changing) sensory input pattern for the maintenance of normal, intelligent, adaptive behavior.
- The withdrawal of pattern or the presence of excessive or conflicting patterns causes an increase in neurocognitive entropy (disorganization) and initiates behaviors designed to counterbalance that entropy (pattern-, order-, and structure-seeking behaviors).
- While it is clear that prolonged exposure to an excess of or an absence of pattern can be profoundly damaging, *brief* exposure appears to be a fundamental key to change.

In essence, human beings are pattern-detecting, pattern-producing, pattern-consuming, and pattern-dependent organisms: consumers of information ... informivores.

REFERENCES

Ashby, W. R. (1952). *Design for a Brain*. New York: John Wiley & Sons.

Baldwin, J. and Baldwin, J. (1981). *Behavior Principles in Everyday Life*, 2nd ed. Englewood Cliffs, NJ: Prentice-Hall.

Bateson, G. (1979). *Mind and Nature: A Necessary Unity*. New York: Bantam Books.

Benyus, J. (1997). *Biomimicry: Innovation Inspired by Nature*. New York: William Morrow and Company, Inc.

Brownfield, C. (1965). *The Brain Benders: A Study of the Effects of Isolation*, 2nd ed. New York: Exposition Press.

Cairns-Smith, A. G. (1996). *Evolving the Mind on the Nature of Matter and the Origin of Consciousness*. New York: Cambridge University Press.

Calvin, W. (1996). *The Cerebral Code: Thinking a Thought in the Mosaics of the Mind*. Cambridge, MA: MIT Press.

Conway, F. and Siegelman, J. (1978). *Snapping: America's Epidemic of Sudden Personality Change*, 2nd ed. New York: Stillpoint Press.

Csikszentmihalyi, M. (1990). *Flow: The Psychology of Optimal Experience*. New York: Harper Perennial.

Hameroff, S. (1987). *Ultimate Computing: Biomolecular Consciousness and Nanotechnology*. New York: North-Holland.

Hebb, D. O. (1949). *The Organization of Behavior: A Neuropsychological Theory*. New York: John Wiley & Sons.

Hunter, E. (1951). *Brain-Washing in Red China: The Calculated Destruction of Men's Minds*. New York: Vanguard Press.

Korzybski, A. (1933). *Science and Sanity*. Englewood, NJ: Institute of General Semantics.

Lifton, R. J. (1961). *Thought Reform and the Psychology of Totalism: A Study of "Brain-washing" in China*. New York: W. W. Norton & Company.

Luria, A. R. (1932). *The Nature of Human Conflicts: An Objective Study of Disorganization and Control of Human Behavior*. New York: Grove Press.

Penrose, R. (1994). *Shadows of the Mind: A Search for the Missing Science of Consciousness*. New York: Oxford University Press.

Sargant, W. (1957). *Battle for the Mind: A Physiology of Conversion and Brain-Washing*. New York: Doubleday & Company.

Schrodinger, E. (1944). *What is Life? With Mind and Matter and Autobiographical Sketches*. Cambridge, MA: Cambridge University Press.

Singer, M. T. and Lalich, J. (1995). *Cults in Our Midst: The Hidden Menace in our Everyday Lives*. San Francisco: Jossey-Bass Publishers.

Vernon, J. (1963). *Inside the Black Room: Studies of Sensory Deprivation*. New York: Clakson N. Potter.

3 Hidden Order and the Origins of Pattern: Thermodynamics, Pattern, and the Human Brain

CONTENTS

In order to construct a sound theoretical foundation for a cognitive science of intervention (a scientific psychology), it is requisite of that theory to describe the relationship between the behavior of matter, energy, information, brain, and mind in such a way that it may reveal cause–effect relationships between these systems; mind and brain must come to be understood in physical terms. It is with such an understanding that psychological intervention may take its place as a branch of science proper.

It is important to note here that cause–effect relationships may not exist in objective reality (whatever that means), but our brains make it so. Just as the brain's design allows us to perceive and respond to difference (change in a stimulus pattern), cause–effect perception is made possible by a property of brain function we refer to as *parallel-sequential compression* of sensory experience. Specifically, in order for conscious awareness to occur with such clarity and "wholeness" that cause–effect relationships can be discerned, the brain's simultaneous incorporation of billions of pieces of sensory experience (stimulus patterns) must be organized and re-presented in a sequential linear format. While we are well aware that this reorganizing process must cause the loss of immediate conscious accessibility of a significant amount of external experience, it may not be so obvious that our very ability to perceive cause and effect is dependent upon such compression and reorganization of sensory information.

What would it be like if instead of being consciously aware of the sequential aspect of this process, we were instead aware of the parallel (simultaneous) part of this process? Such an experiential state is actually quite common to human beings. During periods of dreaming, therapeutic trance, and meditation, one becomes increasingly aware of the parallel nature of information processing. As representations become more loosely connected in space and time, feelings of disorientation and disorganization, as well as a feeling of expanding conscious awareness, are commonly reported. During periods such as these, the brain undergoes a phase transition from a more ordered state (sequential, cause–effect) to a less ordered state, as it experiences an increase in the degrees of freedom in which its aggregates can be coupled (interconnected). This loss in cause–effect perception is one of the many cognitive resultants of the property called entropy.

PATTERN

Somewhere between matter and energy lies the most fundamental principle of interaction between matter particles and force particles, invariant of scale, giving rise to all emergent phenomena found occurring in man and nature. Between matter and energy lies pattern and entropy.

In order to understand brain, mind, behavior, and information and the emergent phenomena arising out of their interaction, such as cause–effect perception, our experience of time, meaning, etc., we must first ask ourselves: What is pattern? Why does pattern exist? What causes it? Why and how does it change? Why is the absence, excess, or presence of conflicting pattern so damaging to neurocognitive stability and biological existence? And how does our brain construct a model of the world that we call "mind," our very sense of reality itself, so that we may behave appropriately and adaptively correspond with, and navigate through, our external environment? These are some of the important questions we shall attempt to answer. Once we have provided a more detailed understanding of the ubiquitous nature of pattern, its origins and implications, we shall then move on to an exposition on predictable ways in which these patterns and their relationships can be represented, understood, and influenced in order to improve the quality of neurocognitive functioning, our resulting internal subjective experience, and life itself.

Pattern is most easily defined, described, and understood by making use of existing tools and terminology from four branches of statistical physics. For a definition of pattern with the greatest utility, we turn to the branch of statistical physics known as thermodynamics. Entropy is defined by the second law of thermodynamics as the degree to which relations among the components of any aggregate are mixed up, unsorted, undifferentiated, unpredictable, unorganized, and random — a state of negative information, a measure of information loss. Pattern can be defined as the state in which the relations between the components of any aggregate are ordered, sorted, differentiated, predictable, and organized in space and time as perceived by our organs of perception and limited by our ability to detect difference, what Schrodinger referred to as negentropy. The existence of pattern, the presence of information, and our ability to detect ever-finer degrees of difference are all inextricably linked together. In nature, pattern *is* the presence of information. Atoms, proteins, and DNA are just a few of nature's eloquent examples. Information is news of difference: a change in stimulus pattern perceived by our sense organs and even our internal cell receptor sites. As Schrodinger pointed out, our very ability to perceive pattern would cease to exist if our sensory organs were tuned too finely with respect to atomic and molecular level elements. It is only through this coarse tuning that we enjoy, as organisms, the benefit of fairly accurate statistical laws (i.e., mathematical descriptions of pattern).

As an example of this relationship of pattern and information, imagine that you are standing at the foot of a quiet pond. As your eyes glance across the pond, you see complete stillness, symmetry. While your eyes are closed, we pick up from the ground a handful of pebbles and rocks varying in size and weight and toss them out into the pond. Once you hear them make contact with the water, you open your eyes and see a pattern formed by the water in response to the pebbles and rocks which made contact with the surface of the water. What was before, a still symmetrical system as far as your eyes could perceive, has now been interrupted by a pattern of falling stones interfering with the stable bonds of the water molecules forming the surface. In effect, what has occurred is an interruption of water molecule pattern-integrity by the introduction of a pattern of stones, resulting in a change of molecular interconnectivity and motion (a new propogating pattern-integrity displacing matter and energy) evolving over space and time. Having only opened your eyes after the pebbles and rocks disappeared below the surface, your eyes inform you of their existence via the remaining water pattern. By looking carefully at the evolving pattern, you are privy to important information about the stones. The first piece of information you gain from the behavior of the water is the number of stones that were thrown into it. You can see this by the number of areas of water displacement that contain radiating, circular rippling patterns. In addition to the number of stones, you can also determine the position of entry. If you look carefully enough, you may also be able to determine their relative sizes and velocities by the height of the ripples and the speed at which they are traveling.

So as you can see, you can acquire much information about the stones simply by analyzing the behavioral patterns of the medium through which they passed. In this case, the medium was the water. We can also go as far as to say that the behavior of information is influenced by the medium that must process it or that it must pass through. The importance of this will become clear in later chapters, when we study the behavior of information patterns — the objective of the field of memetics — and also in Section II, where we discuss in great detail how to use NeuroPrint to analyze the neurocognitive system (medium) and how it is affected by information patterns that pass through it. This is also important in the study of how information changes as it passes through or is translated from one medium to another, as it must when crossing numerous boundaries of biological differentiation in the human body. We can also see from this example how the abstraction called memory is possible. Since you are able to garner information such as number, size, position, and velocity of the stones after the point in which they disappeared from the surface of the water, the water, as a medium, can be said to have or encode memory: in essence, a record of the pattern created by the stimulus — the stones. The most important properties of the abstraction we call memory are speed and accuracy of pattern reproduction and the duration or durability of that pattern. All of these properties are dependent upon another abstraction called stability. "Stability" is a relative measure of the difficulty with which a pattern-integrity displaces the matter and energy matrix of the encoding medium. In nature, the stability of the medium, which information passes through and organizes into pattern, will determine the quality of memory encoded by that medium. It is important here to recognize that the memory of a pattern is dependent upon the stability of the encoding medium and the stability of that medium is affected by both the state or phase of that medium, as well as the information pattern passing through it. That is, memory is state- or phase-dependent and information-dependent.

In the example of the pond, the memory exhibited by the water can be said to be short-term memory, since minutes, even seconds, later the once distinct pattern becomes unsorted, unorganized, undifferentiated, unpredictable, and once again symmetrical. We can also say that the life or duration of such a pattern was shortened by the force of entropy acting on the encoding medium, the interconnected molecules of water. It also follows then that the more stable the bonds between aggregates in the encoding medium, in this case water molecules, the slower will be the effect of entropy, and thus the longer the duration of memory. The reader may have already anticipated the relationship between the molecular medium we call water and the biological medium we call brain.

Water can be found in any of three conformational states or phases: solid, liquid, or gas. The stability of the hydrogen bonds between water molecules weakens as it moves from solid to gas; and the effects of entropy naturally increase as the temperature of the water rises, increasing the number and random nature of collisions between the water molecules. Imagine now what would happen if you stood in front of the same pond, this time frozen from the effects of winter weather. At the physical level of molecular organization and bond stability, we can begin to draw a fascinating parallel between human behavior and the behavior of simpler systems in nature. Noticing how difficult it is to effect change in people who are "set in their ways" by introducing new information, you can appreciate this parallel in nature's deep design. The colder the temperature, the more stable the bonds between water molecules and the thicker the ice. Imagine now that the pond has ice thick enough to support your body weight. What would happen if we took the same pebbles and rocks that we used during the time the water was in liquid phase and tossed them into the pond? Naturally, most, if not all, of the stones would bounce right off the encoding medium, having no effect whatsoever, leaving no impression. The stones are unable to significantly displace the matter–energy matrix of the ice. In human terms, phrases such as "thick skinned" and "hard headed" make perfect sense from this perspective.

One important thing we can learn from this is that as an encoding medium becomes more stable, it more accurately preserves existing patterns and more effectively resists the introduction or interruption from new patterns of information. In order to influence the encoding medium of water in this case, we have two choices: we can make the information more intense (bigger rocks thrown with greater velocity) or we can influence the medium to transition to a less stable phase (liquid). In some cases, the information pattern alone accomplishes both.

Instead of the stones in the last example, imagine that the sheet of ice was impacted by several pieces of molten rock, like that which would emanate from a volcano or minute fragments of a meteorite. In this case, the information pattern accomplishes both. It is not only quite a bit more intense in its ability to interrupt the existing pattern of the water because of the size and velocity of the fragments of rock, but its heat (thermal energy) also simultaneously changes the phase of the area surrounding impact from solid to liquid as it slowly burns its way through the 6 inches of ice. What is left on the surface of the ice is a very clear and stable record of the interrupting pattern. The resulting pattern of holes in the ice gives us much the same information about the rocks that have now disappeared below the surface. We can again clearly count their number, size, relative position of impact, and velocity (from surrounding damage). What is different here is the accuracy and duration of the memory encoded by the medium of ice. In its solid phase, less susceptible to the forces of entropy, the record of the stones' influence will remain not for seconds or minutes, but for days to possibly weeks. We can say that in the solid phase, the water possesses a property we call *long-term memory*. However, this distinction of long-term and short-term is a relative one. We could produce extremely short-term memory simply by heating the pond, thus influencing it to move from liquid to gas. Imagine how great the effects of entropy and how short term the stability of the encoded pattern would be if the water were in this state, such as might be found if we visited the "hot springs" near a recently active volcano.

This example of the information intensely affecting the encoding medium and influencing a change in phase or state of a portion of that medium has a parallel in the design of the human brain. Compelling evidence suggests that information at different intensity thresholds is capable of initiating a higher frequency of synapto-dendridic activity resulting in the triggering of neuropeptide release, which, in turn, is capable of spontaneously altering the state or phase of a small or large subsystem of the brain, responsible for encoding that information (Black, 1991; Shepherd, 1994; Kandel et al., 1991, 1995). This finding is so vital to the advancement of neurocognitive intervention that we would like to quote from a specific study.

> Frequency of stimulation regulates the chemical nature of the transmitter released in a wide variety of neural systems, centrally and peripherally. In the majority of classical transmitter-peptide neurons, release of peptide is elicited by higher frequencies of stimulation than that required for the release of the classical transmitter alone or in combination. Generally a frequency greater than 2 Hz is required for neuropeptide release, whereas classical transmitter release is evoked by lower rates of stimulation. (Bartfai et al., 1986, pp. 321–330).

From this, Black (1994) concludes the following:

> [A]t very low rates of stimulation, a neuron may act as a purely classical transmitter-releasing cell, at very high rates as a purely peptidergic cell, and in a varying combinatorial fashion at intermediate frequencies. (All of these frequencies fall within the physiologic range.) Consequently by altering impulse frequency and pattern, environmental stimuli may directly alter the nature of the chemical messages released by a neuron and, as a result, the nature of the information communicated. (pp. 79–81)

This insight into nature's deep design, that information can alter the phase of its medium to provide for its own more efficient encoding, has profound implications. At the very least it helps to explain the durable nature of our brain's encoding of phobias and traumas and sheds light on the complex mechanisms found beneath the brain's property of state-dependent learning, memory, and behavior (SDLMB).

Continuing with our pond example, imagine now that we have the ability to instantaneously change the phase of the water to any of the three conformational states — solid, liquid, or gas. Prior to throwing the stones, we prepare the water for their acceptance by changing its phase to liquid. As the stones release from our hands and disappear below the surface of the water, leaving a distinct pattern of radiating ripples, we instantaneously change the phase of the water to solid, thus freezing the encoded pattern into a more stable, durable medium. In this way, the water maintains the memory of the information introduced in a stable ordered pattern while simultaneously resisting the influence of the impact of information of similar intensity.

Again, the reader should be able to anticipate parallels in nature applicable to previously observed human behavior and brain function. While water, as a system of interconnected molecules, has only three conformational phases or states, the human brain is capable of producing an incomprehensible number of states or phases in which it may encode information patterns via complex differentiation of cell structure, neuro-peptide systems, hormone systems, and so on, each state or phase possessing its own relative stability as an encoding medium for information patterns. This makes the human brain and nervous system, as well as its connected and highly differentiated biological systems, capable of transferring an encoded information pattern from less stable encoding mediums such as the systems of brain that subserve what we call short-term memory, to more stable mediums such as the manufacture of new proteins. These new proteins, a form of longer-term, patterned-based or shape-based memory, are capable of transferring shorter-term, less stable patterns to longer-term, more stable encoding mediums by acting as signaling molecules (Black, 1991; Kandel et al., 1991, 1995; Shepherd, 1994). In this way nature uses pattern to translate, transfer, encode, and stabilize information contained within, and used by, biological systems (i.e., memory).

WHY PATTERN EXISTS AND WHAT CAUSES IT

Pattern owes its existence to at least three things. First, and most elusive, is our ability as organisms to perceive it. If we did not have the ability to perceive, adapt to, and influence pattern and order, aside from the fact that it would be impossible to do so, a discussion of pattern would be absurd. It is so interesting to see how we, as human beings, have concluded that pattern, order, and structure should be equated with life and that the lack of pattern, order, and structure should be equated with death. For those things without pattern do not exist in human perception. Recalling that pattern is both a phenomenon of space and time and four-dimensional in nature, our use of pattern as a measure for life and death is quite evident.

For many years prior to the development of modern medical machinery, we used heartbeat — a regular pattern reoccurring in time — to draw the dividing line between life and death. As science extended the range and scope of human perception, we turned to what was once invisible to us to again draw a new dividing line between life and death, namely, brain waves. While our extended perception dramatically changed what system of aggregates we viewed to determine the difference between life and death, we still trusted pattern to give us that answer. In this case, the pattern was the repetitive electromagnetic activity measured by an EEG. With both heartbeat and brain waves, we learned that there is a fine balance between regularity (order, pattern) and irregularity (disorder, entropy), and that radical irregularity of either heartbeat or brain waves tends to signal danger of death for the human organism. This type of distinction seems ubiquitous to human existence.

From day to day, we also tend to take notice of the regularity of movement of the people in our lives. If we see even a slight decrease in their anticipated frequency of movement during a given day, we tend to think something is wrong. If we detect a dramatic decrease in frequency of movement, we assume illness or drastic disorganization of normal functioning. We make these types of pattern distinctions not only with other humans, but even with our pets, vehicles, and appliances. How do you know when you have just gotten a punctured tire? You notice a change in rhythm as the tire continues to return to the point of puncture or continues to lose air. How do you know when the clothing in your washing machine is distributed unevenly? Perhaps when you hear a new knocking pattern on each revolution of the tub during the spin cycle.

How far away from our sensory organs must a fog be to be called a cloud? The farther away it gets, the more organized it appears to our visual receptors. At some point of visual threshold, a cloud comes into existence. As we move too close, the same organization of aggregates appears random and disorganized; and at that point, the cloud no longer exists for us. We're now in a fog. You may have had this experience, driving or flying through a low lying cloud. While the organization of aggregates may not have changed much, our perception of its existence has, simply due to our detection of pattern or its absence.

Given the fact that we are organized to detect pattern and order, there are at least two other factors responsible for the existence of pattern, order, and the resulting structure of things in the biological world: attraction and repulsion. These two opposing tendencies set all things, big and small, somewhere on a dynamic continuum between order and disorder: the tendency of attracting aggregates to organize into pattern and structure, and the tendency for their attractive bonds to weaken, wear out, repel, and become disordered and random (entropy). Since these tendencies keep aggregates of all types and kinds in perpetual fluctuation, they give rise to the emergent property of differentiation that is also responsible for the existence of pattern as perceived by our perceptual organs.

THE UBIQUITOUS NATURE OF PATTERN AT ALL LEVELS, BIG AND SMALL

Where does pattern start? We believe science will find that pattern, structure, and organization will continue to exist to the far reaches of human perception. As we continue to expand the range and scope of our perception, we will continue to discover the presence of pattern at both the very small and the very large. Let us now take a few moments to discuss some of the levels at which human beings have discovered pattern to exist.

PATTERN DYNAMICS AND THE BEHAVIOR OF ATOMS

As we write these words in 2000, many decades have passed since the unveiling of compelling evidence of the existence of atoms: *Brownian motion*, the theory explaining the erratic patterned activity of molecules of liquid (Bothamley, 1993). Since then, the field of physics has traveled far in extending the range and scope of that which it can describe. We now understand that the attracting and repelling forces which we spoke of previously exist at the level of subatomic particles. Responsible for unveiling some of the beauty and the complexity of this relationship, theoretical physicist Richard Feynman, developer of *The Extended Theory of Quantum Electro-Dynamics*, equated the existence of positively and negatively charged particles in an electromagnetic field to the emission and absorption of photons by electrons.* He believed that this property was responsible for the attracting and repelling forces between nonnuclear, subatomic particles creating what we know as electrostatic attraction (Parker, 1993). Like microscopic solar

* A photon is a force particle which mediates interactions between matter particles such as electrons. At the right energy levels or frequencies, photons are experienced by the human nervous system as visible light.

systems, these subatomic particles must maintain a delicate dynamic balance between energy and entropy. When their energy is too great, their behavior becomes more random and unstable. When their energy becomes too low, they become very stable but unable to bond with other atoms into new and larger elements called molecules.

This finding gave way to the *least energy principle*: the notion that subatomic particles must attempt to maintain a state of balance between maximum *combinatory flexibility* (requiring higher energy) and optimum stability, by giving way to its attractive forces and expending the least amount of energy possible (Parker, 1993). If too much energy is expended, entropy will rapidly increase; pattern, structure, and information would be lost and the atom would become dangerously unstable. Since their electron relationship is so dynamic, early statistical chemists decided that all known elements should be defined and classified by the number of protons in the nuclei of the atoms. A complimentary number of electrons are presupposed to exist in the atom's outer shell. Here, again, we looked for the most stable pattern existing at that level of description in order to define and classify, to name the existence of a thing. It turns out that there are 92 stable, long-lived nuclei existing in nature, described by the arrangement (pattern) of their protons and neutrons. These are the nuclei of the atoms of the 92 chemical elements. All elements listed on the periodic table having more than 92 protons are called transuranium elements and are quite unstable.

It is interesting to note here that pattern formation principles, such as the least energy principle, appear to operate at all levels of aggregation; for human beings themselves, complex and diversified aggregations of atoms and molecules also exhibit the least energy principle in their unified behavior. It is a well-recognized observation of the behavioral sciences that human beings naturally seek a minimum energy state while attempting to maintain optimum control of their environment. This has been referred to as *maximum gain for minimum effort* (Dichter, 1971; Wilkie, 1994). This is an example of the *fractal* nature of patterns — self-similarly repetitive at all levels of organization. Have you ever wondered why human beings prefer to buy television sets with remote controls to ones without? If you have ever wondered why the sale of just-add-water pancake mixes seems to continually sky-rocket and why people in general prefer cars equipped with power windows, power steering, and automatic door locks, the least energy principle of atomic and molecular behavior and the maximum gain for minimum effort principle of human systems behavior possess great explanatory power.

Formation of Pattern at the Atomic Level

What influences the formation of pattern at the atomic level? Toward an explanation, we must start with one of the most fundamental units of interaction currently known: an electric charge. Electricity or electric current can be said to be a flow of any charged particles. In actuality, a charge is mediated or transmitted by a force particle. In the case of electromagnetic fields, this force particle is a photon. A charge can be either positive or negative (unless we are describing quantum level interactions, where there can be more than two types of charges).

When charges are oppositely matched, they attract. When they are matched, such as two positive charges or two negative charges, they repel. The deep mystery of why such a property should exist is the domain of Feynman's Quantum Electrodynamics Theory, but for the purposes of describing the origin of pattern, we need not delve so deep.

When charges are balanced, they produce no obvious effects. When they are unbalanced, such as when a charged body has an excess or a deficiency of one type of charge, there is a strong tendency for such imbalance or instability to seek balance and stability. If you have ever rubbed a pen on your sleeve and then picked up little pieces of paper with it, or rubbed a balloon against your hair and then stuck it firmly against the wall, you have seen evidence of static electricity or electrostatic attraction. In these cases, we are using friction to interrupt or disrupt the pattern of positive and negative charges, which was balanced before our intervention. Here again we encounter the phenomenon of pattern interruption. It is important to note that forces between these charges will diminish with the square of the distance between them (Parker, 1993). In such a case, if you double the distance, the force is reduced to a quarter.

However, friction is not the only way of interrupting balanced patterns of charges (the stuff pattern, order, and structure are made of). By the early 1800s, it was found that water could be decomposed with electricity into hydrogen atoms and oxygen atoms. This process is called electrolysis and was used extensively to experiment with the decomposition of other elements such as fused salts. With enough electricity, these salts could be transformed into the highly reactive metals of sodium and potassium.

It should be noted here that sodium and potassium are two charged elements vital to impulse propagation in the neuron and entire nervous system, without which neurocognitive functioning, as we know it, would cease to exist. It is also useful to note that the mass of a product liberated in electrolysis is proportional to the quantity of electricity that has passed through it. With these things in mind, the atomic and molecular mechanics behind the highly controversial intervention called electroconvulsive therapy (ECT) becomes apparent. The effectiveness of ECT in treating psychological disorders such as recalcitrant depression can be attributed to the mild to massive pattern interruption of atomic-molecular information processing pathways in the brain. The target of this pattern interruption is one of the most fundamental levels of pattern formation in the brain: atomic level bonds. The most extreme demonstration of the power of electrolysis to liberate biological bonds and destroy organismic pattern is death by electrocution.

A property of the flow of electricity or electric current is the production of a magnetic field, which emanates perpendicular to the flow of electric charges. As far back as the 1800s, it was known that one changing electric current could produce another via a changing magnetic field (Parker, 1993). We mention this here because this understanding was the source of another set of pattern interruption-based interventions called *magnetic therapy.* It was reasoned that the placement or movement of a magnetic field perpendicular to the flow of electrical current in the human body would produce a change in that electric current and a corresponding biological change in the surrounding tissues, such as an imbalance of charged particles and increased blood flow. Thus, magnetic therapy has been widely used for pain relief, an interruption of pain pathway information transmission. Another example of the phenomenon of the magnetic field properties of electric current can be demonstrated at the annual science fair, as children replicate the classic experiment where a copper wire is coiled around a nail and connected to a dry cell battery, creating an electromagnet to pick up other metal objects.

Before continuing with our discussion of atoms and the origin of pattern, order, and structure, here is one additional example concerning electric charges. Imagine jerking an electric charge — that is, not just moving it, but quickly accelerating and decelerating — creating an electromagnetic pulse. Now imagine a bioelectric tissue, the human heart, whose cells contract collectively in a wave motion upon electrical impulse from a nerve; hence, the birth of the defibrillator or what is commonly referred to in hospitals as "the paddles." The idea behind this is that the pattern of an erratically contracting heart, unable to effectively pump blood, can be interrupted by such a jerking of electrical current and allowed to reestablish its stable rhythmic pattern of collective contraction. In some extreme cases, a heart which ceases to exhibit contractile pattern altogether could have its electrical current and rhythmic contractile pattern reestablished by the presence of another changing current, another effective use for the paddles in a code cart.

THERMODYNAMICS: THE SCIENCE OF ENERGY

The origins of pattern and its importance to neurocognitive functioning become increasingly apparent as we apprehend the relationship among brain, mind, behavior, and information and the laws governing matter and energy. Mind must be understood in physical terms. Let us discuss here a little more about the science of energy, thermodynamics.

The first law of thermodynamics states that all forms of energy are, in principle, interconvertible. Thus, energy and mass are conserved. However, it is the second law of thermodynamics, the property called entropy, that is most important to our present discussion. The property of entropy answers many important questions about the existence of matter and the course of biological life. The first and most basic is the question of why heat flows from a higher temperature body to a lower temperature body.

Ludwig Boltzmann, in 1877, accounted for this effect in terms of molecules and kinetic motion based on the idea of distributions of energies between molecules. His critical realization was that although the molecule-by-molecule situation for a gas will be hopelessly unpredictable, with an incomprehensible number of energy exchanges occurring constantly, the overall distribution of energies (i.e., the number of molecules that have a given energy level) can be predicted statistically. The theory predicted that gas molecules haphazardly colliding with one another will soon shuffle their energies so that the normal distribution of energies will be like that of a shuffled deck of cards, the most probable arrangement. Given the numbers involved, the tendency to the shuffled state must be powerful, this state of shuffledness being statistically more probable than any other more arranged state.

Since the kinetic motion of molecules is inescapable and will naturally redistribute the energies of molecules, heat will always appear to flow from the higher-temperature body (high kinetic motion) to a lower-temperature body (low kinetic motion) until the fully shuffled state of thermodynamic equilibrium is reached in both bodies. This is referred to as the state of maximum entropy: a state of mechanical, chemical, and thermal equilibrium. It is the strong tendency toward entropy, the ubiquitous obedience to the second law of thermodynamics, that prevents ice cubes from growing larger in a cup of hot tea.

THE STRENGTH OF ENTROPY

The strength of this tendency toward entropy depends upon how much more probable the more fully shuffled, disorganized, random state is than the more arranged state that you start with — the initial conditions — for any set of aggregates. Entropy can be said to be the degree of shuffledness. Entropy is not, in itself, energy. Yet, the tendency toward entropy can be increased in the presence of an increase in energy, since this increase in energy is in effect an increase in kinetic motion between atoms and molecules which speeds the process of disorganization: loss of pattern, information, and stability.

Boltzmann studied the effects of entropy extensively and showed how to calculate the thermodynamic probability of different physical states of systems (Bothamley, 1993). He considered this the number of conceivable arrangements for the energies of the molecules in different states; the chances of getting a particular kind of distribution would be directly related to the number of ways in which that kind of distribution could be realized.

To visualize the concept of thermodynamic probability, consider the following metaphor: imagine that instead of arrangements of molecules, we were built from arrangements of books. Now in this example, in order for us to realize a state of organization and order, either of us writers simply needs to be in a neat stack. Any sequence of books in the stack will do to allow us to attain the state of order. You, on the other hand, must be in a neat stack, separated by topic and arranged within each topic in alphabetical order according to authors. All of this is necessary for you to attain a similar state of organization and order. So, taking all of this into consideration, which of us, in our optimum state of order, contains or encodes more information? Obviously you do, because if someone needed to find a single book in your stack, it could be located immediately by recognizing the pattern and searching accordingly. A single glance at you would also reveal how many topics you contain.

Which of us will feel the effects of entropy first? To answer this, you must ask yourself: which of us has more ways in which our state of organization can be realized? Well, since we merely have to be arranged in a stack, we have more ways of realizing a state of order because any random shuffling of sequence will still result in an ordered stack. This means that you will feel the effects of entropy first, since you have only one way in which a state of order can be realized. Let's assume someone removed a book from your stack and returned it in the wrong sequence. You would immediately begin to feel the effects of entropy, and therefore the loss of information. Your pattern will have become more random. We, on the other hand, would not. In this way, nature's encoding mediums must maintain a delicate balance between pattern and entropy in order to preserve information. This will become especially important in Section II when we discuss NeuroPrint and begin to plot probability distributions for commonly appearing neurocognitive states of human subjective experience.

What is important to realize at this point is that there are at least two fundamental factors which influence the creation and disruption of pattern at the atomic level: the attractive, organizing forces between atoms and molecules, and the shuffling, randomizing force (or property) of energy redistribution called entropy. Attractive, organizing forces have the advantage when atoms and molecules move slower (i.e., low temperatures resulting in fewer collisions between atoms and molecules), while the disorganizing tendency called entropy has the upper hand when atoms and molecules move faster (i.e., high temperatures resulting in a greater number of collisions between atoms and molecules: kinetic motion). As we shall see, all aggregates of the physical world, big and small, must maintain a dynamic balance between these two forces. If tendencies toward entropy did not exist, diversification of elements, molecules, and hence all life forms would not exist, and behavioral adaptation of larger organisms would be impossible. In fact, larger organisms themselves would be impossible, since atoms unable to attract other atoms would not form larger elements called molecules. Molecules would not form proteins or any other larger biomachinery. If attractive forces did not exist, such as those within and between atoms, then pattern, order, structure, form, and hence function could not exist. Life is truly a dynamic balance between pattern and entropy, between *Yin* and *Yang*.

FORCES MEDIATING ATOMIC PATTERN FORMATION

Until recently, it was widely agreed that all atoms were made up of negatively charged electrons surrounding a tiny nucleus consisting of positively charged protons and uncharged neutrons packed tightly together. The atom's electrons could be found oscillating in a standing wave pattern like that of a plucked guitar string. This forms the so-called outer shell of the atom. Although historically atoms have been classified by the different patterns of arrangement of these three so-called fundamental particles, as early as the 1960s, scientists began to uncover an organizing pattern beneath that pattern. It is now understood that protons and neutrons are held together in a nuclei of an atom by a force far greater than the electromagnetic force. Ironically, this force is referred to as the *weak force* or sometimes the *nuclear force*. The difference between these forces can be demonstrated by experiencing the difference between an explosion caused by dynamite and one caused by a nuclear bomb. While protons and neutrons were also once believed to be "fundamental" particles of matter, particle physicists have clearly demonstrated that they, too, are made up of smaller arrangements of particles called *quarks*. These quarks are bound together by an even more powerful force referred to as the *strong force*. This force is mediated by a particle known as a *gluon*. Patterns of matter created by these forces are extremely stable and require incredible amounts of energy to disrupt. Particle physicists have accomplished this feat by employing the use of high-speed particle accelerators and bubble chambers, which produce and display high-energy collisions between particles in order to discover new particle building blocks. The process of smashing particles into each other at great speeds does in fact create new particles. The type of particle created depends upon the amount of energy used. However, as Einstein predicted, there are *no* "fundamental building blocks" to be found, "matter" itself *is energy* in different states of motion defined by the famous equation $E = mc^2$. The creation of different "particles of matter" is a property resulting from this motion — each type of motion permits the formation of different types of structure.

In this quantum mechanical view of nature, particles appear to be "created" at local points of interaction between interfering waves, where energy is highly concentrated in space and time. Subatomic particles react to confinement by moving faster — the more confined they are, the faster they move. As a result, protons and neutrons in the nucleus move much faster than electrons do. This is the reason for their great force when we attempt to separate them as we have seen demonstrated with the atomic bomb. Particles gain their "matter-like" appearance from the speed of their motion in the same way that the blades of a fan appear as a solid disk when the fan is turned on high speed. In a never-ending dynamic dance, particles (matter) can appear out of nowhere, transform into one another, and finally annihilate each other, disappearing in a sudden flash of light.

This is a picture of reality not easily grasped by mere sensory experience. Since patterns of subatomic energy form such stable macroscopic patterns (matter) at the temperature range that living organisims can be found, a detailed discussion of quantum-mechanical behavior is not necessary for the purpose of this book. It is only necessary that we be aware of the ever-present nature of pattern, structure, and organization occurring at multiple levels in nature, and most importantly that each level of pattern formation gives rise to its own intrinsic level of stability. Nature translates information from less stable pattern-forming mediums to more stable pattern-forming mediums in order to preserve the "memory" of the information encoded. Conversely, adaptability, diversity, evolution, and change in information are attained by transferring that information from more stable pattern-forming mediums to less stable pattern-forming/copying mediums. Entropy is vital for adaptation and change of any set of aggregates.

FROM ATOMS TO MOLECULES

As we cross the bridge from atoms to molecules, we also enter the world of chemistry. There are two important kinds of chemical bonding that result from the effects we previously described. Salts, such as sodium chloride, exist as a compound because of positively and negatively charged ions packed together via mutual electrostatic attraction. Outer electrons, stuck in their highest-energy standing wave modes, are the most likely to interact with other atoms. Although atoms are electrically neutral, their positive and negative charges are not in the same place. This causes a residual electrical field outside the atom, and hence their ability to attract one another to form infinitely complex patterns of arrangement called molecules. Since atoms can stabilize their electron organization by adding electrons or by attracting each other, they have a strong tendency to do so and therefore make extremely good building blocks for these larger units of molecules. However, not all atoms and atomic bonding capacities are created equal. Atoms differ greatly in their ability to attract other atoms and, therefore, in their ability as an aggregate to resist entropy. This property is called electronegativity.

For purposes of our discussion, the more important type of bonding that occurs in biochemistry is *covalent bonding*. These are the forces responsible for the assembly of molecules, simply accomplished by the joining of atoms together in unique patterns. Covalent bonds depend upon *electron sharing* (Abeles et al., 1992). A simple example can be found in the water molecule. The two hydrogen atoms can create a more stable, lower-energy standing wave pattern if they share their electrons, such that a pair of electrons is being held by both hydrogen nuclei, which are themselves held together. This sharing of a pair of electrons is a covalent bond. Humans are carbon-based life forms. Carbon atoms have four outer electrons available to make four covalent bonds with other atoms. This makes carbon a highly useful molecule for building complex patterns, especially since carbon atoms can form bonds with other carbon atoms without any known limit. Atoms having more limited bonding flexibility can also join with carbon atoms to attain an endless variety of molecules. It is here that we enter the world of organic chemistry.

In the domain of organic chemistry, far greater pattern formation complexity can be realized. At this level, molecules can form covalent bonds with other molecules to construct infinitely complex and diverse patterns of organization that are amazingly dynamic, as covalent bonds can form and reform with no known limit. Both atomic and molecular level bonding make pattern formation potential in man and nature endless.

One of the most fascinating and stable of all molecular patterns is the human genome, which is made up of 3 billion base-pairs of autocatalytic nucleotides (a four-symbol molecular alphabet). Like the stack of books that must be separated by topic and arranged in alphabetical order within each topic, DNA arranges into genes (approximately 100,000 of them) and within each gene the precise sequence of nucleotides, codes for the assembly of amino acids (a 20-symbol molecular alphabet) into specific proteins. These proteins are then arranged into patterns, forming everything from our hormones, enzymes, and neurotransmitters to our heart cells and brain cells: an entire living, functioning human being from a four-symbol molecular alphabet (adenine, guanine, cytosine, and thymine). If the force of entropy becomes too great during the copying or translating processes of these sequences of nucleotides and genes, it could result in devastating macroscopic consequences ranging from deformities to cancers, depending upon what stage in the pattern-copying, pattern-translating process these patterns of nucleotides were interrupted (Kendrew and Lawrence, 1994).

To preserve the information that describes a human being, nature had to create the most stable pattern-encoding medium possible, while still allowing intervention from the force of entropy to allow for copying, translating, adaptation, and evolution. This requires a delicate balance.

Pattern arises from the dynamic interaction of matter and energy in the physical world. The field of physics can predict much of nature's pattern-making behavior and has organized such observations into the laws of physics and chemistry. With a better understanding of these laws and principles, the behavior of larger aggregations of these basic elements, such as humans, can be better understood.

At the molecular level of pattern formation, a suspected law is finally confirmed by observation. Specifically, the pattern of an aggregate determines the function (behavior) of that aggregate. Research scientists working in the field of biomimicry have learned that bioorganisms perform a type of shape-based computing or information transfer. While we now have a clearer idea of how this can be accomplished, since the pattern/function relationship is so crucial to understanding pattern dynamics, let us momentarily explore some additional ways in which this relationship is realized in nature.

Once DNA has been translated from the 4-symbol molecular alphabet (adenine, guanine, cytosine, and thymine) to the 20-symbol molecular alphabet, a protein is born, a process many molecular biologists refer to as *gene expression*. This protein is not just an aimless molecule floating around, waiting to be attracted to something, but it is rather a fascinating and extraordinarily complex bio-machine created to perform a specific function.

Proteins are said to have three structures: primary, secondary and tertiary. The primary structure/pattern of a protein is simply the two-dimensional sequence in which their 20 molecular symbols (amino acids) are arranged. These sequences can be likened to sequences of words on a page containing a particular instruction or directing a specific act or acts for that page to perform. Once the correct molecules are bonded together in a string-like fashioned molecular sentence, the sequence (pattern in two-dimensional space), in conjunction with the environment the protein was designed to interact in, will determine what kind of tiny machine the protein will *fold up* into. This looks like molecular origami in dynamic motion. To visualize this, imagine several flat sheets of paper in two-dimensional space spontaneously folding themselves up into three-dimensional patterns, each into a different, specialized working machine, in response to the instructions that are contained on their pages. In this way, proteins are said to develop their secondary and tertiary structures and are able to translate their linear pattern of two-dimensional sequence (instructions for folding and movement) into function (the three-dimensional behavior of pattern through time). In other words, we can say that a protein is a tiny molecular machine possessing a four-dimensional space-time pattern called function: a pattern of shape and a pattern of movement (Kleinsmith and Kish, 1995). Today molecular biologists understand enough about protein folding to be able to predict some of their pattern-making behavior and corresponding function. Through proper questioning, nature has revealed how it translates two-dimensional pattern (information) into four-dimensional pattern (function/behavior).

Proteins have two types of secondary structure, both of which result from hydrogen bonding. These folding modes are usually determined by the environmental conditions in which they perform their task. There are two such folding modes commonly found in protein behavior, arbitrarily called α and β structures. When a protein exists in its α mode/pattern, the string of amino acids can be found coiled into a spiral called a *helix*. When the protein is in its β mode/pattern, hydrogen bonds are formed between different parts of the chain, which are lying somewhat parallel to each other forming a *sheet*-like structure. Combinations of α helix pattern and β sheet pattern tend to occur frequently in different kinds of proteins. These combinatory units of secondary structure are called *supersecondary structures,* consisting of small segments of α helix and β sheet connected to one another by looped regions. Nature uses the secondary structures of protein primarily for the building of structural biological materials. Only a tiny fraction of all proteins exist in their secondary structure. Instead, most proteins are organized into a tertiary structure created by folding their polypeptide chain into a compact shape that is organized into *domains*, each of which may carry out a different function. In this way, an enzyme may contain one domain that catalyzes a particular chemical reaction and another domain that binds some *ligand** that is required for the action to occur. It is also probable that a domain in one protein designed to carry out a particular function can also be found in other proteins performing a similar function. Unlike secondary structure, which depends upon hydrogen bonds, tertiary structure depends in great part upon environmental conditions such as electrostatic (ionic) bonding, and *hydrophobic* and *hydrophilic* behavior.

While the reader is now familiar with electrostatic bonding, the phenomena of hydrophobic and hydrophilic require some explaining. Hydrophobic means water fearing or water repelling. A hydrophobic region can be predicted to fold away from water when present in its environment and nestle itself safely within the protein's structure, radically determining its shape and function. Hydrophilic means water loving/attracting. These regions of a protein can be predicted to fold in such a way where they are present in the protein's outside structure when water is present in the protein's functional environment, also drastically altering the shape and function of the protein as it passes through different environments. In this way, the presence of water moving in and out of a protein's functional environment tends to determine the conformational folding state or pattern of the protein, creating a rather unstable dynamic pattern shifting from one conformational state to another. *Protein conformation* is the final three-dimensional shape generated by the confluence of all interacting forces.

A very important example of this exists in the human brain. Nerve cells are able to transmit electrical impulses over long distances of its axon by virtue of a specialized molecular protein machine called an *ion-gated protein channel*. While performing this task, a protein can certainly be viewed in its full beauty, complexity, and splendor. Its function is to control the passage of sodium and potassium ions in and out of the cell membrane, maintaining proper electrical resistance (membrane potential) needed to propagate an electrical impulse down the entire length of the axon. This results in chemical conversion of the electrical impulse at the synapse (i.e., *neurosynaptic transmission*). This incredibly complex process, occurring hundreds of billions of times a second throughout the brain, relies on the collective and coordinated behavior of trillions of these ion-gated protein channels switching back and forth between their two conformational states or folding modes in the presence of the proper ions. If we could be reduced to the size of a nerve cell to allow us to watch the process, this quantum mechanical

* An ion, or molecular group that binds to another entity, forming a larger complex.

deformation of a protein channel would appear similar to a camera shutter. However, instead of allowing or preventing the passage of light, we would be observing the passage of sodium and potassium atoms.

As far as electrostatic (ionic) bonding is concerned, those chemical groups carrying a positive charge are electrically attracted to those groups containing a negative charge for the same reasons individual atoms are attracted to each other. The type of bonding that results from the electrostatic effect is far less powerful than covalent bonds, which depend upon electron sharing between atoms. Proteins can be expected to fold so as to leave charged groups on the outside to facilitate attraction and the possibility of assembling into even larger units.

Larger proteins consisting of multiple polypeptide chains connect together to form a *quaternary structure**. Such proteins are called *multisubunit proteins* and each polypeptide chain is referred to as a subunit. The forces that hold subunits together are the same as discussed previously. The net effect of all these folding tendencies or probabilities is a momentary decision depending upon the protein molecule's given amino acid sequence, its specialized regions or domains, and its immediate and dynamically changing environment. The protein performs a delicate balancing act between the appropriate energy state needed to maintain stability and the consequences of the effects of entropy for each of those energy states. Its goal is to keep energy as low as possible, giving in to its attractive forces to attain stability, avoiding repulsive forces as a consequence of poor packing, while at the same time keeping entropy great enough to make adaptation to a changing environment possible. The result of this sensitivity to its surrounding environment is a continual dynamic shifting between slightly different folding patterns (conformations), each in turn facilitating a different function/behavior of the tiny protein machine. This is another example of the infinitely complex pattern-forming potential of nature.

The most important of all lessons we have learned from the molecular level of pattern formation is that pattern determines function (behavior) of an aggregate and a change in pattern results in a change in function (behavior), even at the microscopic level of tiny protein machines. This most fundamental principle of nature can be found operating in all larger, macroscopic aggregates assembled from these constituents, from organs to organisms, to the human brain and the human being.

While proper protein conformation is essential for protein function and the function of the entire organism, entropy — resulting from even a moderate increase in temperature (increased kinetic motion of atoms), such as when fighting an invading virus, or a significant change in local chemical influences, such as the presence of salt, altered pH, or urea — is capable of interrupting or disrupting all of the weak bonds responsible for the assembly, pattern, stability, and hence the function of these tiny molecular machines, which ubiquitously determine the quality of our lives and even life itself. Such conditions cause protein molecules to lose their pattern and information along with their ability to carry out normal function, even though the protein's primary structure has not been altered (Kleinsmith and Kish, 1995). Such conditions could be devastating to an organism depending upon the type of protein

* Quaternary structure is a term used in molecular biology, which refers to the formation of a multisubunit protein from more than one polypeptide chain.

affected and the importance of its function. Such proteins could be involved in the generation of a particular physiological/emotional state or behavior of a gland, organ, or the entire organism. Imagine the possible consequences if the affected protein was the ion-gated protein channel in the brain.

This loss of function resulting from disruption of a protein's conformation, three-dimensional pattern is called *denaturation*. This may be permanent or temporary, since spontaneous refolding is possible under certain conditions. Spontaneous refolding indicates to molecular biologists that all the information required for proper protein folding is inherent in the amino acid sequence (primary structure/pattern) of the polypeptide chain, even though its switching is influenced by environmental conditions.

The reader needs only to ponder briefly to anticipate the enormous and pervasive implications of such a fundamental principle of nature. This understanding helps us to gain insight into the underlying principles and mechanisms behind ECT, psychopharmacology, and the myriad of "brief therapies" (rapid psychotherapeutic interventions) which rely on pattern interruption and pattern disruption as a doorway to change.

PROTEINS, PATTERN, AND COMMUNICATION

An understanding and appreciation of nature's ubiquitous use of pattern, borne out of the interaction between matter and energy, helps us to elucidate many of the anomalies present in the world around us. We have explored the forces responsible for the existence of pattern, how nature uses pattern to encode information, and some of the neurocognitive consequences of the absence or excess of pattern and the presence of conflicting pattern. We have viewed various ways in which nature employs pattern to create memory — the preservation of encoded information by transference of pattern to a more stable pattern-encoding medium or by altering the phase of a medium. We have also noted how nature mediates change, adaptation, diversity, and evolution by the opposite process: transference of pattern to a less stable encoding medium and by balancing the forces of energy and entropy to influence phase and, consequently, pattern (information) stability. We have seen how a two-dimensional pattern, a sequence of nucleotides, may be used to specify a three-dimensional structure and, in turn, translate itself into a four-dimensional pattern called function or behavior (i.e., three-dimensional pattern in motion, through time). It is interesting to note here that the interruption or destruction of pattern at every level has been given a different name, as the study of this phenomenon has been relegated to different fields of research, employing different descriptive vocabularies. In great part, this has disabled previous attempts at unifying observed phenomena.

NATURE'S USE OF PATTERN TO RECOGNIZE, IDENTIFY, AND COMMUNICATE

Nature uses pattern in another ingenious way: to recognize, identify, and communicate information in and between biological aggregates. Due to their dynamic pattern-making and self-assembly capacities, proteins can be arranged in such a way as to be able to recognize more or less any molecule present in its environment. In a sense, a protein possesses a unique signature that has been defined by its primary, secondary, tertiary, and quaternary structures. This property of pattern recognition lies at the center of most of the communication systems within and between cells, glands, and organs, and is responsible for the ability of our diverse biological systems to translate and exchange information essential to collective, coordinated activity (four-dimensional pattern). Without such a system, an organism as large and complex as a human being could not exist. And when such a communication system begins to break down due to the forces of entropy, the list of options are uncoordinated activity, malfunction, sickness, disease, aging, and death.

The same principles apply to how the immune system protects us from viruses and bacteria — via the mechanisms of pattern-entropy dynamics. The immune system must utilize all of the properties of pattern that we have reviewed so far. When a virus enters the body, it too contains a signature that can be recognized. So the first step is for the white blood cell pattern recognition system to identify alien signatures. This recognition and identification process is performed by *B-lymphocytes*. The immune system must then employ its information encoding properties to encode the signature of the invader into a stable-enough medium as to allow for accurate copying. Once B-lymphocytes identify the foreign protein, they initiate the copy process by signaling B-cells to multiply. In this way, the pattern-encoding medium — white blood cells — communicates the invader's signature to a large number of colleagues who collectively track down copies of the invading virus throughout the body. Once captured, the white blood cells manufacture another protein, a marker (chemical tag), which acts as a locator signal for another type of white blood cell, a tiny protein warrior: a *killer T-cell*. Once annihilated, the signature of a virus must be remembered to facilitate a more rapid response of the immune system in the event of a reencounter. To accomplish this, the immune system performs a final step: the transference of the invader's molecular signature to a more stable pattern-encoding medium — a *memory T-cell* or *memory B-cell* capable of encoding and preserving an invader's signature for many years (Stites et al., 1994).

While the actual communication process of molecular information transmission is even more complex — involving signals to increase body temperature, shut down digestion, slow energy expenditure through reduction of musculoskeletal movement, mediate cognitive desires for food, and much more — our intention here is essentially to provide insight into how the properties of pattern mediate complex functions within us.

PATTERN-BASED IDENTITY AND THE STABILITY OF MATTER

Nature's use of pattern to determine identity is not limited to the world of macromolecules; it starts at the atomic level. At the atomic level, both identity and the stability of matter can be attributed to the same principle: the *Pauli exclusion principle*, which states that no pair of identical particles can simultaneously occupy the same quantum energy state. Therefore, electrons occupying the same orbital must have opposite spins. This principle is fundamental to the structure of complex atoms and molecules, without which matter would become unstable and collapse in on itself, and the formation of solid bodies would be impossible.

Fundamental to quantum mechanics is the notion that two systems, such as two atoms in the same quantum state (position/configuration and velocity/momentum in phase space), cannot be distinguished even in principle if they are interchanged physically. This principle is also fundamental to what philosophers refer to as the *pattern identity theory*, which roughly states that two organisms that are identical in their genes and their mind programs are one and the same. In principle, this means that if the information/pattern embodied by two encoding substrates we call organisms is identical, then the organisms themselves are identical (Bothamley, 1993; Parker, 1993; Tipler, 1994). Hence, even at the atomic level, nature has developed a way for aggregates to discern the identity of other aggregates by their quantum state (i.e., their pattern of configuration and momentum). This principle has deep and profound implications for the field of cognitive neurophysics and the study of human behavioral patterns and personality.

PATTERN-BASED FILTERS AND THE FORCES OF ENTROPY

Let us digress for a moment back to the immune system illustration used earlier. In this example, another function of pattern has revealed itself. Nature uses pattern to compare, filter, and categorize information. In this case, the B-lymphocyte is utilized to compare, filter, and categorize incoming information, discerning what is self from what is foreign protein. When the forces of entropy cause pattern to lose its filtering property, the result of such a failure is autoimmune disease. There are many different names given to this class of malfunctions, usually categorized by the specific protein signature that the immune system has confused. Whatever the classification given, the mechanics are basically the same. If a B-lymphocyte confuses one of the body's naturally occurring proteins with a foreign invader, one of our vital systems erroneously becomes the target for attack, resulting in devastating malfunction, disease, and possibly death. Imagine if the target of attack is the entire human nervous system, and the protein identified as foreign is the one that not only insulates electrical impulses within and between nerve cells, but also assists in maintaining maximum speed of impulse propagation over long distances and between vital subsystems and organs. The disease is multiple sclerosis and the protein is myelin. The prognosis: death in up to 27% of its victims (Stites et al., 1994). All this results from the forces of entropy acting upon just one of the body's essential pattern-based filters.

PATTERN-BASED CELLULAR COMMUNICATION

Immune system cells are not the only cells possessing pattern-based communication systems. All systems of the body requiring collective coordinated activity within and between other biosystems depend upon pattern-based communication. Cells are extremely large aggregates of proteins. They are like bio-factories encapsulating large numbers of tiny protein machines, each coded for a specialized task. All processes found deep inside these bio-factories can be mediated by pattern-based signaling or what is sometimes referred to as *shape-based signaling*, occurring both between and within cells. The signaling molecule is sometimes referred to as a ligand, and the receiving molecule is generally referred to as a receptor. These are specialized proteins embedded in the cell's outer membrane or in one of its internal organelles. *Receptors* are pattern-selective (both shape- and time-selective) proteins, each of which initiates specific internal cellular functions upon binding with a ligand with a complimentary shape (three-dimensional pattern) at the precise time (four-dimensional pattern).

One way to visualize this process is to imagine the electrical outlets in your home. Different outlets contain differently shaped sockets. If you attempt to plug a three-pronged toaster plug into a two-pronged socket, the result is no toast. If we take a two-pronged toaster plug and attempt to plug it into a three-pronged socket, the result is some nice toast with some danger of electrical shock due to lack of a grounded socket. If we take a three-pronged toaster plug and place it in a three-pronged socket, we get a perfect fit and perfect function. Try the right shape at the wrong time (i.e., during a power failure) and you end up with cold bread. Ligand binding to receptor sites works in a similar way and forms the very foundation of the field of pharmacology. The most important fact that must be added to grasp the significance of this model is that receptor sockets are not three-dimensional "static" patterns. They are four-dimensional dynamic patterns. Due to the continued shifting of conformational state, the socket can spontaneously change its shape, selectively calling for a different ligand whose shape corresponds with it only at that moment. Imagine attempting to plug a three-pronged toaster plug into a socket, which was dynamically oscillating back and forth between a two-pronged receptor surface and a three-pronged receptor surface. Now your toast is dependent upon expert timing. Bad timing, no toast. Now imagine having a key that you can stick in the outlet whose function is to stabilize either the two-pronged socket mode or the three-pronged socket mode. Such a molecular key is called a *control molecule*. Pattern-based communication systems like this exist throughout the human body. In this way, the brain can communicate with vital organs, muscles, glands, immune system cells, and itself by manufacturing and releasing a pattern-based instruction targeted for a specific system or cells of a system, resulting in stupefyingly complex, collective, coordinated activity of its aggregates (Kleinsmith and Kish, 1995).

Why do ligands bind with receptors in the first place? And how do they know when they have bound with the right receptor? Why do proteins bind with anything for that matter? Ligands bind for the same reasons that atoms do. When a protein binds a molecule that it was designed by natural selection to bind, the combination becomes more stable than it was before the binding. Proteins bind in order to stabilize their secondary and tertiary structures in the same way that atoms bind in order to stabilize the charges in their outer electron shells. Hence, proteins are given greater resistance to entropy due to the matter-organizing forces. In this way, a receptor —

a binding protein — is said to "know" their ligand. Another protein molecule that becomes more stable during the binding process is an enzyme. Enzymes appear to become more stable when binding their substrates, the molecule or molecules that are to react. In this way an enzyme can catalyze a chemical reaction.

Ligands are given many names depending upon their pattern, structure, size, the systems they communicate with, and the distances they travel. Neurotransmitters act within the brain. Neuropeptides mediate communication between the brain and the endocrine system. Hormones mediate communication within the endocrine system via long-distance bloodstream travel. Others acting over short distances are called local mediators, mediating collective, coordinated activity between cells in small regions; this is called *paracrine* control, meaning "beside." Some of these pattern-based signals, called second messengers, operate over even smaller distances, confined only within a single cell. Regardless of the name given to the family of molecules, they all possess the same property — pattern-based communication.

Cellular division/multiplication is just one of many functions mediated by cell receptors and second messengers. This process is important for both growth and repair. How does the process of cell generation know when to stop? It does so through pattern-based communication and feedback. What happens when the forces of entropy destabilize the pattern of the receptor whose function is to signal the normal shutdown of the cell duplicating process such that it can no longer recognize its feedback ligand? The result is cancer.

Let us consider one example of this from molecular biology. *Receptor protein-tyrosine kinases* are transmembrane proteins that contain a growth factor receptor domain, which is exposed on the outer surface of the plasma membrane, in addition to a tyrosine kinase catalytic domain found at the inner surface of the membrane. In a normal receptor, the protein-tyrosine kinases domain is activated upon binding of the appropriate growth factor protein to the receptor site. This binding triggers a series of events that lead to the stimulation of cell growth and division. If an oncogene codes for an abnormal receptor in which the growth factor binding site is disrupted, it will result in unregulated growth and activity of the cell. Disruption of pattern-based recognition and communication at a receptor binding site is just one of several points in a normal growth control pathway whose breakdown can result in cancer. Accurate pattern recognition and precise pattern-based communication are essential for normal growth and development. While it is unnecessary to discuss all of the points in a growth control pathway that are pattern-dependent, it should be noted that there are at least five ways in which pattern-based communication, recognition, and memory can be disrupted resulting in cancer.

The first and the simplest way is called a *point mutation*. An oncogene is a gene whose presence can cause a cell to become malignant. They arise by mutation from normal cellular genes called proto-oncogenes. As mentioned earlier, the genetic code responsible for the architecture and function of the human body is contained within 3 billion base pairs of autocatalytic nucleotides called DNA. The accuracy of the translation of these codes to amino acid sequences is so critical that an oncogene can arise from a single base pair substitution. This substitution is a point mutation, which, in turn, creates an oncogene, which codes for a protein that differs in only a single amino acid.

The second type of pattern-based disruption is *local DNA rearrangement.* It involves the creation of oncogenes, resulting from DNA rearrangements that cause either deletions or base-sequence exchanges between proto-oncogenes and surrounding genes.

The third type is *insertional mutagenesis*, which arises when a virus integrates a DNA copy of its own genetic information into a host chromosome in a region where a proto-oncogene is located. This insertion of the virus's pattern-based memory results in the disruption of pattern of the host proto-oncogene, resulting in cancer.

The fourth type of pattern disruption is *gene amplification.* This simply results from an increased number of copies (overproduction) of a particular proto-oncogene.

The fifth type is *chromosomal translocation*, which results when a portion of one chromosome becomes physically detached and rejoined to another chromosome. One such cancer resulting from this mechanism is called Burkitt's lymphoma and arises when a piece of chromosome 8 is translocated to chromosome 14. In short, at the level of DNA, nature can tolerate very little entropy.

PATTERN STABILITY AND ENERGY BARRIERS

Numerous examples have been shown in which pattern-based properties of nature depend upon the movement of information to more stable pattern-encoding mediums in order to create the property called memory. One of the emergent properties of the brain and its cellular organization is memory, and both the fields of psychology and cognitive science have made a distinction between long- and short-term memory. This process operates even at the level of the immune system, where the signature of a virus can be transferred from a cell possessing a short-term memory-encoding medium to one that possesses a longer-term memory-encoding medium.

Nature can make memory possible via pattern dynamics in at least three ways. First, information can be transferred to a different location or translated to a different system, which is a more stable pattern-encoding medium than the one in which the information was originally encoded. Such a function is used by the immune system when it relegates its responsibility of remembering a foreign protein to a T-memory cell or a B-memory cell.

Second, nature can adjust energy distribution to allow or prevent change in the phase of the original information/pattern-encoding medium by the forces of entropy. The metaphor of the pond illustrated information preservation facilitated by a phase transition of the encoding medium of water. A decrease in energy allows the water molecules to give in to their attractive forces, forming a more solid structure and increasing resistance to change. The stability of an encoding medium makes the long-term memory of a pattern possible. In this instance, nature accomplishes pattern-based memory by initiating a phase transition of an encoding medium (water) from a less stable (liquid) to a more stable phase (ice), thus more effectively resisting the forces of entropy. This method is utilized by nature in the full range of physical and biological systems. In the human brain, neuropeptides (signaling ligands), sometimes acting as a paracrine-mediated control system, can accomplish such a feat by initiating a large-scale phase transition, the target of which could be a specific subsystem or region of the brain (Kandel et al., 1991; Black, 1991; Becker et al.,

1992). This is just one way in which phase transition of an encoding medium can be initiated. Another is through shifts in the frequency of electromagnetic activity, accomplished internally by ion- or ligand-gated protein channel pumps, redistributing positive and negative ions in and around nerve cells. Another way is through a change in frequency of an incoming stimulus pattern from the external environment, which results in a change in frequency of the receiving and transmitting neurons. This is a phenomenon referred to as phase-locking or frequency-locking.

The third way nature accomplishes memory is through *energy barriers*. Energy barriers are stable patterns of attractive tension that resist displacement by other pattern-integrities. The vast majority of collisions between molecules do not lead to any reaction. If all collisions did result in a reaction, biological systems would be terribly unstable and life could not exist. Resistance to surrounding environmental forces is a vital property of the aforementioned covalent bond. Part of what keeps them stable is an energy barrier that must be overcome for the bond to be broken. One example of an energy barrier can be found at the level of the atom. An atomic nucleus contains protons and neutrons. Protons are positively charged particles, whereas neutrons, as their name suggests, are neutral. When there is only one proton present in the nucleus, little force is needed to hold the nucleus together. But what happens when there is more than one proton? Remember that chemical elements are defined by the number of protons in their nucleus. A known chemical element can have up to 92 protons in its nucleus. Also recall that "like" charges repel. So, how do you keep 92 positively charged protons from repelling each other and annihilating the nucleus of an atom? Enter the *weak* or *nuclear force*. This force acts as an energy barrier far stronger than the electromagnetic repulsion created by the positively charged protons, hence keeping them together in a neat, little, stable package.

A similar process can be found operating at the level of molecules. Where two larger molecules or aggregates such as proteins might repel when certain regions of electron probability begin to overlap, a covalent bond would overcome this repulsive tendency, keeping the molecules arranged in a neat package or pattern. The two opposing forces reach a collaborative harmony after moving through a transition state and becoming more stable. To cross such a barrier, a new chemical reaction must have enough energy to reach a critical transition state during a collision in order to break an existing bond and form a new one. Patterns endure in biological organisms because within ordinary temperature ranges, the energy needed to make and break the main covalent bonds in biological molecules is rarely achieved. This is one of the main reasons why we are essentially protected from some of the immediate effects of entropy within a normal body temperature range. However, when body temperature reaches approximately 108°F during a severe illness, brain function is in danger of massive breakdown due to interference with phosphorus and other critical molecular bonds, resulting in massive destruction of pattern, function, and life. At temperatures considerably below normal body temperature, or *hypothermia*, death of the organism also results. In the first instance, too many essential bonds between molecules are broken, and in the second case, not enough. Life is a delicate balance between the two. Whenever there is sufficient energy present to overcome the existing energy barriers, the excess energy must be removed in order for the new bond or pattern of bonds to stabilize. Otherwise, the new pattern will be disrupted by the trapped kinetic energy. It is useful to note here that enzymes are one of nature's ingenious molecules capable of removing energy barriers in precise locations of their choosing. In essence, they do this by stabilizing a transition state for their own reaction (Fersht, 1985).

FROM CELLS TO TISSUES, ORGANS, AND ORGANISMS

How does nature use pattern dynamics to build tissues, organ systems, and organisms from cells? Pattern-based architecture and function depend upon a little "biological glue." We say this only half kiddingly since particle physicists have named one of the known force particles just that way. The force particle that holds together *quarks* into particles of matter called protons and neutrons has been jokingly named *gluon*. When we speak about holding patterns of protons and neutrons together in a nucleus, the biological glue is the *weak force*, which is mediated by particles called *W* and *Z bosons*. Nature holds patterns of electrons together with nuclei using the electromagnetic force, which is mediated by photons. In turn, nature builds molecules from patterns of atoms, and proteins and cells from molecules using the covalent bond.

In any multicellular organism, individual cells must become associated in precise patterns to form tissues, organs, organ systems, and hence a complete functioning organism. This type of large-scale, ordered interaction requires each individual cell to be able to recognize, adhere, and directly communicate with each other. All of this is accomplished with another of nature's ingenious biological glues called a *cell adhesion molecule* (CAM). This accounts for why blood and lymph cells circulate freely throughout the body, while heart cells and lung cells stick together forming elaborate four-dimensional structures of shape and function. What would happen if your heart cells and your lung cells confused each other and formed a heart–lung? Imagine what would happen if your skin cells did not stick together tightly. How long would you last as an organism if your insides continually oozed out? Clearly you owe all of your functional splendor and enduring macroscopic structure to nature's cellular-level glue, the CAM. Without pattern-based recognition, adhesion, and communication, CAMs could not exist and organisms would ooze into oblivion under the forces of entropy.

CAMs have been classified into many different families and separated by both the type of cells they "glue" together and whether their adhesion properties are short or long term. Human beings owe their entire brain and nervous system, and all its resulting properties, to the *neural cell adhesion molecule* or N-CAM, without which an entire lifetime of poignant memories, experiences, learning, and behaviors would disintegrate under the force of entropy. Neural network systems — the most complex and vital sets of communication pathways governing the collective, coordinated behavior of larger organisms — are made up of over 100 trillion interconnected, intercommunicating, signaling pathways owing their structure and function to the essential pattern-based properties of the N-CAM. If any of the pattern-based properties of the N-CAM, such as recognition, adhesion, or communication, became disrupted due to forces of entropy, the biological chaos resulting would be unspeakable. Yet, the N-CAM is just a comparatively simple plasma membrane, glycoprotein, that can exist in any of three forms, all of which are encoded by the same gene; two are transmembrane proteins, and the third is covalently attached to the outer membrane surface. Whichever the case may be, the larger part of the N-CAM molecule protrudes beyond the outer cell surface, and that protruding region contains pattern-based binding sites facilitating the cell-to-cell recognition and adhesion.

While it is critically important that cells that should stick together, such as those that make up organs or cellular communication systems, have structural and functional patterns that endure, the opposite is also true. Cells that should flow freely must "know" not to stick together, except under certain circumstances such as in the case of blood cells. Biological organisms produce antibodies that both block and assist cell adhesion. One of the most important reasons for determining human blood types in preparation for transfusion is that such transfusion involving different blood types could result in antibodies in the bloodstream recognizing and binding the terminal sugars in glycophorin molecules of the red cell membrane, causing blood cells to coagulate. While blood cell coagulation is an essential process for preventing blood loss and promoting repair due to an injury, it could have a quite deadly effect if spontaneous coagulation occurred in key places in the cardiovascular network just after a transfusion. CAMs must not only recognize what to bind together but also when to bind together.

CELL JUNCTIONS AND MACROSCOPIC PATTERN ENDURANCE

Whereas cell adhesion molecules play an important role in cell-to-cell recognition, adherence, and communication, the establishment of enduring, long-term connections between cells usually requires slightly more elaborate structures called *cell junctions*. There are three types of junctions that are commonly recognized, all performing different functions in the development of an organism's architecture and resulting function. Nature uses *tight junctions* for sealing, *plaque-bearing junctions* for adhesion, and *gap junctions* for communication.

Tight junctions create a permeability barrier across a given layer of cells. In addition to preventing substances in the extracellular fluid from passing from one side of the cell layer to the other, the resulting structures of tight junction bonding reveal another way in which nature creates pattern-based filtering. The ability of tight junctions to restrict the movement of molecules across cell layers is especially useful in organs like the bladder, which prevent seepage of urine back into body tissues. When tight junctions exist, molecules may only move from one side of these cellular barriers to the other by passing directly through the cells themselves, which is a process that is precisely controlled by *plasma membrane transport proteins*. In this way, nature creates highly specialized pattern-based filtering systems on ever-increasing macroscopic levels.

Since organisms are constantly in dynamic motion, tissues, organs, and organ systems built from elaborate cellular patterns must endure constant stretching and distortion in order to maintain function. For instance, normal inhalation and exhalation do not cause lung cells to burst apart and go their individual ways. We owe the type of pattern resiliency that prevents this to *plaque-bearing junctions*, which stabilize cells in particular patterns of arrangement against mechanical stress. Junctions of this classification all share one important feature. Their points of connection are the internal cytoskeleton of each cell; and the biological glue joining these points of connection are plaques. By connecting the cytoskeleton of one cell to that of its neighboring cells, such plaque-bearing junctions establish an interconnected cytoskeleton network once thought only to maintain tissue integrity under great mechanical stress. This is one of nature's great accomplishments, which was not fully appreciated until the early 1990s, when quantum mechanical theory was applied to molecular neurobiology. New scientific questioning revealed that nature not only uses cytoskeleton interconnectivity to create the enduring structure for large organisms, but also as the most elaborate information processing and *molecular transport network* ever uncovered by the biological sciences, hence dwarfing in comparative complexity the human macroscopic information processing network called the human *neural-net*.

Thus far we have discussed extracellular signaling pathways and have also noted that second messengers form another set of signaling pathways located within cells. Cells also employ a third communication signaling system facilitated by *gap junctions*, that involve the direct movement of molecules from one cell to another without having to pass through extracellular space. Gap junctions occur in nearly every cell type found in vertebrates and invertebrates, facilitating rapid communication between cells, such as between nerve and muscle cells. The flow of electric current responsible for heartbeat owes its passage to such gap junctions, facilitating collective, coordinated movement of amazing complexity. Gap junctions found in the brain, among many other functions, assist in coordinating rapid muscular activities, which must be mediated between the cerebellum and its intercommunicating musculoskeletal system. Compelling evidence suggests that gap junctions are employed by nature to speed up communication, where mere chemical information transmission across synapses just won't do (Kleinsmith and Kish, 1995).

Now that we have seen how nature's ubiquitous use of pattern-entropy dynamics (the interaction of matter and energy) creates complex, coordinated organisms from atomic and subatomic activity, let us focus on a special class of patterns occurring in man and nature. Since evidence suggests that nature reveals its emergent anomalies through the deep structure of pattern-entropy dynamics, it follows that nature would be prudent to develop dynamics capable of streamlining the process of organism building and matter assembly. What we are suggesting here is a permutation of the least energy principle, applied ubiquitously to all of nature's pattern dynamics. Such a property exists in nature's deep design. These four-dimensional pattern dynamics and their resulting patterns are called *fractals*.

FRACTALS AND PATTERN-ENTROPY DYNAMICS

A special class of pattern occurring in nature, fractals are scale-invariant, self-similar, self-replicating patterns that make it possible to design highly complex structures and function from very simple arrangements of matter. Fractals are ubiquitous in nature's deep design.

The next time you visit a grocery store, purchase a head of broccoli and examine it carefully with a magnifying glass in hand. If you break off the smallest branching area nearest the bud at the top of the head and examine it closely, you will find that the branching pattern, greatly magnified, could structurally pass for the entire head of broccoli. It is evident that all nature had to do to create the elaborate architecture resulting in a head of broccoli was to create a single pattern on one scale and simply duplicate it repeatedly, invariant of that scale.

As science examines nature more closely, fractal principles of pattern construction and organization appear to exist not only invariant to scale, but also invariant to the building materials used. Fractals are ever-present in natural systems throughout the known universe and occur in structure and in process. Fractal organization is employed in many systems throughout the human body. Two such types of fractal organization are *branching patterns* and *folding patterns*. Fractal branching can be found in the circulatory system occurring throughout arteries and veins, the lymphatic system, the cytoskeletal network within all cells, nerve cells and neural networks, and more discretely in muscle and connective tissue. Fractal patterns of organization also occur in the design of bile ducts and calyx filters within the kidneys. This is just one more way in which nature employs pattern to create filters. Fractal folding patterns also occur ubiquitously throughout the human body. The two most obvious examples are the neocortex and the intestines.

There are certain classes of fractal pattern that are much more discrete. Certain stable arrangements of atoms and molecules occur repeatedly in nature invariant of scale, context, and particles of matter utilized. Because of the stability of certain arrangements of matter, natural selection appears to have endowed these stable arrangements with similar functions, also invariant of scale and building materials utilized and existing across contexts. What do neurons, cytoskeletons, veins, arteries, capillaries, the lymphatic system, and the intestines all have in common? They are all designed from one of nature's most stable forms: *tubes*. Nature employs tubes to facilitate communication and transport invariant of scale and context (Volk, 1995). Human beings are tubes within tubes. It is also no accident of nature that human beings are tube makers. Human beings incorporate pattern and order from their extraorganismic environment, as Schrodinger believed, and iterate those patterns first in mind, making cognitive models of their external world. Then these patterns are iterated back into their material, external world whenever human beings want to perform similar functions. Hence humans create tunnels, underground or underwater tubes joining two bodies of land in order to facilitate communication and transport between those two bodies. Such man-made tubes occur everywhere: rocket ships to transport humans throughout space, pipes to transport water from a reservoir to your home, and tubes to transport money at the bank drive-through. Is such a phenomenon an accident? We don't believe so.

If we slice a tube long ways up the middle and flatten it out, we uncover *sheets*, another fractal pattern that occurs stably in nature. Without sheets, tubes would not be possible. Yet sheets also stand alone as fractals in nature, having their own inherent function. Sheets such as cell membranes and skin can help complex arrangements of biological matter resist the forces of entropy. If it were not for sheets occurring in nature, what would keep in your organs and organ systems? While sheets help to keep things inside, they also act as barriers to keep other things outside. Without skin, humans would be immediately susceptible to infection, resulting in death. Sheets can also help to transmit energy (Volk, 1995) . Our eardrums, which transmit patterns of gas movement and compression, provide vital information about our environment. Nature also uses sheets to collect light. One such sheet occurring in nature is the leaf, whose structure allows for maximum collection of light for photosynthesis. But, as nature would have it, what works in one place should work in another. Another biological sheet designed for the collection of light patterns is the retina of our eyes, which collects patterns of light in order to help us navigate in our environment, as well as regulating internal behaviors such as sleep and waking via the production of melatonin. Pattern-based navigation is another important emergent property of the interaction between matter and energy. Looking closely at leaves, we find tubes within sheets. Looking closely at the retina, we find the same, namely, *rods* and *cones*.

As discussed earlier, theoretical physicist Schrodinger (1944) pointed out that life extracts and incorporates order (pattern) from its external environment in order to counterbalance the effects of entropy and maintain stable existence. However, at the time he had this insight, he did not have at his disposal the descriptive tools we have today. Consider this: humans, as life forms, consume plants. The atomic and molecular aggregates that assemble into stable geometric forms of tubes and sheets to make plants are first broken down by protein biochemical catalysts — enzymes — which break their covalent bonds. This is accomplished by a digestive system that nature also fashioned from tubes and sheets (esophagus, intestines, stomach lining, etc.). Once broken down into their basic components and assimilated, these same constituents — whose stable forms are tubes and sheets — are used to make little protein machines that build, repair, and regenerate our own tubes and sheets. Is it any accident that atomic arrangements that form tubes and sheets outside of us also form tubes and sheets inside of us, once incorporated? Not if that is the most stable arrangement of the incorporated matter particles.

Before we move on, let us mention *spheres*, one more stable fractal pattern occurring in nature. When sheets form tubes for transport and communication, nature selects spheres as the vehicle of transport (Volk, 1995). Hence, blood cells are spheres that travel within veins and arteries; neurotransmitters travel through axons and dendrites in spheres called vesicles. Resulting from stable organizations of matter, spheres occur everywhere in nature, invariant of scale and context: eyeballs, soap bubbles, grapes, grapefruits, skulls, planets, and stars. Is it accidental that we have incorporated these patterns of sheets, tubes, and spheres from nature into mind itself, and then used them to build mental models to understand and interact with the world in which we live? One such complex mental model employed by the field of chemistry to understand the atomic arrangement of chemical compounds is the sphere and tube model (sometimes called the ball and stick model) of atomic-chemical organization. How about games such as baseball? What are a bat, a ball, and a base, if not a tube, a sphere, and a sheet?

All of these systems mentioned, and many more, reveal this irregular scale-invariant, self-similar, pattern replication process (Briggs, 1992; Kaandorp, 1994; Mandelbrot, 1983; Nonnenmacher et al., 1994). Pattern-entropy dynamics is itself a fractal process, ubiquitous to all scales of interaction between matter and energy, from the very small to the very large.

SEVEN REASONS WHY NATURE USES FRACTALS TO ORGANIZE MATTER AND FUNCTION

While this is a reasonably new field of examination for science, there appear to be at least seven advantages to organizing matter and energy and, hence, structure and function, in this way.

First, nature can create the enormously complex phenomena surrounding us from the very simple. A property such as this could explain why and how it is possible to construct organisms like human beings, planets, solar systems, and galaxies filled with diverse and unique structures, all from a sea of quarks and leptons, believed by particle physicists to contain little useful information in themselves. Such a process is certainly consistent with the least energy principle of atomic and sub-atomic organization.

This notion leads into a second advantage. If extremely complex organisms can be built from relatively simple rules and information, pattern-based information storage (memory) becomes a much simpler, more energy-efficient and space-efficient task for nature. This explains why one of the most complex structures in the known universe, the human brain, can be assembled from the relatively small amount of information contained within DNA. Mathematical physicists have provided compelling evidence that it would be impossible to specify the 100 trillion points of information passage in the human neural-net with only the pattern-based information contained within 3 billion base pairs of nucleotides in a strand of human DNA. Hence, in this way fractals act as information decompressors, allowing for efficient storage and transfer of exceedingly complex information patterns.

Third, this information compression made possible by fractal organization creates another advantage for large organisms. All information storage about structure and function has an energy cost to that organism. If the amount of information that pattern-based storage mediums must encode, transfer, and utilize is reduced, the energy cost is also reduced. In this way, a limited amount of energy can be distributed over a large number of complex functions simultaneously. Reduced energy costs result in a more efficient parallel operation of systems.

Fourth, out of this type of information compression arises the benefit of error tolerance. The fewer number of aggregates involved in the process of encoding, transferring, and utilizing information, the more resistant the aggregates will be to the randomizing effects of entropy.

Fifth, one inescapable source of entropy and its disruptive force is growth itself. As a fully functioning organism grows from an infant to a fully grown adult, it must maintain completely uninterrupted function throughout all of its systems. If the process of growth and development could not make use of fractal pattern dynamics, then the process of growth itself would interrupt the functional patterns necessary for life (Volk, 1995). Imagine if the heart and lungs did not increase in scale in relation to the overall size of the body. Such breakdowns in pattern do occur. A heart that does not grow in proportional scale to the vascular system, through which it must pump blood, would rapidly exhaust itself and cease to function properly. Fractal pattern dynamics help complex organisms to resist such growth-related disruptions.

Sixth, fractal pattern dynamics allow nature to optimize whatever space is available. Both fractal folding and fractal branching result in space maximization (Volk, 1995). The most dramatic example can be found in the human brain. The neocortex has a macroscopic fractal folding pattern evident by examining its convolutions. If we were to surgically remove a human neocortex and spread it out totally flat on a table, its network would be distributed over the entire size of a tabloid newspaper. Imagine having to walk around with a skull that big in order to store and utilize your most valuable memories, learning, and behavioral strategies. In the same set of tissues, microscopic fractal branching can be observed in nerve cells called neurons, which form the elaborate human neural-net. This fractal branching dynamic allows the ever-diminishing scales of axon and dendrite proliferation to fill up every available nook and cranny of the skull, in and between critical tissues and cells. How would it be possible to construct a human brain from so little information without such a process?

Finally, our example of fractal folding and fractal branching in the brain leads us to a seventh critical advantage. Elaborate fractal systems such as the human neural-net possess a functional property called *degeneracy.* This property allows a single class of information patterns to be represented in a large number of different neural systems at different locations. Such a redundancy of information representation facilitates recovery from damage and helps to prevent devastating losses of function. Medical science has known for some time that when an area of the neocortex is damaged by stroke or trauma, the functions once dependent upon the now damaged area can be relegated to another region of the neural-net. In this way, some lost functions and capacities of brain and mind can be regained. Fractal pattern organization creates organismic flexibility, allowing information and function transfer and translation to more stable systems, hence minimizing potential damage and loss.

One such network that exhibits and utilizes all of the principles mentioned is the recently revealed *subneural, cytoskeletal-net.* Since our ability to extend the range and scope of perception has increased over the last decade, it has become evident that an elaborate signaling and molecular transport network, fractally self-similar to the human synapto-dendritic neural-net, exists deep within not only all of the 100 billion cells making up the human brain, but also within nearly every functional cell of a human organism, completely interconnected for high-speed communication by the plaque-bearing junctions already mentioned. Evidence suggests that science has unveiled a scale-invariant information processing aggregate similar in architectural pattern and function to, and intercommunicating with, the human neural network at the subneural level: a brain within a brain cell, or what mathematical physicist Roger Penrose (1994) refers to as a *shadow of the mind* (Benyus, 1997; Kleinsmith and Kish, 1995; Penrose, 1994; Pribram, 1993, 1994).

THE BRAIN: NATURE'S INTERFACE BETWEEN ORGANISM AND ENVIRONMENT

Just as cells have membranes that separate them from the external environment and receptor sites that perceive, encode, and translate information from the external environments to their internal environments, such is the case for organisms, nature's larger aggregates. As organisms, we possess an arrangement of cells that are functionally self-similar to our smaller constituents and responsible for a similar process: a membrane called skin and receptor sites organized into various classes, the five senses or sense organs.

Each of these receptor sites have been engineered by natural selection and the properties of pattern and entropy to perceive, encode, and translate a portion of the available information in our external environment, referred to as stimuli or information patterns. The differences they report — these information patterns — are changes in pattern occurring in the external environment. Our eyes are tuned to encode and translate changes in visible light, a portion of the electromagnetic field. One class of our skin's receptors, thermal receptors, encodes another portion of that spectrum — infrared or heat. Our ears perceive, encode, and translate changes in patterns of gas movement in our atmosphere, referred to as sound waves. We classify some of these encoded differences as changes in frequency (pitch) and intensity (volume). Our skin encodes a different portion of the same field with our mechano-receptors called vibration. Receptors on our tongues (gustatory taste buds), in our noses (olfactory bulb), and some in our skin (chemo-receptors) encode changes in chemical pattern (atomic arrangement) in our external environment. Together, these receptors are designed to report news of difference to the internal environment of the organism.

Since the aggregation of the organism's cells that need to be informed of these changes in the environment and respond collectively to them is so mammoth, a complex interface is needed to encode, translate, and transfer information patterns to the appropriate regions and systems to initiate a coordinated collective response. The interface is the human brain and nervous system. The pattern-processing medium employed to accomplish this task is the neural-net; the subneural cytoskeletal-net; and the cells, proteins, molecules, and atoms to which it owes its structure and function. From the modest beginnings of a neural tube, neo-cortical sheets, and spheres called ganglion and nuclei, complemented by fractal folding and branching patterns, one of nature's greatest wonders, the brain, is born. Its purpose is the incorporation and translation of the portion of the organism's external environment necessary for the initiation of coordinated collective response to changing information patterns, thereby constructing an internal model of the world essential to organismic navigation within that external environment: the mind.

PATTERN-BASED NAVIGATION, MAP BUILDING, AND MODELS OF THE WORLD

Just as cells incorporate pattern-based information from their environments (through receptor sites) needed to regulate internal behavior and navigate movement within their external environments, organisms do likewise. Without pattern-based communication and navigation guided by signaling proteins, immune system cells would not be able to mediate a collective response against the invasion of foreign proteins. Without pattern-based navigation, how would white blood cells locate an invader?

Neurons in the brain must also navigate. How do 100 billion nerve cells know where to go and in which direction to grow in order to provide the highly complex information-processing pathways between receptor organs, internal regulating systems, and effectors that mediate unified, collective behavior? One important way that this is accomplished is through a pattern-based navigation system, guided by the signaling protein *nerve growth factor* (NGF). In this way, nerve axon and dendrite fractal branching patterns can follow a chemical pathway that specifies the location of the target cells and organs that must be connected. From that point, connections are continually modified by stimulation from internal and external pattern-based information or lack thereof, making use of another of nature's innovations, the *activity-dependent synapses*. Those pathways not used are retracted and those supporting the continued traffic from electrochemical impulses (spike trains) are strengthened (Edelman, 1987; Hebb, 1949; Kandel et al., 1991, 1995; Kleinsmith and Kish, 1995). Through continued cytoskeletal pattern change and neurosynaptic-level modification, the elaborate pattern-processing medium, the human neural-net, is able to change structure and function in the presence of changing internal and external information patterns, thus maintaining continued correspondence with an ever-changing external and internal world. In this way organisms, like human beings, can maintain a relatively accurate and dynamic map or model of their world, thus producing the collective, coordinated behavior referred to as adaptation. The malfunction of such a system due to entropy can result in devastating consequences, ranging from uncoordinated and inappropriate movement to hallucination and schizophrenia.

THE NEURAL-NET AND SUBNEURAL-NET

How does this pattern-processing medium change in the presence of information? Let us consider a few ways in which this is accomplished. While signaling proteins from one cell can specify which type of proteins are manufactured by another cell, the selection and transport of those new signaling molecules down a neuron's axon and to the synapse, the gap between two communicating neurons, is dependent on the cell's cytoskeletal-net. As the pattern of the cytoskeletal-net changes in response to information patterns, so does the selection of neurotransmitters and their pathway of travel.

The subneural cytoskeletal-net is also responsible for the manufacture and placement of structural proteins called *dendridic spines*, which facilitate specific and complex connectivity patterns between neurons, capable of increasing or decreasing the number of potential input pathways to a neuron. Neurons are known to connect with between 1000 and 200,000 other neurons in this way. Mounting evidence suggests that the cytoskeletal-net is capable of spontaneous pattern and pathway change in the presence of the appropriate signaling molecules, thus radically changing connectivity patterns in the neurosynaptic-net of the neocortex (Benyus, 1997; Penrose, 1994; Pribram, 1993, 1994).

The subneural cytoskeletal-net itself is made up of much smaller pattern-based building blocks called *tubulin*, a protein capable of snapping together to form elaborate *microtubules*, tubular networks for communication and transport. These elaborate communication and transport networks hook together in complex and dynamically changing patterns with the help of *microtubule associated proteins* (MAPs), a "molecular glue." Within the context of the cytoskeletal-net, tubulin is known to oscillate between two conformational states (different shapes or patterns) caused by the *dipole oscillation* of an electron located in its elbow, the joint formed when α-tubulin and β-tubulin are reconnected. Research suggests that continual movement of these electrons is implicated in the emergent phenomenon of consciousness, for if this oscillation is restricted by the presence of a large molecule wedged within the tubulin dimer elbow, the organism loses consciousness. When the restriction is removed, consciousness is regained. Such molecules found to behave in this way are classified as anesthetics (Hameroff, 1987).

Hi-speed *biophoton emission* (signaling) throughout the cytoskeletal network is believed to cause this electron oscillation, as well as the resulting conformational change of the tubulin dimer, in response to changing information patterns (electromagnetic waves, quantum microwaves, etc.). Cytoskeletal polymers are known to undergo coherent, collective conformational oscillations in the nanosecond time scale (one billionth of a second). Hence, new patterns of biophoton activity can influence ion-gated protein channels in neurons resulting in changes in neurosynaptic level signaling and connectivity. The resulting conformational changes in tubulin also give rise to *bio-mechanical* communication and transport in the form of *solition waves* (Benyus, 1997; Penrose, 1994; Pribram, 1993, 1994). These waves are caused by pattern-integrities that propogate through the tubulin network by displacing matter and energy.

Neurons are also capable of a myriad of other types of pattern-based processing. More than 60 different neuro-active proteins — neurotransmitters, neurohormones, and neuropeptides — can arise due to signal-induced protein production and folding patterns, and combine to make complex molecular signaling profiles that are responsible for emotions, motor movement, perception, memory, thoughts, etc. Temporal coding patterns can also be accomplished by adjusting the frequency and grouping (interspike intervals) of electrochemical impulses (spike trains) on the neurosynaptic level (Rieke et al., 1997).

Considered together, there is considerable evidence suggesting that these multi-leveled and elaborately interconnected pattern-processing mediums give the human neural-net and its interconnecting pattern-making systems the ability to fully incorporate, represent, translate, and transfer all inputs to a given neuron or neural network, and hence the entire organism, rather than the once closely held belief that inputs are in some way just summed or averaged to facilitate a computation-based decision or behavior of an organism (Benyus, 1997; Penrose, 1994; Pribram, 1993, 1994).

NATURE'S INTERFACE BETWEEN INFORMATION AND ORGANISM AND THE EMERGENCE OF MIND

The human brain is one of nature's newest interfaces between information available in the biosphere and the organism, the biological contents of which betray its modest origins. The human brain is the result of 4000 million years of nature's pattern-based engineering, evolving out of the continued pattern–entropy-assisted aggregation of elements from the biosphere. From the modest beginnings of sulfur-breathing bacteria — the first unit of dissipative, autopoietic, self-replicating life — the human brain is a fractally imbedded history of life on earth, a recapitulation of nature's first 4 billion years of evolutionary accomplishment.

From the earliest ocean salts that now mediate transmission of nerve impulses within our brains, to the mitochondria of our cells, a once free-living bacteria that taught us how to breathe a catastrophically poisonous gas called oxygen, nature has not discarded a single valuable invention from her past. More than two decades of concentrated research in microbiology and paleontology has revealed that nature has employed the most elusive of all pattern-making mechanisms from the very inception of life itself — the fractal embedding of each of its best pattern-based inventions through *symbiosis*, a complete incorporation of one living four-dimensional biological pattern by another. Microbiological evidence suggests that all life has evolved by this process (Fortey, 1977; Margulis and Sagan, 1997).

Schrodinger's deep insight was correct; life incorporates pattern and order to counterbalance the incessant tug of entropy. Without a deep understanding and appreciation of this, one of nature's most fundamental properties, the phenomenon of mind cannot be understood. The faculties of sentient beings can only be misapprehended when viewed through the paradigm-induced expectation that all emergent phenomena of mind arise merely out of the electrochemical summing or averaging of inputs to neurons (computing) and the resulting movement of biochemicals to the appropriate receptors. Nature is far too sophisticated for that.

PATTERN: THE UNIVERSAL LANGUAGE OF NATURE

Nature has been evolving the capacity for mind and information exchange with every pattern-based information encoding medium she has ever invented over the last 4 billion years on earth. If she had not, global networks of cyanobacteria would not have been able to continue to stabilize the earth's delicate balance of life-giving atmospheric oxygen at 21%, and living organisms would have either spontaneously combusted or asphyxiated (Margulis and Sagan, 1997). In the next two chapters and throughout the remainder of this book, we discuss and illustrate in detail how information patterns become transmitted, incorporated, replicated, cleaved, and recombined through the fundamental interaction between matter and energy — pattern and entropy.

Information exchange via incorporation (encoding and symbiosis), replication (reactivation and translation), cleaving (cutting and separating), recombination (arranging and reassociation), and transmission (transfer between and among encoding mediums) is a pattern-based property of the entire biosphere, from the quarks, leptons, and atoms that formed the universe, to the autocatalytic nucleotides that continually translate two-dimensional patterns of nature's past accomplishments into the four-dimensional patterns of all evolving life on earth. When nature created the first semipermeable membrane, she created for the first time an apparent distinction between self and nonself, and an ill-conceived perception of the separation between that which is not really separate in nature. It is this misapprehension of her most unique and complex sentient life forms ever to emerge from the biosphere that has prevented us from understanding nature's deep design and the fundamental unity between brain, mind, behavior, and information.

As continued rigorous scientific inquiry has resulted in a unification of our understanding of electricity, magnetism, and finally light, we are confident that it will also unify our understanding of matter, energy, and information. Einstein's discovery that energy and matter are actually two different forms of the same entity, and that matter could be made from energy, yielded the famous equation $E = mc^2$. We suspect that this equation will have to be extended to account for the hypothesis that information, too, can be converted into energy and matter through the resulting property of their continued interaction — pattern — and that the absence of such pattern, eroded by entropy, is tantamount to the absence of information. If we wish to apprehend the mind of nature, then we must continue to seek understanding of this fundamental unity from a nature's-eye view, and not our own. History reveals that nature has evidently endowed us with such capacity.

THE HAND OF NATURE: THE PATTERN-ENTROPY PRINCIPLE OF MATTER AND ENERGY INTERACTION

Is it possible that all known biological and neurocognitive phenomena emerging from the interaction of brain, mind, behavior, and information can be explained, understood, influenced, and unified by a single fundamental property of nature? If so, this property of interaction would itself be a fractal, invariant of scale, context, or the elements of matter which it organizes, ubiquitous throughout nature from the very inception of the universe.

When did this dynamic interplay between pattern and entropy begin? Physicist Steven Weinberg (1977) estimates that during the first million years of expansion after the Big Bang, the universe cooled from 100 billion to approximately 3000 K. It was at this temperature that the first atomic element could be formed by the coalescing of a single electron and a single proton to form hydrogen, the simplest and most abundant element in nature, her first and most enduring pattern-based matter. From that moment forward, an estimated 15,000 million years ago, the universe continued to expand and cool, bringing forth, out of the conversion from energy to matter, all the stars, galaxies, planets, and our own earth. Fifteen billion years ago, at 3000 K, nature spoke her first word in the language of pattern.

As physicists become increasingly confident that they have reached the most fundamental particles of force and matter responsible for the assembly of the universe, evidence strongly suggests that the intrinsic tendency of those force and matter particles to counterbalance entropy by making pattern is the most fundamental process governing the interaction between matter and energy, invariant of scale, from the beginning of time. Since Einstein's discovery that time moves in a discontinuous fashion, physicists have become increasingly aware that time itself is an epiphenomenon, arising out of the interaction between matter and energy, and governed by the properties of pattern and entropy. Pattern-entropy dynamics appears to be the very process subserving information exchange throughout the biosphere.

It is quite tempting to conclude that the pattern-entropy principle of matter and energy interaction, as a descriptive model, has the capacity to resolve nature's most elusive, emergent anomalies, which have defied explanation. Somewhere between matter and energy, nature reveals its most fundamental property of interaction between matter particles and force particles; that property is the opposing tendencies toward pattern and entropy, without which there would be no universe to behold, no observer to marvel at its splendor, and no medium of information exchange between the two.

As science begins to apply the same rigor to the development of theory and experiment with the intention of studying the behavior of pattern and entropy as it has previously done with the study of matter and energy, we shall soon unveil the deepest of all principles responsible for the world outside and inside of us — the very hand of nature which guides the needle and thread of nature's great tapestry.

REFERENCES

Abeles, R., Frey, P., and Jencks, W. (1992). *Biochemistry.* Boston: Jones and Bartlett Publishers.

Bartfai, T., Iverfeldt, K., Brodin, E., and Ogren, S. (1986). Functional consequences of coexistence of classical and peptide neurotransmitters. In *Progress in Brain Research*, Vol. 68. Hokfelt, T., Fuxe, K., and Pernow B. (Eds.), New York: Elsevier Science Publishers B.V. (Biomedical Division), 321–330.

Becker, J. B., Breedlove, S. M., and Crews, D. (Eds.). (1992). *Behavioral Endocrinology.* Cambridge, MA: MIT Press.

Benyus, J. M. (1997). *Biomimicry: Innovation Inspired by Nature*. New York: William Morrow and Company.

Black, I. B. (1994). *Information in the Brain: A Molecular Perspective.* Cambridge, MA: MIT Press.

Bothamley, J. (1993). *Dictionary of Theories*. London: Gale Research International Ltd.

Briggs, J. (1992). *Fractals: The Patterns of Chaos*. New York: Simon & Schuster.

Dichter, E. (1971). *Motivating Human Behavior.* New York: McGraw-Hill.

Edelman, G. M. (1987). *Neural Darwinism: The Theory of Neuronal Group Selection*. New York: BasicBooks.

Fersht, A. (1985). *Enzyme Structure and Mechanism*, 2nd ed. New York: Freeman.

Fortey, R. (1977). *Life: A Natural History of the First Four Billion Years of Life on Earth.* New York: Alfred Knopf.

Hameroff, S. R. (1987). *Ultimate Computing: Biomolecular Consciousness and Nanotechnology.* Amsterdam: North-Holland.

Hebb, D. O. (1949). *The Organization of Behavior: A Neuropsychological Theory.* New York: John Wiley & Sons.

Kaandorp, J. A. (1994). *Fractal Modelling Growth and Form in Biology.* Berlin: Springer-Verlag.

Kandel, E. R., Schwartz, J. H., and Jessell, T. M. (1991). *Principles of Neural Science,* 3rd ed. Norwalk, CT: Appleton & Lange.

Kandel, E. R., Schwartz, J. H., and Jessell, T. M. (1995). *Essentials of Neural Science and Behavior.* Norwalk, CT: Appleton & Lange.

Kane, G. (1995). *The Particle Garden: Our Universe as Understood by Particle Physicists.* Reading, MA: Addison-Wesley.

Kendrew, S. J. and Lawrence, E. (Eds.). (1994). *The Encyclopedia of Molecular Biology.* Oxford: Blackwell Science Ltd.

Kleinsmith, L. J. and Kish, V. M. (1995). *Principles of Cell and Molecular Biology.* New York: Harper-Collins College Publishers.

Mandelbrot, B. B. (1983). *The Fractal Geometry of Nature.* New York: W. H. Freeman and Company.

Margulis, L. and Sagan, D. (1997). *Microcosmos: Four Billion Years of Microbial Evolution.* Los Angeles: University of California Press.

Nonnenmacher, T. F., Losa, G. A., and Weibel, E. R. (Eds.). (1994). *Fractals in Biology and Medicine.* Basel, Switzerland: Birkhauser Verlag.

Parker, S. (Ed.). (1993). *McGraw-Hill Encyclopedia of Physics,* 2nd ed. New York: McGraw-Hill.

Penrose, R. (1994). *Shadows of the Mind: A Search for the Missing Science of Consciousness.* Oxford: Oxford University Press.

Pribram, K. H. (Ed.). (1993). *Rethinking Neural Networks: Quantum Fields and Biological Data.* Hillsdale, NJ: Lawrence Erlbaum Associates.

Pribram, K. H. (Ed.). (1994). *Origins: Brain & Self Organization.* Hillsdale, NJ: Lawrence Erlbaum Associates.

Rieke, F., Warland, D., de Ruyter van Steveninck, R., and Bialek, W. (1997). *Spikes: Exploring the Neural Code.* Cambridge, MA: MIT Press.

Schrodinger, E. (1944). *What is Life? With Mind and Matter and Autobiographical Sketches.* New York: Cambridge University Press.

Shepherd, G. M. (1994). *Neurobiology,* 3rd ed. New York: Oxford University Press.

Stites, D. P., Terr, A. I., and Parslow, T. G. (1994). *Basic & Clinical Immunology,* 8th ed. Norwalk, CT: Appleton & Lange.

Tipler, F. J. (1994). *The Physics of Immortality.* New York: Bantam Doubleday Dell Publishing Group.

Volk, T. (1995). *Metapatterns Across Space, Time, and Mind.* New York: Columbia University Press.

Weinberg, S. (1977). *The First Three Minutes.* New York: BasicBooks.

Wilkie, W. L. (1994). *Consumer Behavior,* 3rd ed. New York: John Wiley & Sons.

4 Memetics: The Behavior of Information Patterns

CONTENTS

PATTERN, THE LANGUAGE OF NATURE

When the expanding universe cooled and the coalescing of an electron and a proton resulted in the formation of hydrogen, nature had spoken her first word in the language of pattern. As the universe continued to expand and cool, the conversion of energy into matter provided nature with a continuous flow of four-dimensional ink with which to record her expanding vocabulary. With this, her first 92-word vocabulary (i.e., the 92 stable elements) evolved into the language of pattern. Like all learners of language, however, nature, too, needed an eraser. That eraser was the opposing force of entropy, without which the language of pattern could never have evolved in complexity. With the entropy eraser in hand, nature could begin experimenting with the formation of larger aggregates: combining, dissolving, and recombining her atomic words into short molecular phrases. Her increasing dexterity with ink and eraser resulted in the relentless complexification of her language, the universe and life. Out of the interaction between matter and energy, information was born. Yet, early information was quite fragile. Fossil records reveal the mass extinction of many of Earth's life forms occurring approximately every 25 million years. Some physicists have speculated that this is due to our solar system's cyclical movement through the galaxy. Sometimes the stroke of nature's eraser is very broad.

In order for information to be preserved during the continued complexification of language, all languages demand that their aggregates obey certain rules or laws of combination. Without such laws, the balance between pattern and entropy would be tipped and the random aggregation of elements would result in the loss of information, namely, nonsense. Hence, all languages have a grammar. Pattern, too, has a grammar, a set of rules and laws prescribing and constraining the interaction between matter and energy, which appear to have remained consistent from her infancy. It is through this grammar that the continued complexification of the universe preserved and stabilized enough information to bring forth life. Through exhaustive survey of current scientific knowledge, nature continues to reveal a consistent, enduring influence of five grammatical laws: laws of aggregation and pattern interaction, namely:

- **Incorporation** (the absorption of one element, pattern, or aggregate by another)
- **Replication** (the duplication, reactivation, or copying of an element, aggregate, or process)
- **Cleaving** (the separating or dismembering of an aggregate into two or more parts/elements)
- **Recombination** (the aggregation and rearrangement of elements or larger aggregates)
- **Transmitting*** (the isomorphic propagation and exchange of pattern/information through and between elements, aggregates, or coordinate space)

WORDS, PHRASES, SENTENCES, AND THE LAWS OF GRAMMAR

Wherever we choose to look in nature, we will find one or more of these laws of aggregation and pattern interaction at work. While a complete survey could potentially

* Transmission is the propogation or displacement of a pattern-integrity within coordinate space, where information is conserved (i.e., rotation, inversion, translation, solition wave, expansion and contraction, enfolding, unfolding, retrogradation, vibration, and parity by inside-outing or mirror image reflection).

fill an entire encyclopedia, it is necessary to describe a few examples found at varying scales of interaction for the sake of clarity.

Life as we know it is carbon based. Carbon can form the stable, long, complex chains necessary for the processes of life. The formation of these chains is subserved by the behavior of pattern called recombination. To form carbon, radioactive beryllium (element number 4) must incorporate a nucleus of helium (element number 2). This aggregation of pattern by incorporation results in a carbon atom. Carbon is the fourth most abundant element occurring in nature and, by the same mechanism, is the essential building block in the assembly of 86 natural elements heavier than carbon (Abeles et al., 1992). Electrons (matter particles) within carbon and other atoms absorb (incorporate) and emit (transmit) force particles called photons (Kane, 1995; Parker, 1993).

Radioactivity is the spontaneous disintegration, or cleaving, of a chemical isotope into other kinds of atoms. This process is accompanied by the emission of radiation: energy that travels through space interacting with matter — transmission (Parker, 1993). Rhodopsin is a photon-incorporating pigment of vertebrate photoreceptor cells facilitating vision. RNA editing is the process of altering a base sequence of an mRNA molecule by incorporation, cleaving, and recombination of nucleotides, which takes place inside mitochondria and chloroplast mRNAs (Kendrew and Lawrence, 1994; Kleinsmith and Kish, 1995).

Mitochondria, the oxygen-utilizing organelles inside cells, are believed to have been at one time free-living bacteria in the biosphere, recently incorporated within cells by a process known as symbiosis. Recombination is the transmitting and recombining (exchange) of genetic information between two different DNA molecules occurring between two homologous chromosomes during meiosis or cell division. Meiosis also consists of DNA replication and the cleaving of a single diploid cell into four haploid cells. Memory B-cells in the immune system must incorporate (encode) the pattern of an antigen (foreign protein) through a process of recombination, and rapidly replicate themselves while transmitting (secreting) antibodies in the event of reinvasion (incorporation of a foreign protein). An antibody is a protein produced by lymphocytes that binds selectively (combines and recombines) to a specific antigen (Kendrew and Lawrence, 1994).

An antioxidant is a substance which donates electrons to free radicals (cleaving of an electron from one substance which in turn is incorporated by another — transmission). Neurotransmitters, neuropeptides, and neurohormones are transmitter proteins that are formed by combinations of amino acids, cleaved back into their constituents by enzymes, and then incorporated by reuptake gates and recombined for future transmission. Nerve impulse propagation is dependent on the carefully timed incorporation of ions. Subneural microtubules are elaborate networks of the protein tubulin, which spontaneously cleaves and recombines into different patterns in response to information transmission (Kandel et al., 1991; Pribram, 1993, 1994). Phagocytosis is the incorporation of particulate matter into cells via membrane vesicles, which cleave and separate by pinching off the plasma membrane. A protein is a macromolecule that consists of one or more combinations of polypeptide chains, each of which consists of hundreds or even thousands of recombined amino acids. A single length of DNA can be "read" by different cells from different starting points (cleaving), resulting in the production of completely different combinations of proteins (Kandel et al., 1991, 1995; Edelman, 1987).

On a larger scale, 100 billion neurons forming the human neural-net, with their fractally branching axons and dendrites, continuously cleave and recombine pattern processing pathways in response to stimulation of receptors and ion-gated protein channels by information (news of difference). In turn they transmit pattern-based signals intracellularly and extracellularly. Neural networks, subneural networks, and their interconnected pattern-encoding mediums also replicate patterns which were incorporated (perception) at an earlier time. When this occurs, we refer to it as memory or recall. Entire patterns can be replicated and then transmitted from one system to another, as when we see a new dance step (visual system) and learn the dance (motor system). This process is called translation.

Morphogenesis is a process by which cells become recombined and organized into larger aggregates called tissues and organs. Human digestion begins by the incorporation of food, and is followed by the cleaving of that food into its constituents for use in appropriate systems of the body. Digestion is an elaborate system of cleaving and incorporation, resulting in later recombination necessary to build new proteins and tissues. In every case, as we incorporate pattern, we incorporate order and information. Through the continual confluence of genes, memes, and interaction with matter and energy in our environment, mind evolves by the same rules of grammar. Through these five constraints we construct a model of the world.

Of course nature has not limited the use of these five grammars to events on earth. Some of the largest-scale examples of these pattern behaviors occur in space. One such event is referred to as galactic incorporation, a process whereby the gravitational fields of two or more spiral galaxies interact to pull all the stars into one system resulting in an elliptical galaxy. This incorporation process makes the newly formed galaxy more effective at replicating stars. Star formation is triggered when clouds of dust from both galaxies collide and form shock waves that ripple through the new system, assisting gravitational collapse of dense hydrogen clouds (Gribbin, 1993).

In addition to guiding the interaction of matter, energy, and hence information, these grammatical laws appear to be self-referent, in that they often appear to be operating in combinations (prescribed by the laws) to affect the same aggregates in nature, hence evolving the more complex functions and phenomena observed in physics, biochemistry, molecular biology, neuroscience, and the mind. It is with these five descriptions of interaction between matter and energy, mediated by the fundamental opposing tendencies toward pattern and entropy, that all complex aggregations and processes evolved.

Some evolutionary biologists might contend that DNA is the universal language of nature. Yet, seen from the broader perspective of physics, it is evident that the complex arrangements of these nucleotides are merely words and phrases which have evolved from the language of pattern, and were wholly dependent upon the existence of a grammar to maintain their stability and the information they encode. These are words, which only brought forth living arrangements of matter and do not account for all of the nonliving aggregations which came before it.

Numerous biologists agree that RNA, a less stable arrangement of nucleotides, must have existed before DNA, as nature tinkered with better, more stable methods of preserving information (Margulis and Sagan, 1997). Nature's journey toward complexification is surely far from complete. For just as we have seen an increase in her vocabulary from atomic words to molecular phrases, she will surely evolve sentences and paragraphs of increasing complexity and refinement. We human beings, our earth, and the universe we stand in awe of are the result of the first 15 billion years of nature's evolving, maturing thought process. We are the current expression of nature's thought, the newest arrangement born from the interaction of matter and energy; we are nature's precious diary.

Brain, mind, behavior, the information encoded by this four-dimensional aggregate of matter and energy, and all its emerging functions, have evolved from and continue to obey nature's language and inherent rules of grammar. The mind of man and the mind of nature are one and the same. It is an ill-conceived notion, born from limited perspective, that our skin somehow separates us from the forces and materials with which we were assembled. By developing a deep understanding and affinity for nature's grammar — her rules of interaction between matter and energy — we can finally begin to understand ourselves. It is with this intention that we expound on the five laws of nature's grammar and further elucidate the inseparable relationship between brain, mind, behavior, and information. To accomplish this, we shall continue to alter the reader's perceptual vantage point, from that of our own to that of nature's: matter, energy, and information. It is from this broader vantage point that the unification of brain, mind, behavior, and information can be apprehended and a scientific field of intervention evolved.

WHAT IS A MEME?

The term *meme* (rhyming with theme) was first coined by Richard Dawkins in 1976 in order to describe a unit of cultural imitation (Dawkins, 1989). A meme is a unit of information, an internal representation (an information pattern) whose existence influences events in the world, such that more copies of itself get created in other minds (Brodie, 1996). The meme's purpose is simply to make as many copies of itself as possible. Its main vehicle for making those copies of itself and reorganizing the very matter that makes up our physical world is the human brain. The meme is the newest breed of autoreplicator to evolve on earth (Dawkins, 1989); and its ability to rearrange matter, organic or otherwise, is far more rapid than that of its predecessor, the gene. What the gene has accomplished over millions of years, the meme can accomplish in a few centuries. Memes include beliefs, thoughts, images, ideologies, fads, slogans, ways of behaving, as well as cultural, religious, political and educational practices, traditions, etc. Memes are patterns of information, big and small, whose one purpose is to spread symbiotically or parasitically, co-opting the minds and bodies of human beings. Homeopathy and allopathy are memes. Christianity and Judaism are memes. Outcome-based education and the voucher system are memes (Brodie, 1996; Lynch, 1996; Plotkin, 1993; Taylor, 1996; Westoby, 1995).

In this chapter we shall explore the life cycle of a meme and the forces that influence it. We shall answer paradoxical questions such as: "Why do beliefs and practices that help people wither and die, while beliefs and practices that destroy people, gain followers and overtake them?" "How do people adopt erroneous belief systems that end in their own destruction?" "What hidden forces assist in the international spread of beliefs, religions, political views, healing practices?" "How do conflicting ideas and beliefs compete for the limited neural resources of the human brain?" "Why are some ideas so powerful that people are willing to kill or die for them, and how do they differ from other ideas that we can so easily discard as nature's equivalent of a paper plate?"

MANUFACTURING REALITY: UNCOVERING THE HIDDEN FORCES THAT CONTROL THE BEHAVIOR OF INFORMATION PATTERNS

Imagine a world in which information patterns behave as living organisms, controlled by forces hidden from the human eye, as well as the brain that stores and transmits them. Imagine a world where ideas, beliefs, religions, and scientific practices competed with each other for the limited space and copying resources of the human brain. Imagine a world where thoughts design themselves through evolutionary mutation, to become "fit" for survival and transmission to other brains. Imagine a world where you can clearly explain why destructive beliefs and practices survive and spread uncontrollably, while beneficial beliefs and practices and their transmitters are trampled by skepticism and ridicule. Imagine a world where a thought or idea, good or bad, safe or deadly, could be intentionally designed by its creators to co-opt the information, copying resources of the human brain in order to spread itself like an epidemic throughout a culture. Welcome — you have just entered the world of memetics.

Language is the process of associating internal and external sensory experience with arbitrary symbols (stimuli), such that we can reorganize our conscious experience into more complex and constantly evolving higher thought forms (Barthes, 1985; Whitehead, 1927). There is no absolute reality apart from our perception. The map is not the territory. From minute to minute, day to day, and year to year, we do not experience reality directly. Instead, we experience reality relatively. We experience reality through the model-making machinery of the human brain. Both our choice of language spoken and our arrangement of words within that chosen language arrange and destroy living information processing networks in the human brain by strengthening and weakening connective patterns within all its encoding mediums. It is from these constantly evolving connectivity patterns that we derive the models of reality we experience.

Numerous times in our history, the U.S. has been goaded into war and suffered the loss of hundreds of thousands of human lives over the televised public burning of an American flag. How is it possible that the public opinion of an entire country can be unanimously focused like a superorganism in the service of destroying human life from another country simply because of the burning of a piece of colored fabric seen on television? How is it possible for a burnt piece of colored fabric to symbolically enslave the massively complex and supposedly logical machinery of the human brain in such a way that a lifetime of religious training, societal laws, and cultural beliefs about the value of a human life can be abandoned long enough to convert that brain into a blood-thirsty murder machine? Why do healthy diets couched in sound medical reasoning go unused, while hundreds of new-age fad diets proliferate uncontrollably throughout the world, leaving a wake of malnutrition and susceptibility to disease? Why is it so much more difficult for good news to find its way into newspapers and television than it is for bad news? Why is it possible for some incompetent and unscrupulous practitioners in certain fields to earn enormous incomes, hiding behind credentials earned decades earlier, while the world abounds with many competent and trustworthy individuals who lack credentials and end up virtually unnoticed and broke? To answer perplexing questions like these, it will be necessary to illuminate some of the hidden forces governing the behavior of information patterns. Let us begin by defining a few terms and rules.

MEMES: SELF-REPLICATING INFORMATION PATTERNS

A meme is a unit of information whose existence influences events in the world, such that more copies of itself get created in other minds. The meme's purpose, from a meme's eye view, is simply to make as many copies of itself as possible. The meme's replication factory resides in the organic information incorporating tissue of the human brain. Thus, the human brain is the meme's primary vehicle for making copies of itself and reorganizing the very matter that makes up our physical world. To understand how information is capable of reorganizing matter and energy in the physical world, we must first recall from the standard theory of pattern-entropy dynamics that information is inseparable from matter and energy. Without pattern and a pattern-encoding medium made from the interaction of matter and energy, there is no information. We must also make a paradigm shift and see the human brain and body from the point of view of the meme — a pattern-incorporating medium born out of the continued interaction of matter and energy, capable of replicating, recombining, and transmitting pattern and the information encoded within. The human brain is nature's ideal "vehicle" for carrying out meme-mediated, organismic responses capable of physically altering the biosphere. To the meme, a human brain and body are like a computer peripheral whose soul purpose is to act out instructions contained within information patterns. To deepen our understanding of the hidden forces that influence the spread and competition of such memes, we have to begin by defining several types of memes.

DISTINCTION MEME

What is so astonishing about memes is their ability to manufacture the reality experienced by a human brain, while remaining completely invisible and inaccessible to daily human awareness. A *distinction meme* is the most pernicious of all, being abstractions of sensory experience constructed from language designed to focus attention on selected sections of reality (Brodie, 1996). The function of a distinction meme is to cleave and separate patterns of information in preparation for later recombination (Spencer-Brown, 1994). This can be biologically equated to one of the same important functions of an enzyme. Distinction memes come tightly packed with presuppositions difficult for the conscious human mind to unravel or even notice. An interesting example is that of "learning disabilities."

Throughout recorded history, literature has abounded with presuppositional relationships between mentor and student. As far back as ancient Greece, we can see clearly stated presuppositions relating to these relationships. The act of educating (to rear or the transmitting of knowledge from one mind to another) was a responsibility that resided with the educator. This distinction held true for more than 2300 years, until the accidental mutation or intentional design of a distinction meme called "learning disabilities." This term began to receive wide acceptance throughout the 1970s and 1980s, and finally became a household term in the 1990s. The evolving of the learning disabilities meme is quite pernicious in that few parents have ever questioned its purpose for existence or its validity. This linguistic abstraction not only goes unquestioned, but at the same time it often invisibly and immediately transfers the responsibility of education (transfer of knowledge) from a credentialed and experienced educator to an innocent 6- or 7-year-old learner. It accomplishes this transfer of responsibility so quickly, completely, and invisibly that it is somewhat unsettling to the educated mind. The intentional or accidental transmission of this distinction meme has radically rearranged and reorganized matter and energy, eroded the educational system, altered family interaction, created new jobs and tests, and has redistributed wealth so profoundly that it has affected five enormous sectors of our culture: education, medicine, psychology, family, and business. And it is far from finished.*

Today there is hardly a school left in the U.S. that does not have special classes for learning-disabled students. Government provides for special funding whenever a class of learning-disabled students can be filled. Thus, educational institutions realize an increase in financial resources simply by accepting, embracing, and spreading to parents and children the notion (distinction meme) of learning disabilities. Because of this financial reward for "believing" (i.e., acting as if it's true), as memetic engineers we would say that the term learning disabilities is memetically advantaged over competing ideas such as poor teaching and overcrowded classes, which would cost more money to correct. Proponents of opposed ideas will find themselves memetically disadvantaged, as the acceptance of their ideas does not come packaged with a financial reward, but instead incurs financial punishment. The dogged persistence of such a child-damaging idea is memetically advantaged for one fundamental reason: an organismic permutation of the least energy principle of atomic arrangement of elements — maximum gain for min-

* While we argue against the dangerous proliferation of the learning disability meme, we also recognize that there are many children and adults who have not incorporated the necessary cognitive strategies that facilitate the learning of certain kinds of subject matter. An educational system is impaired if it rigidly adheres to certain types of programming that cater to students with learning styles within a certain range, disadvantaging those who have less prevalent styles. Also, a system that chooses to label rather than teach in the students' style and assist students in developing effective learning strategies is employing the learning disability meme in a manner which concerns us.

imum effort. Memes that make use of this fundamental principle of interaction between matter and energy, especially that of the human brain, will be memetically advantaged for transmission and incorporation. In other words, if you pay somebody to remember an idea and believe it, the idea becomes memetically superior in its race for limited brain resources.

Another event that memetically advantaged learning disabilities (utilizing the least energy principle) was the invention of testing and screening — *strategy memes* performed by psychologists, who can also receive an income for believing in learning disabilities. Many psychologists make several hundred dollars per testing battery. So what would be the advantage of not believing in learning disabilities? Testing and screening are easier than educating effectively, and they produce additional income (maximum gain for minimum effort).*

Once psychologists are rewarded for believing in and locating children with learning disabilities, and the educational institutions are rewarded for segregating learning-disabled children into special classes, then the pharmaceutical industry gets its turn to choose whether or not to believe that learning disabilities exist. If pharmaceutical companies believe it, they get to design drugs like Ritalin that are purported to control learning disabilities, resulting in a lifetime annuity from generations of drug-dependent students for the companies.** If the pharmaceutical companies choose not to believe in learning disabilities, they would lose billions of dollars. Here, again, the epidemic-like spread of the learning disabilities information pattern proliferates a third sector of our culture, leaving behind it a massive wake of economic redistribution. So if educational institutions, pharmaceutical companies, and psychological examiners get paid to believe in learning disabilities, there are now three credible, unquestioned cultural institutions that can rapidly influence the continued spread of this distinction meme to the general population: parents and children. How do they do this? For the most part, parents and children are not financially rewarded for believing in learning disabilities. Yet their belief in them is essential for the other three institutions to receive their reward. The answer: the strategic placement and removal of guilt and fear by use of *association memes.* In addition, more children today have two working parents with little time to assist their children in learning and study at home, and they feel guilty if they maintain the belief that returning to full-time parenting could correct the problem. Since the call of mounting bills is the more urgent of the two behavioral choices, perceiving the locus of learning disabilities as being within the children, while crediting a quick-fix pharmaceutical with the solution to the problem, seems the easiest alternative. Here again we see the least energy principle making conditions ripe for the continued incorporation and transmission of the meme.

---

* The authors recognize that psychological testing can provide a valuable service when properly used. Certain standardized instruments can help to define strengths and weaknesses, assisting in the development of an individualized plan for the student.

** While Ritalin and other psychostimulants are often used to reduce hyperactivity and improve attention, these compounds are no substitute for assisting learners in incorporating effective learning strategies. Additionally, there is considerable controversy about the utilization of these drugs, since improved attention and reduction in hyperactivity often have been achieved through nutritional means, as well as through treatment for allergies and other substance sensitivities.

Here are some of the most important things to remember about distinction memes. They are generally abstractions of the sensory world, a cleaving of a sensorially incorporated pattern made possible by language. A distinction meme is like a knife that slices up reality into organized packages, which are endowed with tightly packed presuppositions that channel, assist, and sometimes disable a human brain's ability to create voluntary recombination (reaggregations that subserve thinking) of incorporated pattern. Hence, they restrict perception and model building, while prescribing certain sets of "appropriate" feelings and behaviors (appropriate for the sake of the meme's own replication). Distinction memes are among the fundamental building blocks of human perceptual filters (Whorf, 1995; Korzybski, 1933, 1994).

There is an interesting relationship between presuppositional language patterns and subjective experience. The more presuppositional language patterns you can pack into a brief communication (i.e., paragraph, phrase, sentence, or word), the more difficult it is for the human mind to think critically about the statement. Some linguists have held that if you construct three sentences in a row which contain the same basic, hidden presupposition, that presupposition enters the mind outside of awareness and unquestioned by critical faculties of consciousness (Rushkoff, 1994). Through experimentation, we have learned that the two most important factors in the construction of unquestioned, easily incorporated presuppositions are *presuppositional density* and *confusion.*

PRESUPPOSITIONAL DENSITY

Presuppositional density (Furman, 1998) is simply the number of undetected presuppositions contained per given unit of information pattern (presuppositions per length of linguistic phrase). Naturally, these presuppositions must be directing attentional processes in the same direction in order to be effective. An example would be a salesman who says, "When you take delivery of this dining room set on Tuesday, you will need to give the driver the C.O.D. in a check because they can't carry cash." If this sentence is timed correctly in a presentation, during a point of high-emotional involvement with the dining room set, your mind automatically accepts the fact that you have made the decision to buy it. Now it is just a matter of whether you will be home on Tuesday to receive it and remembering to write the check out in advance for the C.O.D.

The interesting fact about presuppositions is that the higher the density, the more difficult it is to critically evaluate them. High density causes presuppositions to evade conscious awareness. While many educated consumers today could critically see through the last example and keep the decision to buy under their conscious control, this ability to think critically breaks down as we further increase the density of presuppositions. When a single word or term contains sets of presuppositions, simply hearing, incorporating, and recalling the term embed the presuppositions contained, without the brain's awareness. For this reason distinction memes can become the most disruptive to autopoietic functions of mind and brain.

The term "learning disability" contains several effective presuppositions. Just learning the word and accepting it as real abstracts sensory experience (cleaves internal representations) and reorganizes reality in such a way as to place the responsibility of knowledge transfer in a classroom setting on the learner, instead of the teacher, even though for over 2300 years cultural beliefs and practices have dictated otherwise (Barthes, 1985; Spencer-Brown, 1994). The term "learning disability" also contains another presupposition that the brain must accept simply by understanding the term (Burke, 1989; Eco, 1976). That is, when transfer of knowledge does not occur between teacher and a student, not only is it the student's fault, but that student is also the not-so-proud owner of a defective, disabled brain.

COGNITIVE CONFUSION

The second factor facilitating the undetected incorporation of a harmful distinction meme is confusion. Through the study of Ericksonian hypnosis, practitioners have learned that an *embedded command* can be transmitted to, and incorporated in, the mind of the listener, going unnoticed as long as the mind has been adequately prepared with linguistic confusion just prior to the embedded command (Erickson, 1989). Listening to transcripts of hypnotists using this technique, you will find this process normally stretching over two or more sentences (low density). Again, what makes it almost impossible to recognize is high density. When a phrase is designed as an oxymoron, it creates immediate confusion — an increase in entropy facilitated by conflicting information patterns — and prepares the brain for acceptance of the phrase. This forms a conceptual slot — a tension to return to order — which assists the brain in the construction of a new category of reality—a return to pattern and order — from that point forward. A conceptual vacuum is created, waiting to be filled with strategy memes and aggregated by association memes (Brodie, 1996; Key, 1989; Rushkoff, 1994; Lutz, 1996). Sweet sorrow, deafening silence, and mournful optimist are examples of oxymoron. While not as readily apparent, learning disability is also such a term.

The memetic advantage of an oxymoron is that it is easy to remember (replicate) and hard to forget. The brain's attentional processes become compulsively directed to resolve the paradoxical ambiguity (causing an increase in neurocognitive entropy) and return its living information-processing mediums to pattern and order. As a result, it can be very "catchy," like a virus. Biological viruses work in a similar way, gaining access to our replication machinery by confusing our pattern-based filters for identifying foreign proteins and incorporating appropriate substances within our cellular membranes. The oxymoronic construction of the distinction meme is like the protein shell of a virus which allows that virus to enter our own cellular structures unnoticed. The advantage of such a protein shell is that the virus, once inside, can coopt our cell's copying machinery in order to replicate itself. When copying commences, many other normal cellular functions are subordinated in the service of "virus replication," resulting in innumerable copies of these chemical information patterns. Distinction memes operate in the same fashion. It is no accident that memetic engineers utilize principles and models from epidemiology in order to predict the spread of information patterns and contagious thought viruses (Lynch, 1996).

STRATEGY MEMES

Strategy memes are another class of information pattern used for model making and our brain's manufacture of reality. Strategy memes are generally rules-of-thumb that tell you what to do when you encounter certain situations and events (stimuli). We might also refer to strategy memes as heuristics. Their basic purpose is to specify or prescribe behaviors to be mediated by the brain (the meme's ambulatory peripheral) in response to certain stimuli (Brodie, 1996). In this way, information patterns are able to sprout "arms and legs," capable of rearranging physical matter and energy in the external environment as well as within the internal environment. It is useful to think of strategy memes as incorporated and transmitted *patterns of response* to internal and external stimuli. Strategy memes also contain undetectable presuppositions about cause and effect. Hence, they also assist in the construction of unquestioned belief systems used by humans to make sense out of their world and organize subjective reality.

Since strategy memes are behavioral manifestations of a belief in a certain cause-and-effect relationship, when you have incorporated a strategy meme, you may unconsciously believe that behaving in a certain way, as prescribed by the meme in the presence of a given stimuli, will produce a predictable effect. To become a successful strategy meme, that behavior must trigger a chain of events that results in the spread of the strategy meme to another mind. Hence, when you are crossing a busy intersection (stimulus #1) and you see a fast-moving vehicle coming toward you (stimulus #2), if you increase the frequency of your movement to the other side of the street, you live to pass on the meme. The strategy meme here specifies a cause-and-effect relationship between an acceleration of walking behavior and physical safety. This strategy meme — a perfect unit of cultural imitation — spreads itself rapidly throughout metropolitan areas. It is important to realize that memes are not inherently good or bad, right or wrong. They are simply information patterns that are either fit for survival or not (Calvin, 1996; Plotkin, 1993; Taylor, 1996). Strategy memes that make the consequences of our actions more predictable, spread quickly. Predictability is another way of saying that we see a pattern or an order in the environment, hence we pattern our response in order to correspond with and navigate through that environment. Predictability is synonymous with the reduction of neurocognitive entropy and, hence, ambiguity, uncertainty, and fear (Csikszentmihalyi, 1993).

One simple example of a strategy meme is "When you come to a red light while driving in Florida, and you want to turn right without causing an accident, come to a full stop first, look to your left, then to your right, and if it's all clear, turn." An even simpler example is "When you see a police car while driving, slow down." This strategy meme is so powerfully ingrained in drivers that you will commonly see them slow down even when they are not exceeding the speed limit. Yet, this strategy meme is not even explicitly taught during driving school. Instead, it is instantaneously copied by drivers who notice other vehicles slowing down in the presence of a police vehicle. Strategy memes are incorporated because they help us predict and control environmental consequences to our actions, thus reducing neurocognitive entropy.

These are simple examples of a strategy meme, and not very dangerous. Yet some of the greatest atrocities in human history can be attributed to these simple memetic building blocks. From birth, many of us incorporate a belief that the taking of a human life is the ultimate sin. We are educated by our parents, educational institutions, religious institutions, and our government, that killing is just plain wrong under any circumstances — until the pernicious building block called a distinction meme is reaggregated with an opposing strategy meme in a single human brain. Consider the political comedy, *Wag the Dog*, for numerous illustrations of how this process can be intentionally engineered.

The innocent meeting of two information patterns in the brain can create an instantaneous conflict in our current model of the world and a tear in the very fabric of our reality. One such distinction meme is war. In the presence of this new event/stimulus, the same institutions that proclaim that killing is sinful not only advocate the taking of human life for a "higher cause" or a "greater good" (distinction memes themselves), but also specify rules-of-thumb (heuristics) to do so effectively (without destroying plants, animals, and buildings). Military personnel go through rigorous training to learn how to kill other human beings effectively, rewiring their nervous systems with new strategy memes and association memes to prepare for the killing fields.

What happens to human brains that have effectively incorporated detailed instructions, behavioral strategies, and appropriate emotions to carry out continual acts of human death and destruction of the "enemy" (also a distinction meme) after the "war" has ended? Answer: Posttraumatic Stress Disorder. Our military-trained brain reenters the "civilian life" distinction meme. In the context of this new perceptual distinction, "civilian life," a brain and body trained rigorously to maim and kill must now incorporate new strategy memes and emotional responses for relating to people and potentially threatening situations. If that military brain is not flexible enough to clearly separate previously incorporated models of reality and reaggregate appropriate distinction memes with complimentary strategy memes, maladaptive behavior will occur (Korzybski, 1933, 1994); and because of the neuromuscular system's previous training as a death and destruction peripheral, society may judge these actions differently. Two movies depicting this paradoxical confusion resulting from conflicting information patterns are *Con Air* and *The Manchurian Candidate*. In such cases new distinction memes like "civilian life" have not effectively cleaved information patterns in preparation for reassociation (recombination).

ASSOCIATION MEMES

While strategy memes are in essence behavioral heuristics formed through the principles of operant conditioning, *association memes* are higher-order stimulus–response bonds or anchors formed through Pavlovian or classical conditioning. Association memes are the third class of memes implicated in the construction of our model of the world (Brodie, 1996). By adding this third building block, the manufacture of almost any reality seems possible for the creative memetic engineer.

Association memes simply perform the task of linking or aggregating two or more other memes in your mind. If a distinction meme cleaves an incorporated sensory pattern, thus activating a new category, perceptual pattern, or internal representation in your mind, and a strategy meme activates sets of appropriate behaviors in the presence of that distinction meme, then an association meme simply functions as "meme glue." Association memes activate arbitrary linkages between information patterns in the same way that an enzyme can catalyze a reaction between two chemical elements. Its extraordinary capability is that of taking any two or more existing memes and aggregating them together as a single pattern, arbitrarily based on the outcome wanted by the memetic engineer (parent, educator, therapist, advertiser, religious or political leader, etc.). In essence, the memetic engineer sits in front of his or her reality editing machine and uses association memes to splice information patterns together. It is unsettling to imagine what could be assembled with such a tool kit. If you live in the U.S., you may remember this one. One of Chevrolet's most famous advertising campaigns went like this:

Baseball, hot dogs, apple pie, and Chevrolet

Since when do these four distinction memes go together? Ever since Pavlov got his dogs to salivate to the sound of a bell, the advertising industry has never been the same.

If you find yourself responding to a stimulus pattern with a somewhat unusual, inappropriate, or out-of-character behavioral pattern, your brain may have incorporated or recombined such an arbitrary linkage in information patterns. Association memes can facilitate the incorporation, replication, and transmitting of strategy memes and distinction memes, and in doing so, replicate themselves. Association memes are commonly utilized intentionally by advertisers to link feelings, that you enjoy experiencing, to internal representations of stimuli, the product they are representing (McCracken, 1990). So if you find yourself at a baseball game eating hot dogs and apple pie and imagining yourself driving home in a brand new Chevrolet when you know you came in a BMW, beware — a memetic engineer has been tinkering with your brain.

If you find yourself in need of an association meme housecleaning, you must begin by asking yourself how your unquestioned and unconscious associations were formed.* Who decided that drinking Budweiser would mean having fun with sexy people? Who decided that Mercedes and BMW mean status (association meme)? What is status (distinction meme) and why does a higher status feel good (association meme)? How are you supposed to behave if you have status or if you don't (strategy meme)? It is through these arbitrary or sometimes intentional linkages in information patterns that our sensorially incorporated model of the world can be enriched or destroyed.

Memes should not be confused with anchors. Memes are far more complex; they are self-replicating, recombinant, transmittable building blocks of mind, brain, and reality, affecting thoughts, feelings, behaviors, and the very meaning of life itself. Memes differ from anchors in at least one very critical way. Memes are

---

* In Section II, the reader will find NeuroPrint useful in modeling the relationships among strategy memes, distinction memes, and association memes and the internal models they create.

information patterns (units of cultural imitation and transmission) incorporated by neurocognitive encoding mediums, whose existence influences events, such that more copies of themselves get created in other minds. Once an effective meme or meme complex has been established, it behaves like a living thing (bio-pattern dependent), traveling through living information processing mediums, independent of its creators. Information patterns, once incorporated into the nervous system, translate into electrochemical, atomic, molecular, cellular, protein, cytoskeletal patterns, etc., that in turn prescribe thoughts, feelings, beliefs, and behaviors, bearing little functional difference from those information patterns encoded within genes, which are sets of chemical instructions prescribing the structure and function of the human brain and body. This is why memetic engineers and researchers use epidemiology models to predict the spread of a meme or meme complex. Once a meme has been given life through a pattern-encoding medium, its primary function is to copy itself. Its spreading behavior is totally independent of its original creators. Memes ensure their own survival and replication by taking advantage of, and adapting to, the idiosyncrasies intrinsic within human brain function: our biological needs and our genetic instructions.

DESIGNER INFORMATION VIRUSES

What would happen if ideas created you instead of you creating ideas? Do people really adopt beliefs or do beliefs adopt people? Do human brains really have control over which memes become dominant and co-opt their information processing resources? Can self-replicating information patterns really become parasitic to their host to the point where they program the host's destruction so that they can go on and continue to replicate from brain to brain? If you earn your living as a memetic engineer, the answers to these questions are quite clear. Human beings cannot control that which they do not know exists (Key, 1993). When we accuse our children of acquiescing to the latest clothing fad, do we really understand the mechanisms by which designer memes and information viruses enslave their free will? When we say their grades in school have dropped because of peer pressure, do we really understand how the very structure of this pressure prepares the brain for penetration and reproduction of meme complexes that program academic failure?

Have you ever known someone who purchased a new car, an expensive one, such as a BMW or a Mercedes? Did you notice drastic changes in their thoughts and behavior? How much did their driving behavior change to avoid getting a dent? How much farther did they walk out of their way in order to park the car safely away from all elements that could mar its new paint? How much more energy did they waste yelling at their children for bringing things into the car that could ruin the new upholstery? How much more money did they spend on premium gas and insurance to protect it from theft? How many times per day during work were their thoughts unexpectedly invaded by images of their car being broken into and stolen? If you think about this from a meme's eye view for a moment, at what point does the car own you? For that matter, at what point do we lose control of any idea or belief such that the balance of control tips in favor of the information pattern? Maybe phobias come to mind. Phobias and compulsions are vivid examples of self-replicating information patterns gone wild. At this point we would like to begin to explore how the intrinsic structure and organization of a human being provides fertile breeding ground where simple memes can grow into pernicious and even deadly information viruses. Let us first review and summarize a few key points:

- A meme is a unit of information whose existence influences events in the world such that more copies of itself get created in other minds. A meme's primary function is simply to make as many stable copies of itself as possible.
- One useful way to classify memes is by their function: distinction, strategy, and association.
- While we explored them separately, it is important to recognize that memes gain their power by splicing themselves together into aggregations that make the most efficient use of our basic biological drives and neurocognitive function.
- Self-replicating information patterns must make use of our biological tissues in order to find the most efficient methods of storage, reproduction, and transmission from brain to brain.
- The first principle determining the efficiency of memes can be stated as follows: information patterns that are most easily and efficiently incorporated, replicated, and transmitted will be memetically advantaged and thus prevail over all other competing memes.
- Distinction memes are simply abstractions of sensory experience constructed from language or other symbol systems like music notation or mathematics, designed to focus attention and perception on selected portions of reality. Their function is to cleave or abstract a portion of an external stimulus field or internal sensory experience (internal representation). Distinction memes allow for the formation of representations in the brain and create categorical separation of sensory stimuli. For the brain to embrace a distinction meme, its sensory receptor cells must first be stimulated; the brain must form an internal representation (incorporate the pattern). Second, the pattern must be recollected (replicatable). When we remember, we literally re-collect representations in specific orders and combinations. The hidden force that controls which representations are recollected together is the association meme.
- Association memes primarily control meaning. While distinction memes cleave and separate patterns into individually replicatable units by defining boundaries around sensory stimuli, association memes splice together these representations (information patterns) via classical conditioning (anchoring), such that different meanings can be apprehended by us when we perceive or remember a given distinction meme. Association memes can splice any two or more memes together into one internal representation. While association memes are able to control meaning, they are also able to splice together distinction memes with behavior patterns (strategy memes). In this way they can specify the function or behavior of the organism they inhabit. Association memes are themselves strings

of distinction memes that facilitate the reaggregations and recombinations (rearrangements) of cleaved information patterns.

- Strategy memes translate strings of distinction memes into four-dimensional patterns called function or behavior (microscopic and macroscopic), within the organism that incorporated them. They can be thought of as instructions, directions, or rules-of-thumb that tell you what to do when you encounter certain external stimuli or internal representations. Strategy memes are heuristics that specify organismic function; they are the meme's instructions for their ambulatory peripheral. In this way, information patterns such as distinction memes, beliefs, values, and ideologies are able to "sprout arms and legs" capable of rearranging physical matter in the external world.

It is through this power of arbitrary linkage that memes become super organisms like belief systems, religions, political ideologies, phobias, compulsions, theories, branches of knowledge, etc. Of course, memes are transmitted from brain to brain through communication. But to truly understand the world from the point of view of an information pattern, we need to broaden our definition of communication. Spoken and written language is only one form of human communication. In essence, communication can be thought of as any calculated stimulation of another conditioned organism. In the world of self-replicating information patterns, things mean what they cause.

Let us now elaborate more on the first principle. Information patterns that are most easily and efficiently incorporated, replicated, and transmitted will be memetically advantaged, and thus prevail over all other competing information patterns (memes). Let's look at incorporation first.

INCORPORATION

For an information pattern to gain entrance to the biological tissue called the brain, and subsequently result in the evolution or devolution of mind, the information must first stimulate collections of pattern-encoding brain cell assemblies and related pattern-encoding mediums strongly enough for existing patterns to change, resulting in the formation of new information-processing pathways. There are at least four ways to accomplish this:

1. Through stimulus intensity.
2. Through repetition. (Both intensity and repetition are encoded by brain cell assemblies in similar ways.)
3. By widening the basin of attraction generated by the stimulus. To accomplish this, the stimulus must be packaged in such a way that it can be sampled simultaneously by as many sensory receptors and representational systems as possible. The more neocortical, subcortical, and subneural systems that carry a representation of the stimulus (incorporate by changing pattern), the more likely and easily the information pattern will be recollected (replicated) without interference from other information patterns.

4. Use of an association meme to splice a new information pattern with one that is already dominating brain activity (easily replicatable). This method can establish an unprecedented efficiency of recollection via information compression.

Let us expound on each of these.

Stimulus Intensity: Why is it possible for a phobia to so powerfully and completely control the thoughts and behaviors of a human being, such that all other desires and goals (also memes) can be subordinated? Think of a person with such a severe flying phobia that all family vacations have to be within driving distance. The occupation choice must allow him/her to remain local. And the unexpected death of a dear relative across the country leaves the flying phobic trembling in conflict while attempting to generate acceptable rationalizations for missing the funeral entirely. One of the most obvious factors at work here, giving fear of flying memetic advantage over all the actions of this organism/human being, is intensity of stimulus. A phobia is thought by many to be a one-time learning (if only mathematics could be learned so quickly).* It is perceived this way in part because we are unaware of all the memetic advantages this information pattern possesses. While it would literally take volumes to model all of these, let us explore a few.

Most flying phobias are caused by either an intense direct experience with flying (incorporated by all available encoding mediums) or an intense representation made as a result of someone describing an intense experience either through a conversation, reading a book, or watching a movie. The intensity is caused in part by the linkage of sensory information (plane in flight) with an intense neuroendocrine response of adrenaline, as well as cascades of numerous other related stress hormones. Of course the effect of this is made even more powerful through an association meme that says "plane crashes result in death." It should be noted that any memes that assist in survival of the organism (protect its host from danger) tend to be memetically advantaged for incorporation and transmission over memes that do not.

Repetition: Another way for information patterns to gain memetic advantage is through repetition. Phobia-producing stimuli, by nature of their intensity, are easily replicatable and result in continuous involuntary reactivation by neurocognitive pattern-encoding mediums. Whenever a phobic subject considers taking a plane flight and giving into logic, spontaneous reactivation of the meme complex strengthens the chemical and atomic bonds, stabilizing the pattern and resulting in the formation of energy barriers that preserve the information processing pathways in the brain and, hence, the information itself. This in turn results in easier, faster, and more efficient future reactivation. An important principle to remember is that whenever two information patterns compete for neurocognitive resources, the one that is reactivated (replicated) faster, with the least amount of energy expenditure, will dominate. Unfortunately for the phobic subject, each repetition (replication) of the internal representation insures domination of that information pattern, making it

* It should be noted that the genesis of many phobias is not consistent with an associative, stimulus–response model. Many appear to be inherited, which does not preclude learning "from a distance."

more resistant to interference from a competing pattern. Replication increases synaptic effectiveness.

Prior to the days of written (literate) culture, oral cultures made use of this knowledge to train people in the principles of their religion. Those religious organizations that required prayer several times per day had a memetic advantage over other religious organizations praying once a day or less. This is simply because those who prayed five times per day were able to remember more easily what they believed and how to act in response to that model of the world (Lynch, 1996). Most of us also remember the multiplication tables and the alphabet as a result of the same process.

Widening the Basin of Attraction: Widening the basin of attraction of an information pattern simply means designing it in such a way that multiple sensory systems can sample it simultaneously, thus encoding a richer representation with greater chances of reactivating without interference. This makes it easy to understand why a movie or a commercial can more easily control recollection of representations than a spoken message. Imagine trying to logically explain the statistical safety of airplane travel over automotive travel to a subject who is phobic of flying. Instead of building an information pattern that can compete memetically against the rich multisensory representation of the bad flying experience, you may inadvertently splice the automotive travel meme neatly to the flying meme and end up with a person who will not leave home to travel anywhere. While this is statistically unlikely, it is not impossible.

Use an Association Meme to Splice a New Information Pattern with One that is Already Dominating Brain Activity: Advertisers use this type of memetic engineering every day in attempts to control mass buying behavior (Key, 1989; McCracken, 1990; Robertson and Kassarjian, 1991). A brand new commercial makes a nice clean example. Here's how it was done. The camera pans in on a traffic jam during morning rush hour. Sequentially, the camera captures the images of three drivers and the cars they are driving in, each driver donning a disgusted expression on his face, clearly upset. The camera frame is open wide enough to show each driver stopped dead in bumper-to-bumper traffic. Then the volume is turned up so that you can literally hear their internal thoughts.

The first person thinks in a fearful tone, "Did I leave the iron on? Will I find my house in flames when I get home?" (unconsciously associating driving in this vehicle with a representation of the loss of the viewer's home).

The second person thinks in a regretful tone, "I think that letter to my boss was too strong. I'm probably going to be fired." (The association meme splices driving in this vehicle with the viewer losing his source of income.)

The third person thinks in a disgusted tone, "I can't believe I wore this, it makes me look so fat!" (The association meme here splices driving in this vehicle with losing the positive regard and respect of the viewer's friends, family, and co-workers.)

Then along comes the camera again and pans into a beautiful new Honda Accord (distinction meme). As the camera pans in toward the driver, you notice he is the only good-looking driver (distinction meme again). Wearing his designer sunglasses (distinction meme), he cranes his neck, not to look at the traffic, but instead to smell the cool, crisp air before he thinks to himself with a grin of delight,

"What a beautiful day. I think I'll take the day off." During all of this, the camera frame never opens up large enough for you to see the distinction meme, "Honda," represented with the distinction meme "traffic jam." Finally, the camera fades back showing the beautiful new Honda, all by itself, driving off the exit ramp to freedom. In this commercial, the distinction meme, "Honda," is associated with looking good, feeling good, being free, and feeling secure enough to take a day off without notice and still keep your job. What does losing your house to fire, losing your job, and being fat and disliked by your friends have to do with the kind of car you drive? Absolutely nothing, until memes are strategically recombined into complexes, designed to be incorporated "whole."

WHY WE INCORPORATE INFORMATION (NEWS OF DIFFERENCE)

The question of why we incorporate any information at all is an important one that can be answered from many different scientific platforms. In Chapter 2, we surveyed a wide variety of studies relating to the cognitive and biological consequences of the absence of information, the excess of information, and the presence of conflicting information. Considering these studies, combined with the brief survey of biological pattern/information-encoding mediums discussed in Chapter 3, it seems incontrovertibly clear that we, as well as the rest of nature's creations, large and small, require a constant and delicately balanced flow of information (pattern/order) in order to counterbalance the dismembering, disordering, cleaving effect of entropy. Without the continued, balanced ordering effect made possible by the incorporation of external pattern, and hence information, through our circuits of interface with the external environment, we would not only fail to correspond appropriately with our external environment, resulting in untold mayhem, but we would also seek the necessary ordering effect of information internally, ultimately resulting in the chaotic iteration or what is known as *positive feedback*.

To get a sense of the damaging, disorganizing effect of positive feedback between interconnected elements, you could take a television camera that is connected to a television monitor, and instead of allowing the camera to incorporate scenes (patterns and information) from the external environment, point it back toward the monitor so it is incorporating from the monitor only the result of the pattern it sends to the monitor. In a surprisingly short time, the result is complete dismemberment of the originally ordered, neatly arranged picture. Now imagine if that monitor were a brain, whose effectors had to coordinate collective action from that information in order to behave appropriately in the world around it. In Section II, when we discuss NeuroPrint, the topological model for representing neurocognitive activity, we shall gain a sense of how readily discrete positive feedback loops such as this can be located and corrected.

Our need to incorporate information becomes most apparent during extreme conditions occurring at opposite ends of the pattern-entropy continuum. The condition described above becomes most apparent during times of sensory restriction and sensory deprivation. Why is this? The second law of thermodynamics clearly warns us of this effect, as the law of entropy states that entropy can only increase in an isolated or closed system. As the brain, our organismic interface with the environment, becomes isolated from or closed to the exchange of information with the environment, disorder must continually increase according to the law of entropy; and, in fact, it does just that. As we shall see in the next chapter, even minor restrictions in information flow can result in serious consequences.

There is a second important reason for why we are compelled to continually incorporate information. Its mechanisms lie on the opposite side of the pattern-entropy continuum. While it is clear that we incorporate information to counterbalance entropy and its disordering effects, we must also incorporate new information in order to counterbalance a condition of too much internal order and regularity.

Theoretical physicist Jacob Bekenstein employed the mathematical apparatus of quantum statistical mechanics to show that quantum systems have only a finite number of possible states (Tipler, 1994). Through exhaustive calculation, Frank J. Tipler, professor of mathematical physics at Tulane University, showed that the human brain is, in fact, a finite state system having $10^{10(45)}$ possible quantum states. While it is true that this is an almost inconceivable number of possible states, it is still finite. Tipler goes to great length to show that information stability requires such systems to have a finite number of states. In order to understand why information stability requires all possible combinations of states to be finite, you must ask yourself how memory would be possible without such conditions. The act of remembering is actually the act of replicating a previous state, originally caused by the agglomeration of sensory patterns during perception. How could we remember information if we could not return to a previous state (people who have sustained hippocampal damage exhibit this inability)? Not only is the total number of possible states finite, but the rate at which information can be processed by the brain is also finite at 4×10^{53} changes of state per second. While both of these numbers are astronomical, in practice macroscopically experienced neurocognitive states never approach this total potential or rate of information processing (Tipler, 1994). Human beings rarely report experiencing more than 45 to 50 discrete neurocognitive states over any given 7-day period. Quite sadly, it appears that our environmental conditions greatly restrict our true quantum mechanical potential. This has very serious implications for the quality of life of a human being. Finite state systems repeat themselves. The force behind this reoccurrence is simply that they are finite. As the range and scope of information available to the brain become restricted, the number of potential states also becomes restricted. As synaptic strengths are modified in the neural-net, attractors* are formed and the range of possible states becomes further restricted. *Experience,*

* Attractors can be represented as regions of the "state space" of a dynamical system toward which trajectories travel as time passes. As long as the parameters are unchanged, if the system passes close enough to the attractor, it will never leave the region. Attractors are *ordered states* of high stability surrounded by instability (apparently random activity). These attractors are surrounded both temporally and spatially by chaotic activity in the brain. Brain cell assemblies are more tightly functionally coupled in attractor regions as a result of the atomic and molecular interactions mentioned earlier. This results in the formation of energy barriers, affording a particular information pattern resistance to change. We will be working extensively with neurocognitive attractors and attractor landscapes in Section II — NeuroPrint.

the interaction between external energy/information fields and our biophysical system, gives rise to relatively stable, reoccurring energy patterns and their barriers (attractors) which become spatiotemporily "enfolded" within the continuous unbroken *holomovement* (holonomic interaction) of our biophysical encoding substrates. At a later time, interaction with similar external energy/information fields can cause a spontaneous "unfolding" of an entire attractor landscape previously "enfolded" by the biophysical system, resulting in the related attractors and trajectories becoming more stable and resistant to change. As these attractors become more stable and the range of potential states more restricted, our organismic interface with the environment becomes less flexible, failing to correspond. This lack of correspondence with the environment results in devastating consequences for the organism.

Significantly limited or restricted incorporation of information tends to force a state vector reduction from the maximum possible quantum states ($10^{10(45)}$), severely restricting the neurocognitive system's potential degrees of freedom. As the number of possible states reduces, the *dwelling time* (i.e., amount of time a system spends in a particular state) for the state vector in each of the existing states tends to increase. This means that as the number of potential states decreases, so does the maximum possible speed of information processing (4×10^{53} changes per second). Observation of human behavior will readily prove that some neurocognitive states can last several minutes to several hours, and in severe conditions, even for several days. States such as depression and fear fit this description quite well.

It should be mentioned here that there are two important types of state reoccurrence. The first is called *Markov reoccurrence* or *Markov chain.* A Markov chain is a system for which the probability of transition to the next state depends only on its present state, not on its past history, except insofar as that history is encoded in its present state. On a macroscopic level, this could be translated as a person having three neurocognitive states and oscillating randomly among them. One example may be a person who experiences worry about finances, fear of health problems, and anxiety about spouse infidelity throughout the day, with random reoccurrence of each discrete state. If this were to continue, with little or no new information coming from the outside and forcing the system into new states, the transition probabilities between each of these three states would begin to be modified based on the history of the system and the strength of the trajectories connecting the three states. In other words, the states would become further organized and ordered into a predictable phase path such that the states reoccur in a predictable sequence (e.g., state two, state one, state three ... two, one, three ... two, one, three ... two, one, three). A *phase path* is one path that is overwhelmingly more likely for the state vector to follow. Phase paths can be easily identified in a *probability density matrix,* as will be elaborated in Section II. This spells serious trouble for a human being.

This increased ordering from a Markov chain to a predictable sequential phase path (the path of transitions between states) results in the second type of state reoccurrence called *quantum reoccurrence*. In this case, the system does not just return to a previous state or attractor; it repeats its entire history again and again between returns over the whole of finite future time, becoming almost periodic.

Thus the repeats of the system are arbitrarily close to one another in detail. We should also mention here that the closer a neurocognitive system comes to collapse, the shorter and more regular will be the time intervals between the reoccurrence of each state. Such a condition can be easily tracked through time using NeuroPrint. With such a topological model, we can see whether a particular intervention has driven the neurocognitive system closer to or further from collapse. In the realm of human behavior, it is not uncommon to find within these behaviors, over time, Markov chains and quantum reoccurrence as periodic and ordered as a swinging pendulum. Such conditions become highly probable as the human brain attempts to counterbalance entropy induced by an excess or severe restriction of information flow. Such restrictions in information flow not only arise from sensory restriction and deprivation, but also from neurocognitive filters (bio-informational architecture) erected by previous information-possessing regularity such as traditions, beliefs, and habits. Sometimes people indicate linguistically when they feel the macroscopic effects of a Markov chain or quantum reoccurrence by saying that they feel "stuck" or "in a rut." Human beings naturally feel compelled to disrupt such a rut by adding new information. This, of course, will be limited to their physical and financial resources. While some will interrupt quantum reoccurrence by reading a new book or turning on the television, others will attempt to completely change their environmental input by going on an exotic vacation in a part of the world where the environment is very different from their own. It is not always this easy to break a Markov chain, as the act of television watching or vacationing in itself can become just another reoccurring state in a collapsing system. It is not uncommon for human beings experiencing such collapse to claim that they can find no meaning or purpose in their life. How could they when they possess an organismic interface initially capable of enormous $10^{10(45)}$ possible states at speeds of 4×10^{53} potential transitions per second, and instead find themselves trapped in an ever-diminishing spiral of quantum reoccurrence?

In summary, we are driven to incorporate information/pattern because it is in the deepest sense our life force. Newly acquired information helps us maintain a balance between a system cycling into disorder on one end of the pattern-entropy continuum, and a system collapsing into reoccurrence and periodicity on the other end of the continuum. Both extremes can result in devastating consequences for the human organism and its neurocognitive interface. It will become clearer in the next chapter that numerous aberrations in mental and physical health point to this very imbalance as their common origin. As information consumers, informavores, we are only just beginning to apprehend the profound implications of a good balanced diet of information. Information processing *is* life, and both overconsumption and underconsumption have their consequences. It follows, then, that the quality of our lives can be influenced by the quality and quantity of the information that we process. Information should be regarded as a living entity, not merely a property of the biosphere.

With a cumulative, profound understanding of this, together with all we have discussed previously in this work, we can now begin to make sense out of the myriad of incomprehensible classifications of neurocognitive, neurobiological malfunction listed in our diagnostic compendiums today. Life is a labyrinth of interconnected, multiply embedded, pattern-based information-processing mediums. These mediums or substrates must consistently and precisely regulate the balance between pattern and entropy and obey the five rules of nature's grammar in order to continue to function and survive. Hence, information can give life or take it away. No physical entity, however stable and impenetrable it seems, can escape the reach of nature's illimitable dominion.

HABITUATION AND LONG-TERM POTENTIATION

Before moving on to information replication, let us discuss one set of opposing mechanisms that nature has developed for nervous systems, our organismic-level interface between information and brain, to incorporate news of difference. These mechanisms are *habituation* and *long-term potentiation* (LTP).

Neurons tend to stop responding to repeated, unchanging stimuli, not just out of temporary exhaustion of available biochemicals, but as part of an intricate cellular strategy of being tuned to respond to change (news of difference). The portion of the axon nearest the cell body, which is responsible for initiating an impulse, contains Ca^{2+}-gated K^{+} channels (i.e., calcium and potassium), which can only be opened from inside. If an axon has been firing repeatedly, the buildup of internal Ca^{2+} triggers the K^{+} channel proteins to open. The subsequent escape of K^{+} from within the nerve cell changes the relationship between the internal (intracellular) and external (extracellular) voltage of the cell, making it progressively more difficult to activate. The result — our neural-net only propagates impulses when there is news of difference (Kandel et al., 1991).

What happens if a human being finds himself in sensory reduction or sensory deprivation conditions where there are few or no changes (information) to incorporate? The brain turns to its internal systems to generate the necessary patterns that maintain normal metabolic activity until positive feedback gives the force of entropy the upper hand and runs the neurocognitive system into chaotic iteration. If such a condition prevails for a prolonged period, neurocognitive activity will transition from boredom-induced day dreaming and fidgeting — commonly referred to as Attention Deficit Disorder (ADD) and Attention Deficit/Hyperactivity Disorder (ADHD) — to vivid hallucination, paranoia, microscopic and macroscopic behavioral malfunction, and eventually disease and death (Conway and Siegelman, 1995).

Accounts of such an understanding being misused in the past can easily be found within literature concerning POW thought reform and cult conversion. The conditions are simple. The subject is placed in sensory deprivation conditions until his behavior indicates that his neurocognitive balance has tipped in favor of entropy. Then the subject is given access to two buttons initiating recorded messages that the subject can listen to over loudspeakers placed in the room or cell. One contains a single unchanging, monotonous message, sometimes relating to the subject's own political ideology. The other button initiates a different recorded message each time it is pressed by the subject. Each message is part of an intricate meme complex of distinction, association, and strategy memes: a new political or religious ideology designed to be incorporated "whole" by the subject (Chakotin, 1940; Hunter, 1951; Lifton, 1989).

One of the cellular mechanisms facilitating incorporation of the new meme complex at this point is LTP. *Synapses* (potential connecting points for information transmission between nerve cells) can become potentiated (making signals easier to pass between nerve cells), resulting in the formation of new information processing pathways. Sometimes this is called a change in synaptic effectiveness. The cellular strategy works like this: special proteins located in the postsynaptic membranes of the receiver cells are double-gated. These special proteins have both ligand-gated ion channels and voltage-gated channels. These gates will open to allow calcium ions in if both their neurotransmitter is bound (which might only require a single impulse) and there is a nearby voltage increase due to a strong train of impulses through at least one nearby synapse of the same cell. The synapse is said to have become potentiated (Kandel et al., 1991, 1995). In this way, new memes can be readily incorporated and fused together as a single aggregate. This is an example of just one way that one pattern-encoding medium has been designed by nature to obey her grammatical laws of incorporation and recombination on ever-increasing scales of aggregation. There are myriads of other ways.

REPLICATION (REPRODUCTION)

The memetic advantage of any self-replicating information pattern also depends upon reproduction or *replicatability*. The easier, quicker, and more accurately an information pattern can be replicated (reproduced) and transmitted, with a minimum expenditure of energy, the faster it can duplicate and spread. In addition, the greater the variety of mediums for reproduction, the faster and wider it will spread. Let's explore a simple example.

Honda has a very easily reproduced (replicated) symbol (distinction meme) attached to everything that Honda stands for by association memes. The symbol for Honda is simply an *H*. Toyota, on the other hand, created a symbol a few years ago in much the same way. It was quite a bit more complicated to reproduce than Honda's symbol if the two had to be drawn. This allows for numerous errors (mutations) to occur in reproduction. During the time television advertising was making efforts to link Toyota's new symbol to their car, Mark was very busy and not able to watch television. Consequently, when the cars bearing this new symbol entered the roadways, he was unable to identify the make of the car and attach the new attributes that he liked or disliked to Toyota. While in the past he had owned both Toyotas and Hondas, needless to say, today he and his wife own a Honda — or, more accurately stated, Honda owns them.

Another powerful example of reproducibility can be illustrated by religious symbols. Judaism is recognized by the Star of David, a six-pointed star. Christianity is identifiable by the symbol of the cross or the crucifix, which can be reproduced by even the smallest child that can hold a crayon, simply by drawing two crossed lines. The simplicity of the cross or crucifix gives Christianity a tremendous memetic advantage over Judaism. While both symbols can be recognizably reproduced on many mediums, the cross can even be invisibly traced by making the sign of the cross on your body during a religious service (a symbol recognizable even to a small child by a mere hand gesture). Try doing this with the Star of David. In the same respect, the grave site of a Christian to be remembered can bear a symbol of his religious persuasion by simply using two sticks. Try doing that with the Star of David. The Star of David never had such memetic advantage. How many religious symbols can be accurately, easily, and quickly reproduced by so many pattern-encoding mediums?

TRANSMITTABILITY

The medium by which a meme is transmitted is a *vector*. This is closely tied to its reproducibility. The number of different vectors capable of transmitting an information pattern will affect its memetic advantage. A simple example of this can be found in ancient religions. Those religions that recorded their beliefs and doctrines in writing had memetic advantage over those religions which evolved and persisted by oral transmission of culture and doctrine, simply because it is easier to remember what you believe and how to behave when you have a written record of it. More importantly, a written record allows for far less accidental mutation of the message (information pattern) during incorporation and transmittance. Written scripture allowed more accurate spatiotemporal replication. Those vectors requiring the least amount of cognitive energy are memetically superior.

If we explore more modern-day vectors, it becomes easy to understand that those information patterns transmittable via television are superior to those transmittable via only radio. Information patterns have recently experienced the ability to traverse several continents simultaneously within minutes via the Worldwide Web. Those information patterns capable of being transmitted and replicated through an Internet vector would have obvious memetic superiority both spatially and temporally as compared to competing memes attempting to replicate through the sale of a paperback book. A recent recording star named Jewel is a perfect example of how vectors for transmittance can bestow a memetic advantage upon a meme complex, and sometimes their originator. With the recent advent of the global Internet, it is only now that we humans are approaching the information exchange capacity that cyanobacteria communities have possessed for over 2 billion years on earth (Margulis and Sagan, 1997; Fortey, 1997).

CLEAVING AND RECOMBINATION

If human beings simply incorporated, replicated, and then transmitted patterns, we would certainly be the consummate automatons, unable to influence any of our own behaviors. However, this is not the end of the story. Cleaving and recombining is the stuff that creativity, innovation, analysis, logic, and the elusive process we call thinking are all made from. It only takes replication to remember, recall, or recollect, but remembering is not thinking.

When Einstein was questioned about his own innovative thinking process, he contended that it was merely the product of recombinant, reproducible elements. He did not have a sense that words even entered into the process. Aristotle believed similarly that all thinking was association. What they missed was quite subtle: the power of an extensive scientific vocabulary of distinction memes to slice up (cleave) incorporated sensory experience, allowing for infinitely more complex possibilities of recombination. Evidence suggests that the extent and inherent complexity of a person's vocabulary and facility with grammar significantly influence his/her capacity to think in the sense we are using it here. But this is still only scratching the surface. All symbols and symbol systems are distinction memes, some of which are associated with extremely complex concepts (Burke, 1989; Eco, 1976). Consider mathematical symbols, music notation, or symbols representing the complex concepts of quantum physics. These symbols, like macromolecules (the result of recombination of thousands of individual amino acids), are highly compressed abstractions of hundreds of interrelated laws, axioms, and hypotheses, which themselves are recombinations of smaller information patterns.

MEME SYSTEMS AND MEME COMPLEXES

Impressive evidence suggests that the evolving complexity of mind and thought, and the corresponding increase in the level of self-determined thought, emotions, and behavior available to a given person, is directly proportional to the number and complexity of symbol systems (distinction *meme systems*) that he or she has incorporated and has facility with (Deacon, 1997; Devlin, 1995; Korzybski, 1933, 1994; Walton, 1995; Werner and Kaplan, 1963). A meme system varies from a *meme complex* in one very important way. A meme system is a system of recombinant elements that allow the user to cleave and recombine incorporated sensory experience, while a meme complex is meant to be incorporated whole with the intention of transmitting an entire model or worldview. A belief system, political ideology, tradition, religion, or commercials are all good examples of meme complexes.

When meme complexes compete for control over a human being's response to the external environment, without the chance of being cleaved and recombined by a meme system, significant malfunction distress can be predicted. Anybody who was raised equally by two parents transmitting quantitatively different models of the world can intimately understand this. If early education experiences are similarly structured, thinking can be seriously crippled. Other examples of this relentless memetic competition can be found in interfaith and interracial marriages, and conflict between political parties.

Symbol systems are not the only way to cleave and recombine inadvertently incorporated information patterns in need of disintegration or reassembly. The field of NLP evolved numerous tools to study the structure of subjective experience, one of which is submodalities. Submodalities are the "atomic building blocks" of sensory incorporation and replication. If used with precision, this meme system possesses the inherent capability to significantly increase one's ability to cleave and recombine internal experience, meaning, and, hence, a model of the world. Anchoring, based on classical (respondent) and operant (instrumental) conditioning, also facilitates cleaving and recombining of experience. Both anchors and submodalities can be used together to facilitate education, as well as neurocognitive and medical intervention.

THE BEHAVIOR OF INFORMATION AND CLUES ABOUT MIND AND CULTURE

Imagine that you are a self-replicating information pattern, a meme. Your primary function is to locate an information copying-transmitting medium (a vector) and duplicate yourself. Self-replication is one of the most powerful forces on earth. It is as basic as propagation of the species. Some beliefs, ideas, traditions, and behaviors dominate human thought obsessively for millennia, while others reach their peak and die out like fads. Why has *Barbie* endured for so long while hundreds of other types of dolls come and go year after year? Is *Ken's* success due to his association with Barbie? Why does bad news travel faster than good news? Why does your child tell the same jokes that you did as a child? (Why did the chicken cross the road? Want to hear a dirty joke? A pig fell in the mud.) Why do miniskirts keep reappearing in our culture approximately every 20 years? When two conflicting information patterns meet in the limited space of the human brain, what determines which information pattern wins control of the human host and its behavior?

Think of American presidents for a minute. Why are assassinated presidents like Abraham Lincoln and John F. Kennedy more famous than many others? What do you remember about Richard Nixon? Oliver North? What causes us to remember George Washington (first U.S. president), Neil Armstrong (first man on moon), and the Wright brothers (first successful air flight)? What was number 2? These are just a few of the areas under exploration by memetic researchers and the reader can already answer some of these questions.

If you were a meme, how would you compete for survival? An effective analogy can be found in the biological world: a virus. A virus is simply a small length of nucleotides (genes) with a protein shell, simply a sequence of chemical information searching for a copy medium. Since human cells possess copy machinery to replicate their own DNA, a virus simply has to insert itself neatly into the DNA (become incorporated by cleaving and reaggregating an existing length of nucleotides) in order to get copied along with it. Biological viruses have no ability to self-replicate. They must find a host, deceive the host's membrane receptors and pattern-based defense system, and keep themselves alive long enough to make sufficient copies, before moving on to the next host. Information pattern memes must do the same.

SYMBIOTIC VS. PARASITIC MEMES

A useful distinction to make here is the distinction between a *symbiotic* meme and a *parasitic* meme. Symbiotic memes are self-replicating information patterns that preserve their host, while parasitic memes are self-replicating information patterns that destroy their host (Westoby, 1995; Lynch, 1996; Brodie, 1996).

Two simple examples of a symbiotic meme are "avoid high cholesterol foods" and "children shouldn't play with guns." Since high cholesterol foods are believed to be a causative factor in heart disease, a meme to avoid high cholesterol foods might preserve its host statistically several years longer. This would allow for several more years in which the host would be alive to continue to spread the meme. The same is true for "children shouldn't play with guns." Since children and guns seem to result in accidents and death more often than children and Frisbees, the "children shouldn't play with guns" meme might preserve its host long enough to have his own children and pass on the meme.

On the other hand, one well-known example of a parasitic meme is the Kamikaze pilot, trained to use his plane and himself as a missile, once he uses up the missiles that he is carrying. This method of eliminating military targets is clearly hazardous to the health of the host. The Kamikaze meme works for the common good of a larger superorganism, rather than its own individual host, since success of this meme will always result in the death of the host. Dead hosts are not good teachers of Kamikaze beliefs and techniques.

All things being relatively equal, symbiotic memes are memetically advantaged for slow, continuous, and stable spread over generations. Symbiotic memes have been abundant in religions such as Judaism, where parents teach their children to avoid pork and other pig by-products. Since the consumption of pork was believed to spread disease and death prior to modern methods of food processing and storage, those religious leaders apprehended by logic also noticed that "dead believers" were not good at spreading the word of God. This was particularly important if you lived at a time when culture and tradition were transmitted orally, and the spread of memes was totally dependent on live hosts. Naturally, this type of symbiotic meme has less influence on the spread of memes in literate societies, where beliefs and traditions can be recorded, keeping them alive and spreading long after one's death.

BAD NEWS AND THE BEHAVIOR OF INFORMATION

Have you ever noticed that bad news spreads faster than good? Some customer service gurus tell us that if someone likes your service, they will tell 1 to 3 people. But if someone has a bad experience with your company, they will tell 11 to 14 people. How many nights per week would you find yourself glued to the evening news if every evening report started with some 7-year-old boy scout who helped an elderly lady across the street, and ended with an interview with one of your neighbors who was just reunited with his lost dog? Could you imagine spending $1 million in advertising as a sponsor of this news program? Who would ever buy your product? While you find your company in receivership, your competitor who invested its $1 million in the network that covered the newest serial killer, school bombing, and hurricane is laughing all the way to the bank. Why do bad diets that don't work permanently prevail over good diets that do? There are many mechanisms at work here. Let's explore a few of them.

The first mechanism here is attention. Information patterns must get attention in order to spread. For one thing, if news were always the same, our brain would habituate to the stimulus. So variability and novelty are important in maintaining attention. The brain is wired to alert us to difference or changes in stimuli. How many times have you taken a young child to a store only to be embarrassed by an event such as her shouting at the top of her lungs and pointing, "Hey, look at the lady with the mustache!" Why is it that your 3 year old always notices the one female out of 500 with a slightly higher density of facial hair and is compelled to shout about it in public? The brain is wired to pay attention to and remember difference. Paying attention to difference allows us to adapt to rapidly changing circumstances and in some cases to avoid danger and preserve life.

Another mechanism at work here is the neuroendocrine system. While we would all like to believe that good ideas and beliefs would prevail over bad, unfortunately this appears to be true only in fairy tales. The neuroendocrine system is wired to keep us alive. Changes in stimuli that signal danger or compromise our safety are always given high priority in a properly working nervous system. Therefore, given limited spatio-temporal resources of both brain and communication vectors, memes that carry survival information will always enslave available neural resources, leaving little space or time for good things to be recognized. You can see examples of this neuroendocrine hierarchy at work everywhere you look.

Let's say that you have money invested in the stock market and you hear of some event that will impact negatively on all those you know who have similar portfolios. Let's also assume that each of these same people love hamburgers and you also have found out that McDonald's has a 59¢ special. Which piece of news would cause you to immediately call all of these people? Which is sure to capture their attention? If you are like most people, it will be the news of the impending stock market doom. Why? While both of these memes are symbiotic in nature, they are not equal. The 59¢ special is a symbiotic meme because food assists in survival, and more food for the same money assists in longer survival or survival of more people. But if you have a full refrigerator at home or if you have just eaten lunch, this news will have little effect in self-organizing the attention circuitry of your brain. However, during conditions of a famine, the 59¢ burger meme would be memetically advantaged and spread like an epidemic.

Now, let's go back and take a look at the impending stock market crash. Notice how the intensity of the disaster naturally increases each time it is discussed. What was originally a minor fluctuation will become an inevitable full-market crash in no time. So why would most people decide to spread this news, rather than news of the 59¢ burgers? Given present conditions, news of a stock market crash can protect people from symbolic danger (economic crisis). It is symbolic danger, of course, because loss of money is an abstraction far removed from actual physical danger, but still associated with it. This type of danger is quite different from spreading news of killer bees or deadly hurricanes. Yet it still has the power to self-organize the neuroendocrine and biomechanical control systems of a human being. Another added benefit above and beyond protection from danger is the reward given to the bearer of bad news. Simply by spreading the news of the stock market crash, the bearer of the news feels important for having access to exclusive information. Second, the bearer feels proud to be able to help all of his or her acquaintances to protect their nest eggs. Third, the bearer is normally rewarded with, at the very least, a sincere "thank you" from the receiver and is possibly even now owed a reciprocal exchange of value in the future. You might say saving someone's financial future would at least be worth a fine dinner. While both of these memes are symbiotic, they are not memetically equal.

What about rubbernecking delays? Have you ever been caught in bumper-to-bumper traffic on the way to work in 7 miles of "parking lot," only to find an hour later that the accident was on the other side of the highway? What happened? Thousands of people slowed down to look at bodies strewn over the highway from the last ten-car collision. Although some might say that this is the morbid curiosity of a sick culture at work, rather it is your healthy neuroendocrine system doing the job that it was wired by nature to do. Painful, gut-wrenching scenes like this attract nearly unanimous attention because the human neuroendocrine system searches the situation for what it can learn to protect itself in the future. Rubbernecking is a symbiotic strategy meme.

WHY WE INCORPORATE PARASITIC INFORMATION PATTERNS

By what mechanisms do parasitic information patterns gain control of human beings? What about bad diets? Why are good healthy diets that create permanent weight loss (a symbiotic meme) memetically disadvantaged to those that create fast temporary (a parasitic meme) weight loss? This seems counterintuitive, doesn't it? This is because no single factor by itself can influence the outcome of memetic proliferation. There is a different mechanism at work here, competing for the same organismic resources. This strain of ineffective diets provides greater opportunity for the spread of their diet meme. Consider this. Let's say you just heard about the super weight-off plan, where you can lose 30 pounds in 5 weeks. First of all, why does this sound so good? Because this meme makes use of a very basic human need — the principle of "maximum gain for minimal effort." It is the very same principle that causes quick-fix drugs, pyramid scams, network marketing, get-rich-quick, and just-add-water cake mix memes to proliferate. How can you resist the idea of losing 30 pounds in 5 weeks, if you feel self-conscious about your weight?

First of all, this kind of meme interrupts logical reasoning faculties in our neurocognitive system. The mistake here is that we start to multiply. If I'm 30 pounds lighter in 5 weeks, then I'll be 60 pounds lighter in 10 weeks. An understanding of physiology reveals this faulty reasoning. The excitement of losing 60 pounds of excessive flab causes you to start talking (transmitting) about the diet, even before it starts to work. Then some interesting things start to happen. As you begin to lose the first 10 pounds of water weight, your family, friends, and office workers begin to detect "difference" — a change in stimuli — in this case, a difference in your bodily proportions. As your clothes start swimming on you, it becomes the talk of the town. As a result, you get a great deal of attention and positive reinforcement. Others who think they are overweight fall prey to the maximum gain for minimal effort principle and also crave the attention they see you are getting.

Once you finally start bottoming out as your metabolism compensates for famine conditions, there is a sharp decrease in the number of people who start noticing you and your miraculous weight change. If you decide to maintain this new weight for too long, you will lose social reinforcement and the dramatic attention you were receiving during your weight plunge. If this hurts too much, the negative reinforcement causes you to stop the diet and begin the gaining process. However, even this does not reestablish the level of social reinforcement and attention that you initially received from the rapid weight loss. While others are quick to notice your rapid weight loss, it is much more difficult for them to notice your slower weight gain. So you commit to starting this process all over again from scratch for the thrill of some more social reinforcement and attention. Pretty soon you find yourself 4 pounds heavier than you were before you started. Enter the next diet fad.

As a result of the confluence of these mechanisms, good diets that promote good health and permanent weight change may be memetically disadvantaged over physiologically destructive diets that create fast temporary change, even though good diets are symbiotic memes. Self-replicating information patterns do not care if they are good or bad. Their primary function is to make copies of themselves and compete for the limited resources of the neurocognitive system and its effectors, which in turn ensures their survival.

CONSTRUCTING A MODEL OF THE WORLD — INCORPORATING THE MIND OF NATURE

The human brain is truly one of the most intricate and beautiful expressions of nature's maturing thought process. Written in the language of pattern obeying nature's laws of grammar, its tentative existence and stable operation are totally dependent on the maintenance of a dynamic balance between pattern and entropy, ink and eraser.

The brain's purpose is to maintain an organismic-level interface between internal and external information necessary to initiate collective, coordinated responses to a changing environment. The brain must insure that the organism continues to correspond with the ever-changing arrangement of the world in which it lives. If it fails to do so, it may soon perish, caught in the inescapable stroke of nature's great eraser. To accomplish and maintain this dynamically shifting correspondence, the brain must construct an internal model of the world. Through its carefully tuned receptors, it must incorporate the evolving mind of nature and continue to revise its internal model. It must do this in lock-step with nature's revisions and its own journey through nature's changing terrain, following as closely as if it were nature's own shadow; for that which does not correspond will eventually be erased.

The brain embarks on this journey bearing just a few key instructions for its own self-organization and function, written in the language of pattern — each step of the way, tuning to external conditions — bound by the constraints of nature's grammar. To keep up with nature's evolving thought, it must maintain a state of operation that perpetually teeters on the knife's edge of instability, maintaining as many degrees of freedom as possible without giving entropy the upper hand. In order to counterbalance entropy during periods of reduced externally changing stimuli, the human brain actively seeks new information to incorporate and evolves continually in complexity under the sheer self-organizing force of its newly incorporated information, giving way to a relentlessly evolving mind.

But at stages along the journey things go wrong. Under the conflicting pressures of genes and memes, repeated replication of previously incorporated pattern begins to stabilize the organism's function, structure, and perception. This in turn gives rise to ridged behavioral habits, perceptual filters, and conceptual prejudice which further reduce the organism's degrees of freedom, causing it to respond in a way that is no longer quite in phase with the world outside. Its rapidly stabilizing sensory pathways begin deleting, distorting, and generalizing newly incorporated information patterns, which are no longer able to overcome the mounting functional energy barriers or cross the bridgeless waterways of unpotentiated synapses. The organism does not fear change; instead it fears entropy. For once the balance shifts too far, the very identity of the organism will be lost in the chaotic dismemberment. In response to entropy's tug, it becomes more patterned, ordered, and brittle, giving birth to a predicable personality with inflexible thoughts, emotions, opinions, skepticism, and beliefs — unable to evolve, fearing and avoiding the incorporation of new knowledge: so to speak, hardening of the categories.

The day-to-day external interplay between matter and energy that once led to corresponding change in the carefully tuned and flexible organism now acts as a cleaving force. Like a sheet of ice cracked in two by a falling tree, carried separate ways forever dismembered, it thinks how things could have been if it had only changed to water in time. In continued response to conflicting information patterns and the forces of pattern and entropy, the brain's once dynamic information-processing pathways now form energy barriers, giving rise to an increase in functional stability and ridged structural boundaries, which represent an internal "model" of a world that no longer exists on the outside. The brain has stabilized an internal "map" for navigation that in no way resembles or corresponds to the current external territory.

Paradoxically, the organism becomes more stable and at the same time more susceptible to the force of entropy, ready to break itself against the very laws of nature that once gave it life. The neurocognitive and biophysical epiphenomena arising from this paradoxical state — from phobias, compulsions, and learning disabilities to the plethora of psychoses and physical diseases that have forever plagued human existence — have confounded human intellect from the inception of scientific inquiry. It is therefore our intention and hope that this text brings deeper pragmatic insight to those practitioners and scientists engaged in the important work of evolving humanity and its culture.

REFERENCES

Abeles, R., Frey, P., and Jencks, W. (1992). *Biochemistry.* Boston, MA: Jones and Bartlett Publishers.

Barthes, R. (1985). *The Semiotic Challenge*. Berkeley: University of California Press.

Becker, J. B., Breedlove, S. M., and Crews, D. (Eds.). (1992). *Behavioral Endocrinology.* Cambridge, MA: MIT Press.

Benyus, J. M. (1997). *Biomimicry: Innovation Inspired by Nature*. New York: William Morrow and Company.

Boole, G. (1958). *An Investigation of the Laws of Thought on Which are Founded the Mathematical Theories of Logic and Probabilities*. New York: Dover Publications.

Brodie, R. (1996). *Virus of the Mind: The New Science of the Meme*. Seattle: Integral Press.

Burke, K. (1989). *On Symbols and Society.* Chicago, IL: University of Chicago Press.

Calvin, W. H. (1996). *The Cerebral Code: Thinking a Thought in the Mosaics of the Mind.* Cambridge, MA: MIT Press.

Carroll, J. B. (Ed.). (1956). *Language Thought and Reality: Selected Writings of Benjamin Lee Whorf.* Cambridge, MA: MIT Press.

Chakotin, S. (1940). *The Rape of the Masses: The Psychology of Totalitarian Political Propaganda.* New York: Fortean Society.

Conway, F. and Siegelman, J. (1995). *Snapping: America's Epidemic of Sudden Personality Change (2nd Edition).* New York: Stillpoint Press.

Csikszentmihalyi, M. (1993). *The Evolving Self: A Psychology for the Third Millennium.* New York: HarperPerennial.

Dawkins, R. (1989). *The Selfish Gene.* Oxford: Oxford University Press.

Deacon, T. W. (1997). *The Symbolic Species: The Co-Evolution of Language and the Brain.* New York: W. W. Norton & Company.

Devlin, K. (1995). *Logic and Information.* Cambridge: Cambridge University Press.

Eco, U. (1976). *A Theory of Semiotics.* Bloomington: Indiana University Press.

Edelman, G. M. (1987). *Neural Darwinism: The Theory of Neuronal Group Selection.* New York: BasicBooks.

Edelman, M. (1964). *The Symbolic Uses of Politics.* Urbana, IL: University of Illinois Press.

Erickson, M. H. (1989). *The Nature of Hypnosis and Suggestion: The Collected Papers of Milton H. Erickson on Hypnosis Volume I.* New York: Irvington.

Fersht, A. (1985). *Enzyme Structure and Mechanism*, 2nd ed. New York: Freeman.

Fortey, R. (1977). *Life: A Natural History of the First Four Billion Years of Life on Earth.* New York: Alfred Knopf.

Furman, M. (1998a). Memetics: The behavior of information. *Anchor Point*, July, 27, 12–17.

Gribbin, J. (1993). *In the Beginning: The Birth of a Living Universe.* New York: Little, Brown & Company.

Hunter, E. (1951). *Brain-Washing in Red China: The Calculated Destruction of Men's Minds.* New York: Vanguard Press.

Kandel, E. R., Schwartz, J. H., and Jessell, T. M. (Eds.). (1991). *Principles of Neural Science*, 3rd ed. Norwalk, CT: Appleton & Lange.

Kandel, E. R., Schwartz, J. H., and Jessell, T. M. (Eds.). (1995). *Essentials of Neural Science and Behavior.* Norwalk, CT: Appleton & Lange.

Kane, G. (1995). *The Particle Garden: Our Universe as Understood by Particle Physicists.* Reading, MA: Addison-Wesley.

Kendrew, S. J. and Lawrence, E. (Eds.). (1994). *The Encyclopedia of Molecular Biology.* Oxford: Blackwell Science Ltd.

Key, W. B. (1989). *The Age of Manipulation: The Con in Confidence. The Sin in Sincere.* Lanham, MD: Madison Books.

Key, W. B. (1993). *The Age of Manipulation: The Con in Confidence. The Sin in Sincere.* Lanham, MD: Madison, Books.

Kleinsmith, L. J. and Kish, V. M. (1995). *Principles of Cell and Molecular Biology.* New York: Harper-Collins College Publishers.

Korzybski, A. (1933). *Science and Sanity.* Englewood, NJ: Institute of General Semantics.

Korzybski, A. (1994). *Science and Sanity: An Introduction to Non-Aristotelian Systems and General Semantics (5th Edition).* Englewood, NJ: Institute of General Semantics.

Lifton, J. (1989). *Thought Reform and the Psychology of Totalism: A Study of "Brainwashing" in China.* Chapel Hill: University of North Carolina Press.

Luria, A. R. (1932). *The Nature of Human Conflicts: An Objective Study of Disorganization and Control of Human Behavior.* New York: Grove Press.

Lutz, W. (1996). *The New Doublespeak: Why No One Knows What Anyone's Saying Anymore.* New York: HarperCollins Publishers.

Lynch, A. (1996). *Thought Contagion: How Belief Spreads Through Society: The New Science of Memes.* New York: BasicBooks.
Margulis, L. and Sagan, D. (1997). *Microcosmos: Four Billion Years of Microbial Evolution.* Los Angeles: University of California Press.
McCracken, G. (1990). *Culture & Consumption.* Bloomington: Indiana University Press.
Parker, S. (Ed.). (1993). *McGraw-Hill Encyclopedia of Physics,* 2nd ed. New York: McGraw-Hill.
Penrose, R. (1994). *Shadows of the Mind: A Search for the Missing Science of Consciousness.* Oxford: Oxford University Press.
Plotkin, H. (1993). *Darwin Machines and the Nature of Knowledge.* Cambridge, MA: Harvard University Press.
Pribram, K. H. (Ed.). (1993). *Rethinking Neural Networks: Quantum Fields and Biological Data.* Hillsdale, NJ: Lawrence Erlbaum Associates.
Pribram, K. H. (Ed.). (1994). *Origins: Brain & Self Organization.* Hillsdale, NJ: Lawrence Erlbaum Associates.
Rieke, F., Warland, D., de Ruyter van Steveninck, R., and Bialek, W. (1997). *Spikes: Exploring the Neural Code.* Cambridge, MA: MIT Press.
Robertson, T. S. and Kassarjian, H. H. (1991). *Handbook of Consumer Behavior.* Englewood Cliffs, NJ: Prentice-Hall.
Rossi, E. L. (Ed.). (1980). *The Nature of Hypnosis and Suggestion: The Collected Papers of Milton H. Erickson on Hypnosis Volume I.* New York: Irvington.
Rushkoff, D. (1994). *Media Virus: Hidden Agendas in Popular Culture.* New York: Ballantine Books.
Shepherd, G. M. (1994). *Neurobiology,* 3rd ed. New York: Oxford University Press.
Singer, M. T. and Lalich, J. (1995). *Cults in Our Midst: The Hidden Menace in Our Everyday Lives.* San Francisco: Jossey-Bass.
Spencer-Brown, G. (1994). *Laws of Form.* London: George Allen and Unwin Ltd.
Stites, D. P., Terr, A. I., and Parslow, T. G. (1994). *Basic & Clinical Immunology,* 8th ed. Norwalk, CT: Appleton & Lange.
Taylor, G. (1996). *Cultural Selection: Why Some Achievements Survive the Test of Time — And Others Don't.* New York: BasicBooks.
Tipler, F. J. (1994). *The Physics of Immortality.* New York: Doubleday.
Walton, D. N. (1995). *Informal Logic: The Handbook for Critical Argumentation.* New York: Cambridge University Press.
Werner, H. and Kaplan, B. (1963). *Symbol Formation: An Organismic-Developmental Approach to Language and the Expression of Thought.* New York: John Wiley & Sons.
Westoby, A. (1995). The Ecology of Intentions: How to Make Memes and Influence People: Culturology. London: unpublished work.
Whitehead, A. N. (1927). *Symbolism: Its Meaning and Effect.* New York: Fordham University Press.
Whorf, B. L. (1995). *Language Thought and Reality.* Cambridge, MA: MIT Press.

5 The Life of the Mind: Infogenesis and Information Translation in the Language of Pattern

CONTENTS

What is life? As all else in science, the answer to this question continues to evolve as we increase the range and scope of our perception. As scientists, we continually return to this question, imbued with an intuitive sense that the answer will reveal how to improve the quality of human life. We instinctively realize that in order for there to be an appreciable difference in the quality of human life, we must make a profound change in the way we understand the act of living. To accomplish a radical paradigm shift, we must see a thing or a process from a completely new perspective, and within that perspective ask entirely new questions. Such a paradigm shift as we are suggesting can be accomplished by regarding information/pattern as living entities or structures, subject to the same physical laws as matter and energy, evolved and shaped by the same selection pressures as all other life in the biosphere. Thoughts, ideas, beliefs, feelings, habits, and traditions (memes) should be viewed as self-replicating living entities capable of moving from one encoding substrate to another with great ease, leaving each substrate significantly altered as a result of its presence and passage. Information is capable of traversing all of nature's boundaries, from the so-called nonliving matter to the world of living matter, from outside an organism to inside, from organ to organ, system to system, cell to cell, protein to protein, molecule to molecule, and atom to atom. To deeply understand and profoundly influence the human brain, mind, and its emergent behavior, we must comprehend information in a distinctly different way — as "informed" matter and energy.

In science, information means something quite different than it does when used in nonscientific contexts. The Latin root of information is *eidos*, which eventually became *informatio*, from *informare*, meaning to "give form," "shape," and "guide" (Loewenstein, 1999). The Latin roots are much closer to the way in which we discuss information in science today. In science, information is a universal measure of order in a system, in the same way that entropy is a universal measure of disorder. The relationship between the two is inextricable. An increase in entropy always implies a decrease in information and vice versa.

Information translation is really just another way of saying that information/pattern is being exchanged between two diverse or disparate substrates, making up a coherent, coordinated system. The language of pattern must necessarily be translated into different substrates (exchanged between substrates) in order for the parts of a system to maintain memory and perform the collective, coordinated behavior we commonly refer to as organismic. In the foregoing chapters, we have provided a myriad of isolated examples of such information exchange. Effective information exchange is wholly dependent upon the five behaviors of pattern mentioned before: incorporation, replication, cleaving, recombination, and transmitting. The fidelity of information exchange is absolutely necessary for coherent, coordinated behavior of an organism or a system of organisms.

As information transitions from substrate to substrate by the process of pattern translation, the behavior of that information will determine the behavior of its substrate. Thus, the information itself is substrate-neutral. So what then is actually living? As information is being processed, we see only its current, animated, and coherent substrate, changing its atomic and molecular configurations, and call it life. Yet information continues to leave one substrate for another, leaving behind a mere living fossil of its former existence until the next exchange. (Cairns-Smith, 1985; Loewenstein, 1999; Maynard Smith and Szathmary, 1999). As a subterranean cave is dynamically shaped by the passage of water, the human brain and its resulting mind and behaviors are shaped by the passage and presence of information, the universal measure of order within a system. In a sense, the human being becomes a living fossil, a dynamic informational bioarchitecture formed from the interaction of information within the agglomeration of matter and energy. In the words of theoretical physicist Roger Penrose:

> Most of the material of our bodies and brains, after all, is being continuously replaced, and it is just its *pattern* that persists. Moreover, matter itself seems to have merely a transient existence since it can be converted from one form to another. Even the *mass* of a material body, which provides a precise physical measure of the quantity of matter that the body contains, can in appropriate circumstances be converted into pure energy (according to Einstein's famous $E = mc^2$) — so even material substance seems to be able to convert itself into something with a mere theoretical mathematical actuality. Furthermore, quantum theory seems to tell us that material particles are merely 'waves' of information … Thus, matter itself is nebulous and transient; and it is not at all unreasonable to suppose that the persistence of 'self' might have more to do with the preservation of *patterns* than of actual material particles (1994, pp. 13–14).

Information clearly manifests itself in both the structure and function of a human being, hence altering behavior microscopically and macroscopically as it takes up residence, replicating, translating, and passing from system to system, crossing all interfaces of a human body. As practitioners, if we wish to change mind, brain, or behavior with greater precision and predictability, we must first make a model of the present structural and functional relationships which are representative of the entire neurocognitive system (informational bioarchitecture), or at least a significantly large portion, to fully comprehend how both structure and function has been organized and constrained by the forces of information, and hence modified the bioarchitecture its passage has left behind. It is only with this broader systemic view that we may effectively and predictably influence and improve the quality of life of the information processing aggregate we call a human being.

Whether we see information as a living entity or life as the act of information processing, nonetheless we have adopted a more useful perspective for understanding human life, as well as for the designing of more effective interventions. The correctness of this approach can only be supported by the extent of its utility. As the reader gains deeper insight into how information is exchanged from substrate to substrate within a human being, the systemic effects of substrate alteration will become clearer.

The Standard Theory of Pattern-Entropy Dynamics postulates that all manner of function and malfunction in a human being can be ascribed to the five behaviors of information/pattern*:

- **Incorporation** (the absorption of one element, pattern, or aggregate by another)
- **Replication** (the duplication, reactivation, or copying of an element, aggregate, or process)
- **Cleaving** (the separating or dismembering of an aggregate into two or more parts/elements)
- **Recombination** (the aggregation and rearrangement of elements or larger aggregates)
- **Transmitting** (the isomorphic propagation and exchange of pattern/information through and between elements, aggregates, or coordinate space)

Therefore, all that we see as malfunction in the human organism can be attributed to the expansion and contraction of natural variation in these five behaviors of pattern, around a central tendency, which is stabilized by local, environmental selection pressures. Psychology refers to this central tendency, stabilized by natural selection, as the normative state of an organism. For ease of understanding, we can begin by parsing the deleterious results of these variations into three major domains:

1. **The Principle of Pattern-Entropy Balance** — An imbalance of information flow across the interface (brain) between environment and organism,

* All complex/compound structure and process (function) in nature (normal and abnormal) can be ascribed to the recursion of the five behaviors of information/pattern. Recursion is the proclivity of patterns to incorporate isomorphic transformations of themselves. (See Appendix 2 in Section II.)

resulting in either increasing entropy (neurocognitive and biophysical disordering) or the formation of Markov Chains and ultimate Quantum Reoccurrence (neurocognitive and biophysical periodicity)

2. **The Principle of Organism-Environment Correspondence** — A lack of correspondence between organism and environment, resulting from either incorrect or incomplete translation of information (information/pattern exchange) across the organism's interface, or the untimely, spontaneous replication of information by the organism, observed outside of the context to which it corresponded, and in which it was originally incorporated
3. **The Principle of Conflict/Competition Between Information Patterns** — A disruption in information translation between subsystems, organs, cellular interfaces, and molecular signaling systems due to the incorporation, replication, and transmission of conflicting information patterns with separate and competing functions, and resulting in action that is both uncoordinated and not collective or coherent in nature

As a practitioner or scientist comes to apprehend this relationship between the behavior of information and its substrate, the previously anomalous structure (form) and function (behavior) of an organism becomes increasingly understandable, predictable, and influence-able.

The ubiquity of pattern that can be found both in humans and nature has been described in great detail. We have discussed pattern behavior from the subatomic level to the highly complex level of a human being. It is an incontrovertible fact that pattern and information are inseparable. As Tipler (1994) argues, pattern is just another name for information. With the foundation laid thus far, it is now possible for cognitive-neurophysics to offer answers to some very difficult questions that will greatly influence both the quality of human life and the scientist-practitioner's depth of understanding of the power of information to influence that life. What is the ultimate source of the biological information that influences the creation and evolution of human mental and physical life? What are the conditions necessary for such complex organisms as human beings to arise? Should we expect human beings to continue to complexify throughout life, or is this just an illusion caused by errors in reasoning and a closely held cultural belief in the idea of progress? Why have human beings created a distinction between living and nonliving matter, and is such a distinction useful or a hindrance to understanding human behavior? While the reader may be sufficiently prepared to anticipate the answers to these questions, let us expand the picture somewhat further to elaborate the connection between human beings and the natural, physical systems from which they are created.

INFOGENESIS: THE ORIGIN OF THE INFORMATION THAT WE PROCESS

The birth of our universe as we know it, now commonly referred to as the "big bang," marked the beginning of time, space, matter, and information. Information is not possible to detect, even if it could exist by itself, without observing its behavior manifested in the organization of matter and energy as it is processed. The presence of information makes itself known by arranging matter and energy in various forms or structures, subject to the known (and even yet-to-be-known) laws of physics. As we understand it, the passage of time is measured in entropy, the tendency of a system to irreversibly dissolve into disorder. Such an understanding seems to be at the very root of our physical organization. For if we were to see the shattered pieces of a teacup rise off the floor, jump onto the kitchen table, and reassemble themselves, it would surely occur to us that time was running backward. We humans are massive agglomerations of atoms, continually arranged and rearranged by the flow of information from our environment. In a very deep sense, these arrangements of matter and energy can be said to "know" the direction of time, which is not only critical to our perception or illusion of cause and effect, but also to the totality of human behavior, much of which, if not all, is time-dependent.

There is an important piece of the story that is critical for a complete understanding of information and its effects. Information, because of its dependence upon matter and energy, can be said to have had its birth contingent upon the appearance of matter in the universe, which provided it with some substrate or element to act on or with which to interact. Yet, at the time of the universe's birth, information was not as complex and varied as it is now. It is most critical to realize that as information in our universe complexified, the very same processes gave birth to the spontaneous appearance of organisms (agglomerations of matter and energy) that were equally complex and capable of exchanging that information. That is, in nature, the sender (transmitter) and the receiver (incorporater) evolved and continue to evolve simultaneously. (A recent example of this is the frightening increase in ADD- and ADHD-designated students in American schools, born of the continual incorporation and replication of environmental patterns such as television, high-speed video games, and Pentium-driven computers.) Let us now detail the story of infogenesis more clearly.

The "big bang" is considered collectively by physicists and cosmologists to mark the birth of our universe and, hence, the beginning of matter, space, time, and information. However, in the very early stages of this massive explosion, this was not exactly the case. The temperatures produced by the big bang were so enormous that hydrogen gas had not yet formed. Hydrogen was the first atomic structure to be formed, and as far as many physicists are concerned, this marked the origin of all life. If a gas were to be heated to such temperatures as were present during the big bang, it would turn to plasma, and matter, time, and information would not exist within that structure. Such was the case in the beginning. As the plasma expanded and cooled, the drop in temperature allowed electrons to be captured by nuclei. This transition of our universe from plasma to the more discrete matter and energy, of which we are so familiar, can be thought of as a bifurcation, a point at which a system must make a choice between alternatives. The choice made at this transition point where electrons were captured by nuclei truly marks the beginning of the distinction between matter and energy. It is at this point in the genesis of a universe that both time and information can arise (Smolin, 1997). Yet, it is important to understand that information did not just spring into existence at its present complexity upon infogenesis. Information, like all else in the universe, had to evolve by processes of natural selection.

When the universe existed at temperatures where only a single atom could be formed, as a system it was too homogeneous to encode complex information. It was only after the universe expanded to the point of allowing temperature differentials that the laws of thermodynamics and the genesis of complex information and pattern could come into being. A temperature differential can be thought of as an asymmetric distribution of energy and matter. The potential for asymmetry in matter and energy is essential for the birth of information and its continued evolution toward the greater complexity necessary for life and mind to arise. While simple symmetric, periodic patterns are information-poor, complex asymmetric, nonperiodic (almost random) patterns can encode tremendous amounts of information. Physicists, information theorists, and evolution theorists have known this notion for some time (Cairns-Smith, 1985; Davies, 1999; Einstein, 1961; Goldstein and Goldstein, 1993; Kauffman, 1993; Maynard Smith and Szathmary, 1999; Prigogine, 1996; Schrodinger, 1944; Smolin, 1997).

At the time that the universe was made up of only homogeneous clouds of hydrogen, there were no amalgamations of matter, or information complex or diverse enough to construct a human being, and hence no information complex enough to be incorporated in order to maintain the existence of a human being, if one had been previously amalgamated. So where then did the complexity of information we know today come from? How did such homogeneous and information-poor beginnings evolve into the range of complex, biological information that amalgamates and evolves the human brain of today? The answer is an ever-widening *entropy gap*. An entropy gap can be understood as the difference between the present entropy of a system and the maximum possible entropy of that system.

Almost a half century of study in the fields of thermodynamics, nonequilibrium physics, and the dynamics of unstable systems has revealed that the further a system is forced from equilibrium (the larger the entropy gap — asymmetry of matter and energy distribution), the more successive its bifurcations, and hence, the more diversified its parts become, leading to innovation, novelty, and new forms, and thus greater complexity of information and potential for information exchange. Information exchange (communication) is absolutely necessary for coherent, collective, macroscopic behavior to arise in a complexifying system. As a system is forced further from equilibrium, bifurcating, diversifying, and complexifying, each diversified part emerging must communicate with all others in order for coherent, macroscopic behavior, and the organisms performing them, to be possible. As diversification and communication-assisted coherence (pattern transmission–incorporation) emerges, so does the complex structure called organism. This deep insight, over the last half century of nonequilibrium physics research, has significantly blurred the distinction previously made by biologists between living (animate) and nonliving (inanimate) matter, and hence life and death itself (Baser et al., 1983; Haken, 1983; Haken, 1988; Kauffman, 1993; Kruger, 1991; Mishra et al., 1994; Prigogine, 1996; Thelen and Smith, 1994).

So, then, how was our homogeneous universe of hydrogen atoms forced so far from equilibrium and homogeneity to produce such complex informational architectures as human beings and the information that we process? The answer is gravity. Gravity is a universal property of our universe, always attractive and operating at infinite distance, proportional to the mass of related structures. The force of gravity is never zero. Recall that death, thermodynamically described, means thermal, chemical, and mechanical equilibrium or maximum entropy. The initial homogeneous environment of our young universe could have easily evolved into a state of thermal and chemical equilibrium possessing little to no information and no chance for complexity, since it only contained a single atomic structure, hydrogen, and was continuing to cool down as it expanded from the initial explosion. However, as physicists now understand, the *positive energy* from the big bang must have produced equal and opposite *negative energy* that was always attractive, counterbalancing the positive energy to result in a net energy of zero for the universe as a whole. That negative energy, equal to and opposite from the enormous energy of the big bang, is known as the *gravitational field* (Davies, 1999; Progogine, 1996; Smolin, 1997). The spontaneous appearance of complexifying information patterns owes its existence to this gravitational field. As Davies (1999) put it:

> This is really just the second law of thermodynamics revisited, because the spontaneous appearance of information in the universe would be equivalent to a reduction of the entropy of the universe ... The conclusion we are led to is that the universe came stocked with information, or negative entropy, from the word 'go.' (p. 62)

HOW A GRAVITATIONAL FIELD PRODUCES INFORMATION FROM HOMOGENEOUS MATTER

While the early universe was evolving toward thermal and chemical equilibrium, the force of gravity prevented the third requirement for death — mechanical equilibrium. Its universal attraction forced the hydrogen to agglomerate into discrete clouds of increasing density. As the mass of the hydrogen clouds increased, so did the attraction of gravity. As this attraction continued and the kinetic motion of hydrogen atoms sped up, temperatures began to rise sharply. At a critical point of temperature and pressure, the hydrogen nucleus became capable of incorporating another proton (a pattern of quarks), thus forming helium. This resulted in the birth of a star. It is the existence of gravity and resulting stars, in a continually expanding and cooling universe, that maintains the necessary *entropy gap* (i.e., the difference between the present disorder and the maximum possible disorder of a system, resulting in the maintenance of an asymmetrical distribution of matter and energy) from which all life and evolving complexification feeds. Without this gap and the gravitational field that produced and maintains it today, self-organizing systems, life, information, and human brains could not exist (Davies, 1999; Haken, 1988; Prigogine, 1996; Schrodinger, 1944). Today physicists are well aware of the fact that all the atomic elements heavier than hydrogen originated in the gravitational processes that formed the stars and galaxies. This process of atomic element manufacture by incorporation is known as *nucleosynthesis*. Carbon-based life would not have been possible if it were not for the fact that carbon atoms, and life's other essential building blocks, were formed by *nuclear fusion* (atomic pattern incorporation) in the process of making stars. In a very real sense, we can say that life and information as we know it arose from stardust.

Both the material building blocks (atoms) necessary to make life and the essential entropy gap (thermal, chemical, and mechanical disequilibrium) needed to sustain it owe their existence to the gravitational field. Therefore, gravity is one of the most important universal forces responsible for increasing the distance from equilibrium in systems from human brains to the entire universe. The profound and far-reaching implications of gravity and the entropy gap are still extant today and make themselves felt in human processes even so subtle as *cortical field activation cues* (Furman, December 1996). These are cues an external observer can use to predict what type of internal sensory representations a subject may be activating when incorporating or replicating a given set of information patterns.

The reader will recall that memory is a phase transition from a less organized (closer to equilibrium state) to a more organized (further from equilibrium) state. Watch closely the next time you ask a person a question that requires him/her to formulate a complicated internal, visual picture in order to give you the answer. Prior to responding, the person's eyes and head will tilt upward, allowing gravitational effects to assist in the sustained increase of regional blood flow to the visual (occipital) cortex located toward the rear of the skull. A PET scan would, in fact, reveal this increase in regional cerebral blood flow, which to neuroscientists signifies a temporary increase in functional complexification (an increase in asymmetrical distribution of matter and energy) of the corresponding region of the neural-net (the visual cortex) necessary for constructing and maintaining a visual image (Roland, 1993; Posner and Raichle, 1994). This simple and discrete head and eye movement seen ubiquitously in human beings is the entropy gap at work. In order for the necessary information to arise in the visual cortex, to produce the answer to your question, that particular region of the brain has to be forced far from mechanical and chemical equilibrium in order to incorporate and replicate the necessary information. As both Davies (1999) and Prigogine (1996) observe, diversification and information increase as any system is forced farther from equilibrium; and, conversely, homogeneous systems near equilibrium are information-poor. It follows, therefore, that in order for the human brain to continue to incorporate and process complex information, it must necessarily be forced farther from equilibrium. The more complex the information to be incorporated and processed, the farther from equilibrium it must be forced. The capacity for the gravitational field to generate mechanical disequilibrium (asymmetric/aperiodic patterns of matter and energy distribution) in human brains, and hence a sustained increase in information incorporation, is clearly observable through PET scan analysis (Roland, 1993; Posner and Raichle, 1994).

BOUNDARIES, BORDERS, AND SEMIPERMEABLE BARRIERS

While the force of gravity alone may be capable of sustaining the necessary entropy gap (nonequilibrium on a galactic scale), its force, and by implication its effectiveness in doing so, diminishes when we consider the smaller agglomerations of matter, energy, and information from which living organisms are built. They, too, must maintain a state of thermodynamic nonequilibrium (asymmetrical distribution of matter and energy) in order to sustain life and mind. Yet, at this small scale the force of gravity is weaker than that of the electromagnetic force. For life and mind to exist, nature had to create a new innovation — the boundary — bidirectional, semipermeable barriers separating living, nonequilibrium systems from their *source* (of matter, energy, and information) and their *sink* (repository for used, disordered matter and energy) — entropy. Without such boundaries, no smaller-scale system (biological organism) can maintain the necessary asymmetrical distribution of matter and energy necessary for its existence.

This can be simply understood by imagining a rectangular container with a partition down the center separating gases of two different temperatures (nonequilibrium). Once the partition is removed, the two gases rapidly begin to mix, and molecular collisions bring the entire container of gas to a uniform temperature (thermodynamic equilibrium — maximum entropy) at which point no further ordered behavior can occur within reasonable statistical probability.

The life of the biosphere as a whole is sustained by the presence of such a semipermeable barrier, an atmosphere. Earth's atmosphere is made up of gases and particles of various masses trapped by gravity. The atmosphere also contains a region called a *magnetosphere*, an electromagnetic field created by tectonic plate movement. The resulting semipermeable barrier regulates the flow of informed matter and energy into the biosphere from its *source* (the sun) as well as the flow of disordered (entropic) matter and energy to its *sink*: space.

In the same regard, nature had to design boundaries for smaller living systems. Thus, cells and organs have semipermeable membranes and organisms have skin as well as other elaborately configured sense organs designed to regulate the flow of matter and energy and maintain internal conditions of thermodynamic nonequilibrium necessary for life. Fatalities resulting from third-degree burns, malaria, gunshot wounds, tetanus, peritonitis, and ulcers have one important thing in common: a barrier at some level was compromised. When barriers are compromised, it is difficult and sometimes impossible for an organism to maintain the asymmetrical distribution of energy and matter necessary for life.

The Standard Theory of Pattern-Entropy Dynamics postulates that all human behavior can be understood as the regulation of informed (ordered) and entropic (disordered) matter and energy across the barrier(s) which separate the system (organism) and its various subsystems from source and sink. Thus, the purpose of all organismic level behavior is to maintain a delicate, dynamic state of nonequilibrium relative to the environment.

Once single-cell (prokaryotic) and multicellular (eukaryotic) life got started, nature had a new challenge before her — how to exchange information between unending numbers of multiple embedded, diversified boundaries (barriers within barriers).

INFORMATION TRANSLATION: PATTERN EXCHANGE BETWEEN DISPARATE SUBSTRATES

We have discussed where information comes from, why we incorporate information, how nature can use substrates of differential stability to create memory and to forget, how nature uses pattern for communication between disparate parts of a system resulting in collective coherent behavior, and how nature uses pattern to determine identity between discrete elements. We then discussed how nature uses pattern to compare and filter (four-dimensional selectivity) information and navigate aggregates, large and small, through their environment. It is clear that pattern-entropy dynamics is the keystone to a deeper understanding of brain, mind, behavior, information, and thus life itself. Yet a mere understanding of this relationship between humans and nature does not assuage our daunting struggle for an increasing quality of life. We must go beyond understanding the relationship between information and organism to physically and intentionally direct its evolution to the best of our capability. Such is the purpose of NeuroPrint, which is covered in Section II. Beforehand, a few additional questions must be answered and clarifications made.

THE FORCE THAT MEDIATES INFORMATION TRANSLATION BETWEEN SUBSTRATES

While we now understand that gravity is an essential universal force responsible for the existence of the information that biological organisms exchange with their environments, it is not by itself capable of mediating the complex information exchange that gave rise to organisms like ourselves. The most influential of the four known forces (strong, weak, electromagnetic, and gravity) in mediating complex information exchange in our universe is the electromagnetic force. The electromagnetic field of an *emitter molecule* organizes the atoms of the *receiver molecule* by making the atoms of the receiver deploy themselves in an analogue spatial pattern (Loewenstein, 1999).

It is no accident that carbon is an essential building block of the complex organizations of matter and energy referred to as life, and hence it is the most fundamental Tinkertoy® upon which the electromagnetic force acts. Carbon, as an encoding substrate, unlike the other known naturally occurring elements, is particularly efficient at preserving the stability of the complex information needed to assemble the living organism aggregates. It is also essential in preserving the fidelity necessary for replication and exchange of complex information. Carbon atoms can share their electrons to form stable covalent bonds, resulting in greater stability for information patterns encoded by the larger, more complex substrates built from a carbon foundation, such as proteins. The electrostatic fields of carbon constrain the atoms to occupy exactly four points in space around itself. Its quantum energy levels are few and wide apart. All of these intrinsic and unique attributes characteristic of carbon afford larger structures (patterns) built from carbon a greater precision and stability of organization (Lowenstein, 1999). Carbon's ability to bond in this way with atoms such as hydrogen, oxygen, and nitrogen endows it with the ability to be arranged by the electromagnetic force into a tremendously complex variety of asymmetric, aperiodic patterns. Therefore, the carbon atom is the best naturally occurring building material for the construction of a complex, diversified, precise, and stable informational bioarchitecture.

While gravity is the essential force that created the necessary conditions (an entropy gap) for information to exist in our universe, the electromagnetic force is most influential in mediating the exchange of that information between encoding elements and their large substrates, the most successful of which are carbon and the proteins built from them. This is the very keystone of bio-information translation, making complex life possible.

As a result of the confluence of these forces and elements, the human organism is a *critical system* in that a change in information, and hence the function or structure of any level of the system, can effectively initiate a change in every level of the system. Since the human organism operates at points very far from equilibrium, it is highly sensitive to initial conditions. Nature is capable of affecting all of the changes in such a system merely by shuffling the arrangement of the five grammatical rules of the language of pattern. At this time we would like to elucidate the process of information translation in finer detail. But as we do this, it should be borne in mind that this is only one of many ways in which this translation process can take place. It is always dangerous to elaborate such a parochial view, since one may erroneously assume that it is the only way in which the process can occur. Yet it is also essential for the reader to be able to incorporate the finer details of such a process.

As the range and scope of our perception increased in the biomedical sciences less than two decades ago, it gave way to the understanding that the human brain connects, interfaces, and communicates with absolutely every other system in the aggregation of cells and molecules of the human organism. From the muscles of the heart and skeleton to the varied arrays of immune system molecules that initiate collective and coherent defense against foreign invaders, the human organism is an elaborately interconnecting web of circular communication patterns, an information exchange network. While there is no true centralized location for the translation of information patterns in such a decentralized self-organizing system, it is helpful for sake of articulation and illustration to single out one pathway and arbitrarily punctuate it with discrete boundaries. A system well fitted for such an illustration is the limbic–hypothalamic–pituitary system, which can be thought of as a funnel drawing in information from disparate neural networks in the neocortex and translating its neurotransmitter-based encoding of information into hormonal and peptide-based information, which circulates through and between systems of the body. Many specialized neuroendocrine cells found in the brain's hypothalamus can translate information encoded in the form of nerve impulse patterns of the neocortex directly into hormonal messenger patterns that circulate throughout the body. In this way, we are able to translate pictures, sounds, tactile sensations, and even chemical information incorporated from our environment into feelings and emotions. This is made possible because one end of the neuroendocrine cell receives electrochemical impulses from the cerebral cortex via neurotransmitters, and the other end of the neuroendocrine cell discharges a neurohormone or neuropeptide in order to transmit information to some tissue of the body. These information patterns sometimes travel by bloodstream over long distances, or by paracrine mechanisms over short distances.

This is merely one of many information loops, large and small, that nature developed for brain-to-organism communication of environmental information. In this case, information patterns incorporated by the neural and subneural net are transmitted in parallel to another more fluid information processing substrate (pituitary-endocrinal information patterns) capable of less restricted movement throughout the organism. This can be thought of as a circulating nervous system. In this paradigm, the hypothalamus is an interface that translates environmental information patterns from a state of greater restricted activity (along nerve cells) to a state of lesser-restricted activity (greater mobility to circulate from system to system). This circulating nervous system of neuropeptides and hormones can also transmit information back to the neural-net, affecting global changes in brain function, as in the case where desire to eat is inhibited during onset of immune system attack on a foreign invader. This particular feedback loop (circular information exchange) is just one such system (pathway for the travel of proteins) implicated in the multiscale array of circular information exchange systems, ascribed to the establishment and maintenance of a physiological state.

For example, to produce a state referred to as "stress" or "anxiety," neuroendocrine cells must trigger the release of ACTH (adrenocorticotropic hormone), which in concert with catecholamines transmits a message to the endocrine and autonomic nervous systems to ready the body for "fight or flight." When adrenaline and catecholamines reach the next interface (a cell membrane), they exchange their information with secondary messengers inside a cell, one of which is known as cyclic adenosine monophoshate (cAMP). From a biophysical point of view, this transaction can be seen as the translation of information from one substrate to another, in this case between different types of proteins. While transmission of information along this pathway is useful for obviating immediate, environmental danger, the continual replication of such information patterns outside of the original context in which they were incorporated can lead to interference with immune system information processing.

Molecular biologists have reported that the long-term presence of such molecules can interfere with the immune system at the level of DNA by altering the gene expression of several vital proteins (four-dimensional, molecular machines) such as interleukin-2 (IL-2) receptors. Psychoneuroimmunologists have identified IL-2 as an essential protein pattern involved in the prevention and regression of cancer. In order to counteract the dangerous alteration of gene expression for IL-2, switching molecules called *transcription factors* must be sent from the brain directly to the DNA of the involved cells in order to switch back "on" the silenced gene. If this does not occur, ACTH and cortisol will continue to inhibit IL-2 synthesis. This is an example of the principle of internal conflict, or competition between information patterns. Endocrinology research has shown that the macroscopically observable physiological state most likely to resolve this conflict, and result in the necessary counterbalancing neuropeptide profile, is a state that we abstractly refer to as "exhilaration" (Ader et al., 1991; Despopoulos and Silbernagl, 1991; Greenspan and Baxter, 1994; Stites et al., 1994).

There are three important things to extrapolate from this illustration. The communication of an environmental danger across the organism's multiscale information interface requires the incorporation, replication, and transmission of information patterns from one substrate to another. Some of these encoding substrates allow for greater mobility of information (greater degrees of freedom), thus allowing the original pattern to be cleaved and replicated and to hence inform (shape or guide) disparate systems simultaneously. The cleaved patterns are later recombined to produce collective and coherent behavior of "fight or flight" which requires that information originating in the neurocognitive system is exchanged between the neuroendocrine system, cardiovascular system, and the musculoskeletal system. The five behaviors of pattern are at work at all times and in all systems in some combination, in order to mediate this information/pattern exchange.

Second, regardless of the ever-increasing number of names we assign to these encoding substrates, based on their particular molecular pattern and the circuitous paths of intraorganismic and interorganismic travel, the major players in the biological information translation game are atoms and proteins, amalgamated and arranged into a wide variety of asymmetric, aperiodic, information-rich patterns by the major mediating forces of gravity and the electromagnetic force.

Third, when one of these spatiotemporal patterns becomes too stable, it can be replicated with far greater ease, and its replication can be inadvertently triggered outside of appropriate context (the context in which it was incorporated). Such replication outside of appropriate context results in a myriad of abstracted malfunctions, including but not limited to phobias, compulsions, hallucinations, and post-traumatic stress disorder. This leads to the answer of an important question left unstated to this point.

WHY WE CAN'T STOP THINKING

Have you ever noticed how difficult it is to maintain a quiescent mind for more than a few seconds? How long can you sit or lie quietly without a single thought or sensory representation arising spontaneously in your mind? When a representation does arise, is it one of worry or excitement? Fear or pleasure? Want or satisfaction? Work or family? Unless you have had extensive training in silencing this activity through methods such as meditation, you probably cannot stop thinking for very long. Why can't we stop thinking once information has been incorporated? The answer is related to the reason why we incorporate information in the first place. What we call thinking of this sort is primarily the spontaneous replication of information patterns competing for the finite, limited information processing resources of the brain. To maintain the delicate balance between pattern and entropy, our brains are continually either incorporating, replicating, cleaving, recombining, or transmitting information. Therefore we watch television or read books (incorporate), remember and reminisce (replicate), analyze and evaluate (cleave), create and innovate (recombine), and talk on the phone or socialize at parties (transmit), relentlessly performing the five behaviors of information patterns. So when we are lying in bed in the dark, with restricted sensory input to incorporate and few people to talk with (transmit), we are limited to cleaving and recombining the information patterns that we are capable of replicating.

Which patterns will we replicate when incorporation and transmission are suppressed? We will replicate those patterns that have become most stable under the changing selection pressures from our environment (Blackmore, 1999; Calvin, 1996; Darwin, 1993; Edelman, 1987; Plotkin, 1993). (These patterns can be easily identified, modeled, and altered with the assistance of NeuroPrint.) Therefore, our ability to influence our spontaneously arising states of mind in large part depends on what patterns we incorporate and our facility with the essential behaviors of cleaving and recombining of those information patterns. Our formal education systems do not give equal emphasis to these information-processing behaviors. Only a paucity of examples of this type of training exists, in contrast to the emphasis placed on the mere incorporation, replication, and transmission of information patterns. In light of the insights made possible by the Standard Theory of Pattern-Entropy Dynamics, we would at the very least summarily urge a reevaluation of both current educational content and process. The standard theory necessarily postulates that such a resulting restructure of educational content and a rebalancing of training in the five grammars of the language of pattern (information-processing behaviors) could in principle result in not only a greater quality of life for the individual, but also for the whole of humanity. If life itself is information processing, it follows that the information we process and how we process and organize it will necessarily have a profound impact on the life of the mind, and hence the quality of human life.

MODELING THE ORGANIZATION OF INTERNAL COGNITIVE DOMAINS

In Section II, we shall employ the tools of NeuroPrint to elucidate the intricate informational bio-architecture that such stable information patterns tend to sculpt. It will then become easier to apprehend how the confluence of environmental information patterns manifest themselves in the structure and function of our cognitive-neurophysical systems, and in turn mediate the very quality of our mental and physical lives.

Since living beings continuously lose information during the act of living, there is only one way to prevent such a system from prematurely tending toward its state of highest probability, that of thermodynamic equilibrium. The key to preventing this devolution of a living system is to infuse new information so that the organism may maintain its high degree of order, as pointed out by Schrodinger in 1944 and numerous physicists since that time. Of all the pattern substrates we have discussed so far, the most versatile of all in accomplishing this task are the biological substrates called proteins. They are, to date, nature's most elaborate and versatile information-exchanging substrates responsible for the collection of environmental information and the exchange of that information across the many multiscale interfaces that make up a living organism.

In physics, the law of conservation warns us that the environment must necessarily undergo an equivalent increase in thermodynamic entropy; for each and every bit of information an organism gains, the entropy in its environment must rise by a certain amount. Hence, a living organism is continually exchanging information for entropy during the act of living. Sometimes the resulting entropy (disorder) can be reincorporated into the organism if it is sustained too closely within the organism's immediate environment (sink). As organisms, it is not always easy in many cases to remove ourselves from an environment to which we have exchanged entropy for information. Even if we can, we must also realize the deeper fact that we are environments within environments. That is to say that the most immediate, external environment of one organ or system inside our bodies is another organ or system; and the immediate external environment of a cell of one of our organs is the other cells that make up the same organ. When we do not view ourselves systemically, we can erroneously create disorder in one part of us as a result of creating order in another part of us. Thus, interventions must be designed systemically, carefully taking into account the potential effect on other essential, information-exchanging organizations of matter and energy both inside (such as belief systems) and outside (such as family members) of the human being.

FROM PROTEIN PATTERNS TO COGNITIVE NEUROPHYSICAL DOMAINS

As agents of change, our most effective interventions will be designed from macroscopic manipulation of information/pattern domains within a human, cognitive-neurophysical system. To accomplish such a task effectively, we must have a sense of how the behaviors of pattern at the microscopic level arise in homologous patterns at the macroscopic level. As noted, proteins are incredibly well suited for the recognition and identification of other molecules. This ability allows proteins to select appropriate molecules out of many possible choices, and this recognition occurs by direct molecule-to-molecule interaction. In this way, a protein "feels" the shape or pattern of a molecule, and thus the information it has to exchange. Each protein in our bodies, from neurotransmitters to second messengers, is endowed with a certain amount of information inherent within its orderly pattern. If a molecule in the external environment of a protein fits the proper configuration, it will join by way of intermolecular attractive forces and complete the information exchange.

Upon selection of a molecule, a protein undergoes a conformational change in its configuration as it is endowed with new information. In this sense, a protein can be said to have cognition. When the exchange is complete, the protein's configuration must be reset to its original state of information (i.e., clear its memory), to be ready for the next cognitive cycle. Loewenstein (1999) has persuasively argued that the unavoidable thermodynamic price is paid when the protein resets to its original informational state; entropy is exchanged with the protein's external environment during the act of forgetting. If you have ever become physically exhausted, hot, and fidgety after strenuous mental activity such as cramming for an exam or watching an intense action movie, you have undoubtedly experienced the information/entropy exchange macroscopically, between your neurocognitive and biophysical systems. This information/entropy exchange has deleterious consequences for children and teachers in a classroom environment. Learning (information incorporation and replication) is paid for by the conversion of biochemical energy into thermal and biomechanical entropy, causing incessant fidgeting and talking commonly mistaken as ADHD and hyperactivity (Furman, December 1998).

Information is captured and incorporated from our environment (stimuli) and translated from system to system, cleaved and separated, and later recombined and agglomerated, to give us an ongoing, evolving, unbroken experience of cognition. The evolving confluence of these ever-present behaviors of pattern gives rise to our mental life. In this way, the life of the mind is an incorporation of the mind of nature.

In both the fields of psychological intervention and education we arbitrarily cleave this seamless, multiscale act of information/pattern exchange into four macroscopic domains of convenience: stimuli, thoughts, emotions, and behaviors. It is critical to realize that the four domains of pattern overlap extensively and their boundaries blur, for they are no more than arbitrary linguistic parsing of seamless experience. Nevertheless, in order to understand how human behavior is organized by information, we must start here. Section II on NeuroPrint can be utilized to model the cognitive-neurophysical topology of a human being using these four domains.

FOUNDATION FOR A COGNITIVE NEUROPHYSICS

In summation, the Standard Theory of Pattern-Entropy Dynamics postulates the following:

1. Pattern is the universal language of nature. Pattern-Entropy Dynamics is the genesis of information regardless of substrate. Pattern is a substrate-neutral process fundamental to all living and nonliving matter at all scales. Pattern is synonymous with "informed" matter and energy — information.
2. Pattern and entropy are dynamically exchanged between living systems (organisms) and nonliving systems of matter and energy.
3. Pattern-entropy dynamics is the fundamental process driving the development of all microscopic and macroscopic behavior from the simple to the complex, including but not limited to memory, recognition, identification, comparing, filtering, communicating, navigating, coordinated collective action, thought, and mind. All of these behaviors are by-products of pattern-entropy dynamics.
4. All behaviors of greater complexity found in man and nature can be ascribed to the five fundamental grammars (behaviors) of pattern that are substrate and level-neutral: incorporation, replication, cleaving, recombining, and transmitting.

5. All human behavior can be understood as the regulation of informed (ordered) and entropic (disordered) matter and energy across the barrier(s) which separate the system (organism) and its various subsystems from *source* and *sink*. Thus, the purpose of all organismic level behavior is to maintain a delicate, dynamic state of nonequilibrium relative to the environment, which is necessary for healthy function, collective coordinated activity, and life itself. Behavior is a state-bound or state-dependent property of a system.
6. A significant and prolonged absence or excess of pattern, or presence of conflicting pattern, can be damaging or even fatal to the human, biophysical system (organism), while brief, entropy-increasing exposure is the keystone to change. The behavior of a substrate is by necessity the behavior of its cumulative information processing.
7. Brain, mind, behavior, and information are inseparable manifestations (manifest properties) of a single, fundamental, dynamic process — pattern-entropy dynamics. A change in one property necessarily engenders a change in all.
8. Behaviors, thoughts, emotions, and stimuli are all dynamic, spatiotemporal patterns of "informed" matter and energy resulting in a living, dynamic bio-architecture which is totally dependent upon the relative arrangement, order, structure, and probability amplitudes of these patterns, for its continued existence and healthy operation. Hence, the quality of human life can be significantly altered by the internal redesign of this "informed" dynamic bio-architecture. All effective interventions must necessarily be briefly entropy increasing.
9. All manner of malfunction in the human organism can be attributed to the expansion and contraction of natural variation in the five behaviors of pattern and the following three principles:
    a. **The Principle of Pattern-Entropy Balance** — An imbalance of information flow across the interface (brain) between environment and organism, resulting in either increasing entropy (neurocognitive and biophysical disordering) or the formation of Markov chains and ultimate quantum reoccurrence (neurocognitive and biophysical periodicity).
    b. **The Principle of Organism-Environment Correspondence** — A lack of correspondence between organism and environment, resulting from either incorrect or incomplete translation of information (information/pattern exchange) across the organism's interface, or the untimely, spontaneous replication of information by the organism, observed outside of the context to which it corresponded and in which it was originally incorporated.
    c. **The Principle of Conflict/Competition Between Information Patterns** — A disruption in information translation between subsystems, organs, cellular interfaces, and molecular signaling systems due to the incorporation, replication, and transmission of conflicting information

patterns with separate and competing functions, and resulting in action that is both uncoordinated and not collective or coherent in nature.

As a practitioner or scientist completely apprehends this relationship between the behavior of information and the behavior of its substrate, the previously anomalous structure (form) and function (behavior) of an organism becomes increasingly predictable, influenceable, and understandable.

REFERENCES

Ader, R., Felten, D. L., and Cohen, N. (Eds.). (1991). *Psychoneuroimmunology (2nd Edition).* New York: Academic Press.

Basar, E., Flohr, H., Haken, H., and Mandell, A. J. (Eds.). (1983). *Synergetics of the Brain.* New York: Springer-Verlag.

Blackmore, S. (1999). *The Meme Machine*. New York: Oxford University Press.

Cairns-Smith, A. G. (1985). *Seven Clues to the Origin of Life*. New York: Cambridge University Press.

Calvin, W. H. (1996). *The Cerebral Code: Thinking a Thought in the Mosaics of the Mind.* Cambridge, MA: MIT Press.

Darwin, C. (1993). *The Origin of Species*. New York: The Modern Library.

Davies, P. (1999). *The Fifth Miracle: The Search for the Origin and Meaning of Life*. New York: Simon & Schuster.

Despopoulos, A. and Silbernagl, S. (1991). *Color Atlas of Physiology (4th Edition).* New York: Georg Thieme Verlag Stuttgart.

Edelman, G. M. (1987). *Neural Darwinism: The Theory of Neuronal Group Selection*. New York: BasicBooks.

Einstein, A. (1961). *Relativity: The Special and the General Theory.* New York: Three Rivers Press.

Furman, M. (1996b). Foundation of neurocognitive modeling: eye movement — a window to the brain. *Anchor Point*, December, 14, 10–12.

Furman, M. (1997a). NeuroPrint — Human Performance Modeling and Engineering. Unpublished lecture on video. Orlando, FL.

Furman, M. (1998b). Intelligent learning systems. *Anchor Point*, December, 12(12), 21–29.

Goldstein, M. and Goldstein, I. F. (1993). *The Refrigerator and the Universe: Understanding the Laws of Energy.* Cambridge, MA: Harvard University Press.

Greenspan, F. S. and Baxter, J. D. (1994). *Basic & Clinical Endocrinology (4th Edition).* Norwalk, CT: Appleton & Lange.

Haken, H. (1983). *Synergetics: An Introduction Nonequilibrium Phase Transitions and Self-Organization in Physics, Chemistry and Biology (3rd Edition).* New York: Springer-Verlag.

Haken, H. (1988). *Information and Self-Organization: A Macroscopic Approach to Complex Systems*. New York: Springer-Verlag.

Kauffman, S. (1993). *The Origins of Order: Self-Organization and Selection in Evolution.* New York: Oxford University Press.

Kelso, J. A. S. (1995). *Dynamic Patterns: The Self-Organization of Brain and Behavior.* Cambridge, MA: MIT Press.

Kruger, J. (Ed.). (1991). *Neuronal Cooperativity.* New York: Springer-Verlag.

Loewenstein, W. R. (1999). *The Touchstone of Life: Molecular Information, Cell Communication, and the Foundations of Life*. New York: Oxford University Press.

Maynard Smith, J. and Szathmary, E. (1999). *The Origins of Life: From the Birth of Life to the Origin of Language*. New York: Oxford University Press.
Mishra, R. K., Maab, D., and Zwierlein, E. (1994). *On Self-Organization: An Interdisciplinary Search for a Unifying Principle*. New York: Springer-Verlag.
Penrose, R. (1994). *Shadows of the Mind: A Search for the Missing Science of Consciousness*. New York: Oxford University Press.
Plotkin, H. (1993). *Darwin Machines and the Nature of Knowledge*. Cambridge, MA: Harvard University Press.
Posner, M. and Raichle, M. (1994). *Images of Mind*. New York: W. H. Freeman and Company.
Prigogine, I. (1996). *The End of Certainty: Time, Chaos, and the New Laws of Nature*. New York: The Free Press.
Roland, P. E. (1993). *Brain Activation*. New York: Wiley-Liss.
Schrodinger, E. (1944). *What is Life? With Mind and Matter and Autobiographical Sketches*. New York: Cambridge University Press.
Smolin, L. (1997). *The Life of the Cosmos*. New York: Oxford University Press.
Stites, D. P., Terr, A. I., and Parslow, T. G. (1994). *Basic & Clinical Immunology (8th Edition)*. Norwalk, CT: Appleton & Lange.
Thelen, E. and Smith, L. (1994). *A Dynamic Systems Approach to the Development of Cognition and Action*. Cambridge, MA: MIT Press.
Tipler, F. (1994). *The Physics of Immortality*. New York: Anchor Books Doubleday.

Section II

NeuroPrint and the Standard Theory of Pattern-Entropy Dynamics

"Until thought is understood — better yet, more than understood, *perceived* — it will actually control us; but it will create the impression that it is our servant, that it is just doing what we want it to do."

David Bohm

"Knowledge organizes itself geometrically."

R. Buckminster Fuller

6 NeuroPrint — Purpose and Preliminaries

CONTENTS

THEORY GEOMETRIZING: THE EINSTEIN/MINKOWSKI SOLUTION

In 1905, Albert Einstein published four original papers on what seemed to be very disparate branches of physics. The theories presented in these papers were so revolutionary in scope that they became paradigm theories, shifting the way scientists perceived the world and the universe. One of these later became known as the special theory of relativity. However, regardless of the accuracy and applicability of Einstein's theories, they did not begin to gain acceptance among physicists until about the middle of 1907. Herman Minkowski, one of Einstein's former mathematics professors, first noticed the primary reason for this delay. He observed that the special theory of relativity, in the form in which it was originally articulated, defied visualization. Most of the relations between space and time were described in arcane mathematical formulas, understood by only a handful of theoretical physicists at the time. To overcome this problem, Minkowski developed a geometrical topology of expressing space–time relationships. Minkowski's pictorial representation of the implications of Einstein's theory dramatically accelerated its acceptance and utility among the scientific community, while making these revolutionary concepts accessible to the nonspecialist as well. Similarly, NeuroPrint has been designed as a pictorial articulation of the implications of the standard theory of pattern-entropy dynamics. While in Section I we explored thermodynamics, the physics of energy dynamics, in order to help us develop insight into cognitive neurodynamics, in this section we combine knowledge of thermodynamics with dynamical systems physics, the science of change.

A DYNAMICAL BLUEPRINT OF "INFORMED" BIO-ARCHITECTURE

The purpose of NeuroPrint is to provide a simple diagramming technique that can be used to model neurocognitive topology. To deeply understand human behavior, we must first be able to visualize the bio-architecture that results from the organizing force of information and the disorganizing force of entropy. Brain, mind, and behavior should be thought of as spatiotemporal bio-architectures resulting from the flow of information and entropy across the boundaries separating human systems from their environment.

It is essential to understand that the human bio-architecture resulting from the continued flow of information and entropy is dynamic, constantly changing. One of the least stable, most dynamic processing substrates of the brain, the cytoskeletal-net, changes pattern within the nanosecond (billionth of a second) time scale in response to information. However, with regard to the more stable cognitive domains of habitual behavior — thoughts and emotions — we are discussing information patterns relegated to the more stable substrates of the human brain and biophysical system. NeuroPrint will be used in this section to model the structure and organization (arrangement) of and the relationship between these neurocognitive domains and the stimuli (information patterns) that created them. In this way, both scientist and practitioner can easily visualize the system as a whole.

As a NeuroPrint is developed on paper, it becomes an indispensable map that guides intervention and deepens understanding. At the same time, it is information about the neurocognitive system that can be fed back through the very system from which it came. It has been observed that such a feedback process can yield spontaneous change in a person who has for the first time become aware of hidden internal relationships (Furman, 1997a). NeuroPrint can be thought of as a dynamical blueprint of the neurocognitive bio-architecture. The very process of modeling neurocognitive activity necessarily changes that activity.

THE PURPOSE OF NEUROPRINT

Behaviors, thoughts, emotions, and stimuli are dynamic, spatiotemporal patterns of "informed" matter and energy, resulting in a living, dynamic bio-architecture which is dependent upon the relative arrangement, order, structure, and probability amplitudes of these patterns for its continued existence and healthy operation. These complex patterns are the foundation of our daily cognitive neurodynamics. Hence, the quality of human life can be significantly altered by the internal redesign of this dynamic bio-architecture. Bearing this in mind, NeuroPrint has been designed to accomplish the following:

- NeuroPrint makes the relations among reoccurring neurocognitive patterns visualizable within the larger systems of which they are a part and also affect. The resulting picture is referred to as *Hilbert Space* (or *phase space*), a three-dimensional space where all possible states of a system are plotted. Hilbert Space is also a representation of the neurocognitive system's *design space*: a four-dimensional space where all possible states can be realized spatiotemporally. Design space is where information interacts to dynamically create the informational bio-architecture, which results in the states and behaviors of a system. When the neurocognitive system repeatedly tends toward a particular functional organization or pattern in design space or phase space, that pattern is called an *attractor*. For example, within a particular neurocognitive system there may be attractors such as anticipation, distant, "wanting to escape," boredom, satisfaction, joy, etc. Attractors formed by complex systems tend to organize into larger spatiotemporal clusters called *basins of attraction*, which in turn organize into larger agglomerations referred to as *attractor landscapes*. The result is a profound visual representation of hidden dynamic relationships organized by the flow of information through the neurocognitive system.

- NeuroPrint provides a way to predict the behavior of greatest probability for a given context (i.e., the stimulus field of "informed" matter and energy). NeuroPrint is an algorithmic process derived from the Standard Theory of Pattern-Entropy Dynamics that can be used to yield a topological model of any selected domain of neurocognitive design space.
- NeuroPrint provides a map to guide intervention designed to influence bio-architecture and resulting behavior, while allowing the practitioner to predict other probable effects (ecology) of a chosen intervention. Hence, it makes the selection of appropriate intervention and the design of new intervention tools much simpler.
- NeuroPrint modeling reveals areas of high leverage where small changes can be made in order to facilitate large-scale systemic reorganization. Hence, the practitioner diminishes the time and effort necessary to realize the therapeutic objective.
- NeuroPrint reveals system readiness to change as well as indicates the relative degree of stability of each neurocognitive domain and each pattern within a domain modeled (i.e., the more unstable, the more ready to change). The standard theory of pattern-entropy dynamics provides us with the principles necessary to effect system readiness to change. By necessity, all interventions must be temporarily entropy increasing. This increase in entropy can be targeted to the entire system, such as in the case of hypnotic trance induction, or it can be targeted to a selected domain of neurocognitive activity, whether the domain is large (behavior, emotion, or thought) or small (submodality of a sensory representational system). When targeting a selected domain, the scientist/practitioner can choose from a more narrow class of pattern-interrupting interventions.
- NeuroPrint provides a way to categorize and organize intervention tools by the effect they have on the neurocognitive bio-architecture. These effects include stabilizing neurocognitive domains (neural network attractors), destabilizing neurocognitive domains, and rearranging trajectories (phase paths) between (connecting) neurocognitive domains.
- NeuroPrint provides a method to measure the stability of a selected neurocognitive domain relative to any other connected neurocognitive domains, thus indicating the probability of any existing information pattern being activated, and in turn enslaving the biomechanical system in the service of a collective, coordinated behavior. NeuroPrint also guides

the rearrangement of this probability distribution in favor of a selected ecological outcome.

- The same measurement algorithm used to determine the initial stability of a pattern (state, behavior, thought, etc.), its control over the system, and its readiness to change can also be used to measure the present stability and organization of that pattern relative to itself at an earlier time. This type of comparison allows us to precisely gage the degree to which an intervention has produced reliable, stable change in an intended direction, while also providing a way of anticipating future problems through the detailed analysis of present structure.

Behavior is an intrinsic emergent property of our internal informational bio-architecture. It is a macroscopic window to the microorganization within and provides profound clues to disturbances in that organization. With the precise modeling tools of NeuroPrint, we develop a detailed picture of the cognitive neurodynamics that influence the quality of our daily lives. With this new way of seeing we may finally begin to understand the deep structure that lies within.

TOOLS NEEDED FOR THE DEVELOPMENT OF A NEUROPRINT

The following are the only essential tools needed develop a NeuroPrint:

- A large sketch pad
- Colored pencils or thin markers
- A stopwatch or a watch with a second hand
- A basic four-function calculator

7 Developing a NeuroPrint of Cognitive Neurodynamics

CONTENTS

STANDARD METHODS AND TECHNIQUES

NeuroPrint is a content-neutral algorithm for modeling the flow of informed matter and energy and the resulting bio-architecture. The picture it yields is a topology of relations between aggregates (attractors) which determine the structure and function of the larger bio-architecture. While NeuroPrint can be applied to a multitude of human systems, the examples provided here focus on the neurocognitive system of the individual. The principles of NeuroPrint are best learned by initially modeling your own cognitive processes.

STEP 1: LISTING PHYSIOLOGICAL/EMOTIONAL STATES EXPERIENCED WITHIN A GIVEN TIME FRAME

List on a separate smaller piece of paper or at the top right-hand corner of the large sketch pad all of the physiological/emotional states you or your subject have experienced over the last few weeks to 1 month. Use two columns which separate negative (–) states from positive (+) states, placing the symbols at the top of each column. One way to facilitate this process is by looking at a calendar, appointment book, daily planner, or journal for specific notes you have made, thereby triggering memories of each day individually. It is preferable to do this in order, to reveal *transition states* between major emotional attractors, which may be hidden. A transition state is an intermediary state used to get from one emotion to another. Since these states can be quite elusive, close attention will be needed to list them in the order in which they were experienced. Mentally scan through each day, noticing both reoccurring states and new states. If you have difficulty remembering these emotional states, start by scanning your memory by events, conversations, places, clothing worn or foods eaten each day, books read or movies watched during the week, changes in weather, songs heard, smells, etc. As you continue this scan, you will notice that those emotions that are more intense or more frequent will be more easily remembered (replicated by neurocognitive and biophysical system). The emerging picture at this stage reveals the *degrees of freedom* the neurocognitive system has to experience (organize into) discrete states or phases.

The actual degrees of freedom possessed by a mature system is far less than the initial potential of a finite state system of this complexity. Notice Table 7.1. This

TABLE 7.1
Neurocognitive Degrees of Freedom for State/Phase (Total Number of Recurrent States = 31)

(–)	(+)
Boredom	Excitement
Frustration	Curiosity
Anger	Love
Anxiety	Satisfaction
Exhaustion	Contentment
Confusion	Joy
Overwhelmed	Pride
Fat	Loved
Torn/indecisive	Settle
Fear	Anticipation
Trapped/confined →	Escape–Possibility–Planning
Disappointment	Happiness
Stressed	Trust
Embarrassed	Trusted
Annoyed	
Unorganized →	Organized

neurocognitive system possesses 31 recurrent states/phases or degrees of freedom. In this case, negative and positive transition choices are approximately equal in number, which is rarely the case. As neurocognitive systems approach collapse from too much order (periodicity), their total degrees of freedom for state/phase decreases significantly. When they approach collapse from too much entropy, their degrees of freedom increase until macroscopic state differentiation is almost impossible.

While you are cataloging emotional states/phases, there are a few helpful things to bear in mind. First, emotional states/phases should be named with a single word whenever possible. The names of emotion states/phases are distinction memes. In the example above, the subject had a state called "feeling a need to get out" which presupposed perceived options that could be used to "get out." The subject determined that this was qualitatively different than the feeling of being "trapped," which presupposed that there were no perceived options that could be used to "get out." To differentiate the state of "feeling a need to get out" from a state of "trapped," the description of the feeling was shortened to "escape." Whenever the name of a state is altered, it is important to determine that it still reliably triggers the intended state/phase.

Second, in this case, while "trapped" is considered by the subject to be an emotion of negative (–) valence, the feeling of "escape" is considered to be positive (+) as it acts as a transition state that quickly and reliably leads to "possibility" (+), and then to a state of mental "planning" (+). Arrows can be used to delineate states that are chained together in an ongoing trajectory or phase path (transitional bridges linking states). It is advisable to delineate as many of these states as possible at this stage, since these emotional states (attractors) will be grouped closely together when diagramming the NeuroPrint in Step 3.

Third, a clear distinction should be made between states such as feeling "love for" someone and feeling "loved by" someone. Similarly, feeling "trusted by" someone does not necessarily mean that you feel "trust for" someone. These states are qualitatively different and give rise to distinctly different *state-bound behaviors*. To accurately diagram these states, they must occupy different spatial locations on the NeuroPrint, as we will see in Step 7.

Step 2: Determining the Frequency (F) of Replication or Recurrence for Each State

Next we must determine the relative frequency of occurrence/replication of each biophysical state/phase for a given period. Relative frequency is an indicator of the stability of a state/phase, which is essential for predicting the most probable state/phase, and by implication, the most probable behavior for a given context (stimulus field/information field). The frequency (F) of an emotional state/phase is often proportional to the number of stimuli in the stimulus field (internal and external) that can potentially trigger that state (its basin of attraction).

To determine frequency, both daily and weekly time scales should be considered. If an emotion is experienced frequently, it is best to consider the number of times per day and then multiply by seven (7) days to yield a weekly index. The resulting frequency index should be converted to either one or the other (occurrences per day

TABLE 7.2
Frequency Distribution Index for State/Phase (Number of Times per Week)

(F)	(-)	(F)	(+)
2	Boredom	4	Excitement
10	Frustration	3	Curiosity
5	Anger	42	Love
3	Anxiety	14	Satisfaction
10	Exhaustion	7	Contentment
2	Confusion	5	Joy
10	Overwhelmed	5	Pride
3	Fat	14	Loved
10	Torn/indecisive	4	Settle
3	Fear	7	Anticipation
10	Trapped/confined →	10	Escape–Possibility–Planning
4	Disappointment	14	Happiness
14	Stressed	10	Trust
1	Embarrassed	10	Trusted
5	Annoyed		
1	Unorganized →	7	Organized
92		149	

or per week) for all of the emotional states, depending upon which one yields the most useful frequency distribution picture.

The frequency distribution index (Table 7.2) begins to yield a qualitatively different picture. Notice that in Table 7.1 what seemed to be an equal distribution of possible state/phase alternatives for the neurocognitive system is actually an asymmetrical temporal distribution of system alternatives. The most dramatic contrast is that the neurocognitive system phase transitions to the state of "love" 42 times per week, while phase transitioning to the state of "embarrassment" a mere 1 time per week. The state of "stress" occurs 14 times per week, while a state of "curiosity" occurs only 3 times per week.

Totaling both positive and negative columns yields another overall picture. In a given week, this neurocognitive system evidences more positive state/phases than negative ones (a ratio of 149:92). However, this ratio does not necessarily mean that the subject spends more time experiencing positive feelings and less time having negative ones. A frequency distribution index is only one of three measures of a state's stability. In this case, the frequency distribution calls attention to the fact that the neurocognitive system's biochemical and biomechanical architecture is more often influenced by positive rather than negative stimuli.

As we will see in Steps 3 and 4, *phase velocity* and *dwelling time* create an even more asymmetrical distribution of alternatives. For example, dwelling time tells us the duration of time the system spends in a particular state. It may appear initially that "love," with a frequency index of 42, has greater control over the system than "anger," with a frequency index of 5. However, if we determine that the average dwelling time for the state of "love" is 2 minutes and the average dwelling time for "anger" is 2 hours, then the state/phase of "love" would determine the system's behaviors for an average of 1 hour and 22 minutes per week, while the state/phase "anger" would determine the system's behaviors for an average of 10 hours per week. Therefore, all three measures of state (attractor) stability must be taken into account in order to yield an accurate picture.

Step 3: Determining Phase Velocity (V) for Each Recurrent State/Phase

The speed at which a biophysical state can be replicated is called its phase velocity. Phase velocity is measured with a stopwatch or the second hand of a watch, beginning from the time the subject is asked to experience the state and ending when the first physiological indicators are observed that the emotional state has been entered.* Alternatively asking the subject to nod when first beginning to experience the state may facilitate this process. Phase velocity will usually be measured in seconds. Measurements of phase velocity can vary, relative to the stability of the attractor from which the state vector is coming. Because of this, we may sometimes choose to measure phase velocity only after we have diagrammed the connecting states.

Those biophysical patterns that are most quickly and easily replicated (i.e., having the highest phase velocities) during competition for limited spatiotemporal resources of the neurocognitive system will necessarily be most effective in enslaving the biomechanical system in the service of a collective, coordinated behavior. Phase velocity is the second important indicator of attractor stability. The higher the phase velocity (the faster the subject experiences the emotion), the more stable the state/phase. The lower the phase velocity (the longer it takes for the subject to experience the emotion), the less stable the state/phase.

Whether the subject is asked to begin experiencing the state or the state is triggered with a stimulus, this is referred to as *releasing the system (state vector) from its initial conditions*. When the subject has just entered the state/phase, this is referred to as the *trapping zone* of the emotion attractor, the point at which the state vector is *captured*. The neurocognitive system can also be released from its initial conditions by using a stimulus (anchor) that commonly triggers the state/phase. For example, if the subject becomes "angry" whenever insulted, the actual act of insulting would yield a far more accurate index of phase velocity than by simply asking the subject to experience "anger."

Obviously, the practitioner must make a decision as to the ecology of such a technique. When it is not deemed ecological by the practitioner to use the actual stimulus, such as when treating a snake phobia, the subject should simply imagine the stimulus. Imagining a stimulus that reliably triggers an emotional state/phase will still yield a much more accurate index of phase velocity than simply asking the subject to experience the state. Of course, in many instances the subject will lead into the emotional state by doing such imagining.

---

* Notice our culture's habitual use of "container" metaphors when referring to emotional states.

In summary, there are three ways in which phase velocity can be measured:

- Having the subject simply access the emotional state under consideration, without also accessing an internal stimulus such as an image or sound
- Having the subject imagine a stimulus that reliably triggers the state to be measured
- Reconstructing the stimulus field that reliably triggers the state to be measured

In Table 7.3, states having the highest phase velocity (indicated by the short time needed to activate state) are represented by the lowest number of seconds.

In Table 7.3, we have indicated the phase velocity of each state/phase in seconds. Notice that "excitement," "curiosity," "love," "happiness," and feeling "trusted" all have extremely high phase velocities of 1 second. This means that the subject's neurocognitive system only requires an average of 1 second to transition into any of these states. Also notice that "anger," "confusion," "overwhelmed," "fat," "trapped/confined," "stressed," and "unorganized" all take at least 10 times longer and some take up to 40 to 60 times longer for the neurocognitive system to activate. In a competition for the neurocognitive system's limited spatiotemporal resources, any of the positive states just mentioned would overcome the negative states or phases, providing that the system was released from both competing initial conditions at the same time. For example, if the subject entered

TABLE 7.3
Phase Velocity Distribution Index (Time Needed to Activate/Replicate State in Seconds)

(F)	(V)	(–)	(F)	(V)	(+)
2	**3**	Boredom	4	**1**	Excitement
10	**2**	Frustration	3	**1**	Curiosity
5	**15**	Anger	42	**1**	Love
3	**3**	Anxiety	14	**4**	Satisfaction
10	**2**	Exhaustion	7	**2**	Contentment
2	**60**	Confusion	5	**3**	Joy
10	**20**	Overwhelmed	5	**5**	Pride
3	**40**	Fat	14	**3**	Loved
10	**5**	Torn/indecisive	4	**25**	Settle
3	**1**	Fear	7	**2**	Anticipation
10	**20**	Trapped/confined →	10	**21**	Escape–Possibility–Planning
4	**3**	Disappointment	14	**1**	Happiness
14	**10**	Stressed	10	**8**	Trust
1	**2**	Embarrassed	10	**1**	Trusted
5	**1**	Annoyed			
1	**65**	Unorganized →	7	**10**	Organized
92			149		

a particular *information field* (home, office, etc.) where two competing stimuli exist that reliably trigger positive and negative emotional states, respectively, if they were encountered at the same time, we can predict that the neurocognitive system would enter the positive state.

Using the above phase velocity distribution, we simulated several events containing a stimulus (information pattern) that reliably triggers "excitement" (having a phase velocity of 1 second) and a stimulus that reliably triggers "disappointment" (having a phase velocity of 3 seconds); and, as predicted by the Standard Theory of Pattern-Entropy Dynamics, the subject reliably transitioned into the "excitement" attractor (Furman, 1997b). This relationship holds true for all the states in the distribution and has important implications which influence the construction, use, and outcome of interventions designed to release the neurocognitive system from two or more initial conditions simultaneously.

One such intervention used by practitioners of NLP is called "collapsing anchors," which forces two or more attractors (usually states) to compete for the state vector at the same time. While this process has been described as resulting in an "integration" of the two states, it rather appears that collapsing anchors serves to perturb (disrupt) both attractors, leading to a temporary increase in neurocognitive entropy and a *shallowing* (weakening) of both attractors. The body sensations created during this temporary increase in entropy (disordering of both patterns) are experienced by the subject as a hypnotic trance.

When measuring phase velocity, some states/phases are measurable independently, while others always require transition from a previous state/phase. In Table 7.3, the subject could not experience "pride" without first experiencing "joy." In this case, "joy" becomes part of the basin of attraction for "pride." This means that the state of "joy" is a stimulus or set of initial conditions that the neurocognitive system must be released from in order to experience "pride." In order to measure the phase velocity for "pride," we necessarily have to take into account the phase velocity plus the dwelling time (time spent in the state) for "joy." For instance, if "joy" has a phase velocity of 3 seconds and a dwelling time of 2 seconds, "pride" may have a phase velocity of up to 5 seconds.

We encounter the same situation when we measure the phase velocity for "escape." In order to feel the need for "escape," the subject must always initially activate the state of "trapped," which has a phase velocity of 20 seconds and, surprisingly, a dwelling time of only 1 second prior to transition to the state phase of "escape." Therefore, the phase velocity for "escape" is 21 seconds. These intricate relationships between states/phases will become much more apparent when we get to the diagramming steps of NeuroPrint.

It is also valuable to note that when the subject was asked to experience the state of "satisfaction," an internally visualized review of the number of tasks accomplished during a given time period was the necessary initial condition to trigger a transition into "satisfaction." In this case, the phase velocity of 4 seconds was the length of time it took for this internal visual review.

Step 4: Determining Dwelling Time (D) for Each Recurrent State/Phase

Dwelling time is the length of time a particular state or phase (attractor) in design space is activated without interruption — the duration of a state/phase. If a state is represented by phase space, its dwelling time is the length of time the state vector dwells in a particular attractor of an attractor landscape.

Thinking of it this way helps us to develop a dynamic (four-dimensional) representation of an attractor landscape for use in computer simulation or even mental simulation. It is helpful to read a final NeuroPrint by imagining a ball or marble traveling over a bumpy terrain, with hills and valleys representing attractors. Those attractors that are more stable are represented by deeper valleys, and attractors with a low phase velocity are represented by a steep hill with a ditch or drop at the top (like the mouth of a volcano) to capture the ball. The depth of the volcano would represent the stability of the attractor. Dynamic visualization is important for reading a final NeuroPrint and understanding the cognitive neurodynamics it represents.

Dwelling time is the third indicator of stability of an attractor. Dwelling time can be measured with a stopwatch or second hand of a watch for states of short duration. We begin to calculate dwelling time when the subject reaches the trapping zone of the emotional attractor. Measurement stops when subject begins phase transition to a new state. For states of long duration (more than a few minutes), the subject should estimate the relative amount of time each emotional state is experienced simply by reviewing several different contexts in which the state previously occurred.

In some cases sufficient time is not available to precisely time each state's dwelling time. In such cases a reasonably accurate representation of each state's relative stability can be garnered through verbal elicitation. For example, the subject can be asked to mentally represent several instances of the emotional state being measured and then directed as follows: "Once the external stimuli are removed, on average, for how many seconds, minutes, hours do you continue to experience (feel) the state of (X) before transitioning to another state or selecting a behavior that interrupts that state?"

In Table 7.4 dwelling time is represented in minutes, since only one state, contentment, lasted for longer than one hour. The time frame chosen to represent the result must be consistent for all states/phases being compared by their dwelling time.

It should be noted that in this case most of the states/phases have short dwelling times. When the dwelling times for each phase are short (most under 15 minutes), the states/phases of the neurocognitive system are primarily driven by external stimuli. When they are long (hours to days), such as in severe depression, they are primarily driven by internal stimuli, incorporated memes, and internal organization of trajectories in design space. Of course there are always exceptions. A subject may be immobile (severely limited degrees of behavioral freedom), and therefore unable to leave a particularly influential stimulus field. Such conditions can solidify bioarchitecture rapidly.

TABLE 7.4
Dwelling Time Distribution Index (Number of Minutes System Spends in Each State/Phase)

(F)	(D)	(V)	(–)	(F)	(D)	(V)	(+)
2	**20**	3	Boredom	4	**7**	1	Excitement
10	**3**	2	Frustration	3	**3**	1	Curiosity
5	**20**	15	Anger	42	**1**	1	Love
3	**10**	3	Anxiety	14	**10**	4	Satisfaction
10	**30**	2	Exhaustion	7	**120**	2	Contentment
2	**2**	60	Confusion	5	**4**	3	Joy
10	**3**	20	Overwhelmed	5	**5**	5	Pride
3	**2**	40	Fat	14	**1**	3	Loved
10	**5**	5	Torn	4	**1**	25	Settle
3	**5**	1	Fear	7	**5**	2	Anticipation
10	**1**	20	Trapped →	10	**8**	21	Escape–Possibility–Planning
4	**2**	3	Disappointment	14	**2**	1	Happiness
14	**7**	10	Stressed	10	**1**	8	Trust
1	**1**	2	Embarrassed	10	**1**	1	Trusted
5	**5**	1	Annoyed	**2**	**10**	**1**	**Surprise**
1	**1**	65	Unorganized →	7	**3**	10	Organized

States with shorter dwelling times are commonly used by the neurocognitive system as transition bridges to more stable states. Stable state attractors capture many weaker trajectories and make them part of their own attractor basin in order to maintain biochemical and biomechanical control of the system. This is referred to as *enslaving*, described in more detail in a later step.

During the elicitation (Table 7.4), the subject noticed a significant difference in the dwelling time for the state of "pride" from context to context. Sometimes it was 5 minutes and at other times it was 15 minutes. Upon further questioning we realized that when the subject was "surprised" with the accomplishment, the dwelling time was 15 minutes. However, 10 of the 15 minutes was actually spent in the state of "surprise" and only 5 in the state of "pride." This is a very common way of uncovering missing (unrepresented) states. Since "surprise" had not previously been represented, we measured phase velocity and frequency before including it in the dwelling time distribution. Including this new state in our neurocognitive state representation adds one degree of freedom for state selection to the original 31 degrees computed in Table 7.1.

Although not represented here, it is important to note that the particular state-bound behavior chosen by the subject while in the state will necessarily determine the dwelling time of that state. Some behaviors selected will interrupt and weaken the state and others will reinforce and stabilize it. This is where a careful analysis of habitual behaviors becomes important, as they can potentially influence system state selection and stability. If relationships such as this exist in design space, it is much easier to uncover them when state-bound behaviors are diagrammed with NeuroPrint.

Step 5: Determining Stability Distribution Index for Neurocognitive System

While frequency of recurrence is a good measure of the relative stability of a system's state/phase, a much more reliable indicator is the *stability index*. We arrive at a stability index (S) by multiplying frequency (F) by dwelling time (D).

$$S = F \times D$$

This index (Table 7.5) shows us the total time that the neurocognitive system spends in one particular phase/state relative to another and yields a much more accurate picture of the asymmetrical distribution of neurocognitive resources expended.

The stability of an attractor is referred to as its *depth*. The more stable the attractor is, the deeper it is. Stability is also an important indicator of resistance to perturbation (disturbance). Deep attractors are able to survive in the face of disturbance. That is, a deep attractor is better able to recapture the *state vector* after a brief interruption or disturbance and co-opt the limited resources of the neurocognitive system.

This can easily be understood by picturing a small marble (state vector) rolling around on a dinner plate (a shallow attractor). If the plate is balanced on a hand and rocked back and forth slightly, the marble will roll off the plate. If a bowl is being balanced nearby (deeper attractor), the marble (state vector) may be captured by it. If the same perturbation is applied to the bowl it will resist the loss of the state vector.

TABLE 7.5
Stability Distribution Index (Relative Stability of Each State/Phase)

(F) x	(D) =	(S)	(–)	(F) x	(D) =	(S)	(+)
2	20	**40**	Boredom	4	7	**28**	Excitement
10	3	**30**	Frustration	3	3	**9**	Curiosity
5	20	**100**	Anger	42	1	**42**	Love
3	10	**30**	Anxiety	14	10	**140**	Satisfaction
10	30	**300**	Exhaustion	7	120	**840**	Contentment
2	2	**4**	Confusion	5	4	**20**	Joy
10	3	**30**	Overwhelmed	5	5	**25**	Pride
3	2	**6**	Fat	14	1	**14**	Loved
10	5	**50**	Torn	4	1	**4**	Settle
3	5	**15**	Fear	7	5	**35**	Anticipation
10	1	**10**	Trapped →	10	8	**80**	Escape–Possibility–Planning
4	2	**8**	Disappointment	14	2	**28**	Happiness
14	7	**98**	Stressed	10	1	**10**	Trust
1	1	**1**	Embarrassed	10	1	**10**	Trusted
5	5	**25**	Annoyed	2	10	**20**	Surprise
1	1	**1**	Unorganized →	7	3	**21**	Organized

Given a system with just these two states, a negative one represented by the bowl (deep attractor) and a positive one represented by the plate (shallow attractor), it would be very difficult for the state vector to transition back into the plate (positive/shallow attractor). To do this we must perturb the negative state with a competing pattern, enough for its attractor to shallow, giving up its state vector, and at the same time deepen (stabilize) the positive attractor enough to capture the state vector and resist further perturbation.

This is the very essence of intervention and the keystone to modifying the function of a neurocognitive system. To modify the function of the neurocognitive system, necessarily we must modify the system's *stability distribution* or *rearrange its phase path* (the state vector's transition path through design space). The most thorough and ecological interventions do both. To modify the system's stability distribution, we need little more information than we already have. However, to rearrange its phase path, we must represent that path clearly in design space. We shall come to this shortly. This will become clearer when we map the trajectories (phase path) between emotion and behavior attractors.

It should be borne in mind that if a subject is feeling intense depression, the intensity (on a SUD* scale) of the feeling is not a reliable indicator of stability. Many intense states are not stable under perturbation. Rather, multiplying the frequency of recurrence by the dwelling time will give us the attractor depth (stability) relative to other competing states/phases. If the attractor is shallow and the subject is told a very funny joke, he will transition to another state/phase that has laughter near its basin of attraction, regardless of the intensity of the depression. On the other hand, if the attractor is deep, the state/phase of depression will be resistant to the perturbation (the funny joke) and recapture the state vector after the disturbance. In this case, depression will co-opt his biophysical resources.

In the stability distribution (Table 7.5), contentment has an index of 840 and frustration has an index of 30, which means that the contentment attractor is 28 times deeper (more stable) than the frustration attractor. If the phase velocity of frustration were much higher than the phase velocity for contentment and the two states/phases were released from initial conditions at the same time by a stimulus field (particular context) to compete for limited biochemical and biomechanical resources, frustration would win only temporarily. However, the frustration attractor is shallow and easily perturbed. Once it loses the state vector and the contentment attractor captures it in transition, it will be very difficult for "frustration" to get it back. A far more significant perturbation would be necessary for the contentment attractor to destabilize and give up the state vector.

It is important to note that if we were to add the stability index of each state together, the total time spent in recurrent states should not account for the subject's total waking hours. If the periodic oscillations of recurrent states/phases did in fact occupy all of the system's waking hours, it would indicate a very ridged system in the early stages of collapse. In later stages of collapse, the more stable attractors

* SUD stands for subjective units of distress. It is used as a measure of *intensity*, which is experienced subjectively.

will capture (incorporate) or severely destabilize the weaker ones, slowly decreasing the system's total degrees of freedom, until only a few states/phases are left — cyclically repeating themselves (i.e., a Markov chain).

Rather, a healthy system spends much of its time responding fluidly to a richly changing stimulus field (environment), ready to incorporate the current territory and make a continually updating, internal model of it that accurately corresponds in lock–step. The healthy neurocognitive system never dwells too long at any point in its phase path, and rarely comes back to *exactly* the same place.

Step 6: Determining the Probability Distribution for Neurocognitive System

In our last step, we determined the stability distribution index for the neurocognitive system, which yielded a very accurate asymmetrical picture of temporal resources expended for each state/phase of the system. The stability index also gives us each attractor's relative resistance to perturbation (disturbance). When phase velocity is considered alone, it tells us which of any two or more attractors is most likely initially to capture the state vector after perturbation. However, phase velocity alone does not tell us how well the attractor will hold onto the state vector after it has been captured by it. When we consider the stability index and phase velocity together, it gives us the *probability amplitude* of each state/phase, which in turn yields a *probability function* or density matrix for the entire neurocognitive system.

The probability function tells us the relative probability of the state vector's being captured and held by any attractor, hence co-opting the limited biochemical and biomechanical resources of the entire neurocognitive system. Once it has done that, we can turn back to the stability index for that attractor in order to get an estimate of how long it will hold onto it or how often we can expect to find it there. Each of these three measures of stability gives us a slightly different picture of what is happening in the neurocognitive system. The picture we choose to use at any given time and the portion of the system we choose to look at will be determined by the ultimate outcome desired.

When we choose to modify the system's function systemically, we must look at the largest portion of the system possible in order to anticipate the ecological consequences of an intervention. The same is true when we are modeling neurocognitive activity for the purpose of skill or knowledge transfer. However, there are many times when only a small portion of the entire matrix will be needed. This is covered in more detail later.

To compute the probability amplitude (P) for any attractor (state/phase, behavior, etc.), multiply the frequency of occurrence (F) by the dwelling time (D) and then divide the product by the attractor's phase velocity (V). This is the most extensive calculation used when developing a NeuroPrint or measuring the degree of success of a particular intervention.

$$P = \frac{F \times D}{V}$$

For example, "boredom" has a frequency of 2, a dwelling time of 20, and a phase velocity of 3. In order to compute the probability amplitude for "boredom," simply multiply $2 \times 20 = 40$ and then divide 40 by 3. The probability amplitude for "boredom" is 13.33.

In the previous step, when we compared the stability index (Table 7.5) for contentment (840) and frustration (30), we found that the attractor for contentment was 28 times deeper or more stable than the attractor for frustration. Therefore, contentment is more likely to hold onto the state vector once it is captured by it. In Table 7.6, we can see that the phase velocity for these two attractors is equivalent (2 seconds). Applying the formula $F \times D \div V$, the relative attractor depth (stability) difference remains the same. The probability amplitude for contentment is 420 and the probability amplitude for frustration is 15. Obviously, contentment is still 28 times more stable than frustration.

In contrast, looking at the stability distribution index (Table 7.5), the love attractor has a stability index of 42, while the stability index for anger is 100. This tells us that once the state vector is captured by anger, it is 2.38 times more difficult ($100 \div 42$) to perturb the anger attractor than it is to perturb the love attractor. On the other hand, if the love attractor captured the state vector, it would be 2.38 times easier to perturb the love attractor than to perturb the anger attractor. This is an important insight that can be used for couples' intervention.

Notice in Table 7.6 that once we take the phase velocity into consideration for both love (1 second) and anger (15 seconds), the probability amplitude for love remains unchanged at an index of 42, but the probability amplitude for anger falls to 6.67, since the state vector takes 15 seconds to be captured by the anger attractor. This travel time severely limits the anger attractor from co-opting the resources of the neurocognitive system if a particular context (stimulus field) releases the system from both sets of initial conditions simultaneously. In this case, love wins! This is, of course, a healthy probability function for maintaining an intimate relationship. If the probability amplitudes of these two attractors were reversed, the relationships would be short and perhaps quite violent.

In Table 7.5, the stability index for curiosity is 9 and the stability index for confusion is 4. The curiosity attractor is 2.25 times deeper or more stable than confusion, which means that once the state vector is captured by curiosity, a significantly greater perturbation would be necessary to release the state vector from curiosity than vice versa. This proves to be a very resourceful way to deal with problems.

Notice the same two emotions in the probability density matrix (Table 7.6). Since confusion has a phase velocity of 60 and curiosity a phase velocity of 1, if the system is released simultaneously from both initial conditions by a problem (stimulus), the system is 60 times more likely to handle that problem from a state of curiosity than from a state of confusion. This leaves the probability amplitude for curiosity at an index of 9, while the probability amplitude for confusion falls to 0.07. Taking phase velocity into consideration shows us that it is 128 times more likely that the neurocognitive system's resources will be collectively organized by the curiosity attractor than by the confusion attractor. We can assume from this probability function that this particular neurocognitive system is highly effective at

TABLE 7.6
Probability Density Matrix: (Determining the Probability Amplitude for Each State/Phase)

(F) ×	(D) ÷	(V) =	(P)	(–)	(F) ×	(D) ÷	(V) =	(P)	(+)
2	20	3	**13.33**	Boredom	4	7	1	**28.00**	Excitement
10	3	2	**15.00**	Frustration	3	3	1	**9.00**	Curiosity
5	20	15	**6.67**	Anger	42	1	1	**42.00**	Love
3	10	3	**10.00**	Anxiety	14	10	4	**35.00**	Satisfaction
10	30	2	**150.00**	Exhaustion	7	120	2	**420.00**	Contentment
2	2	60	**0.07**	Confusion	5	4	3	**6.67**	Joy
10	3	20	**1.50**	Overwhelmed	5	5	5	**5.00**	Pride
3	2	40	**0.15**	Fat	14	1	3	**4.67**	Loved
10	5	5	**10.00**	Torn	4	1	25	**0.16**	Settle
3	5	1	**15.00**	Fear	7	5	2	**17.50**	Anticipation
10	1	20	**0.50**	Trapped →	10	8	21	**3.81**	Escape: pos/plan
4	2	3	**2.67**	Disappointment	14	2	1	**28.00**	Happiness
14	7	10	**9.80**	Stressed	10	1	8	**1.25**	Trust
1	1	2	**0.50**	Embarrassed	10	1	1	**10.00**	Trusted
5	5	1	**25.00**	Annoyed	2	10	1	**20.00**	Surprise
1	1	65	**0.02**	Unorganized →	7	3	10	**2.10**	Organized

problem solving. As we will see later, when we diagram state-bound behaviors, a far more resourceful class of behaviors is likely to be bound within the curiosity attractor. Thus, when the curiosity attractor captures the state vector, these state-bound behaviors now become available as properties of that particular state/phase.

In Table 7.5, consider the state of stressed with an index of 98 and the state of excitement with an index of 28. It appears from this index that the stressed attractor is 3.5 times deeper or more stable than the excitement attractor. However, this only means that once the state vector is captured, it is 3.5 times more difficult for a perturbation to release it from the stressed attractor than from the excitement attractor. Turning our attention to Table 7.6, the probability amplitude for excitement, with a phase velocity of 1 second, remains the same at an index of 28, while the probability amplitude for stressed falls to 9.80, since its 10-second phase velocity is 10 times slower. Taking phase velocity into consideration, it is actually 2.86 times more likely that we shall find the neurocognitive resources of this system enslaved by the excitement attractor than by the stressed attractor. The probability function is a more accurate predictor of state and behavior than any of the previous indexes considered by themselves. This index is especially valuable in instances where the diagramming steps of the NeuroPrint have not been done and we are unable to see the phase path of the state vector.

The very essence of intervention is the modification of the function of the neurocognitive system by altering the system's attractor stability distribution or by rearranging its phase path, the state vector's transition path through design space. It is critical to have this information in order to make an effective decision about where in the neurocognitive system to target an intervention. It is also equally important to know as accurately as possible the initial values of the neurocognitive system (stability distribution and probability distribution) in order to measure, during intervention, how close one is to achieving the intended outcome at any given time. Once an intervention is believed to be complete, checking the current values for the attractors modified by intervention will yield a current picture of relative stability and probability amplitude of the attractors being considered. This will also reveal whether or not the modifications made by the intervention will be reliable. This makes it possible to design interventions that produce lasting changes in a desired direction.

Last, the stability distribution index and probability density matrix provide us with systemic vision. As we modify the stability of an attractor landscape, we can watch for spontaneous changes in other attractors on the state vector's phase path. As the practitioner develops acuity at noticing these spontaneous systemic changes, they can also be predicted prior to intervention.

Intervention in neurophysical terms simply means perturbing attractors and phase paths in highly specific ways. When we begin perturbing or disturbing a particular attractor, its phase velocity begins to decrease. That is, as we interrupt a pattern that the neurocognitive system has organized its aggregates into, it takes longer for the system to reassemble or replicate that pattern or attractor. This decrease in phase velocity is referred to as *critical slowing*, which is an extremely reliable indicator of a shallowing or weakening attractor.

As the phase velocity of an attractor decreases after perturbation, its stability also decreases. Both frequency of recurrence and dwelling time decrease shortly following the decrease in phase velocity, due to natural competition of other attractors within the neurocognitive system. A rechecking of the stability index or probability function will reveal a dramatic shift in values. In this regard we are able to accurately estimate the degree to which any intervention has been successful in producing an intended change.

Step 7: Frequency Grouping — Diagramming Emotional States/Phases in Design Space

The next step in developing a NeuroPrint is *frequency grouping*, which is the first step in diagramming the NeuroPrint (Figure 7.1). Each state/phase is delineated by a circle drawn in colored pencil or thin colored marker on the sketchpad. Depending on the evaluator's preference, the circles will vary in size in direct proportion to either their stability index or probability amplitude*. The stability index takes into account the state's frequency (F) and dwelling time (D) and the probability amplitude additionally takes into account the phase velocity (V). It is recommended that two different colored pencils be used to represent and differentiate positive and negative states/phases, perhaps green for positive and red for negative.

States with a higher frequency index should be drawn toward the center of the sheet and those with a lower frequency index gradually toward the perimeter. This will properly prepare us for later steps, since states with higher frequency tend to

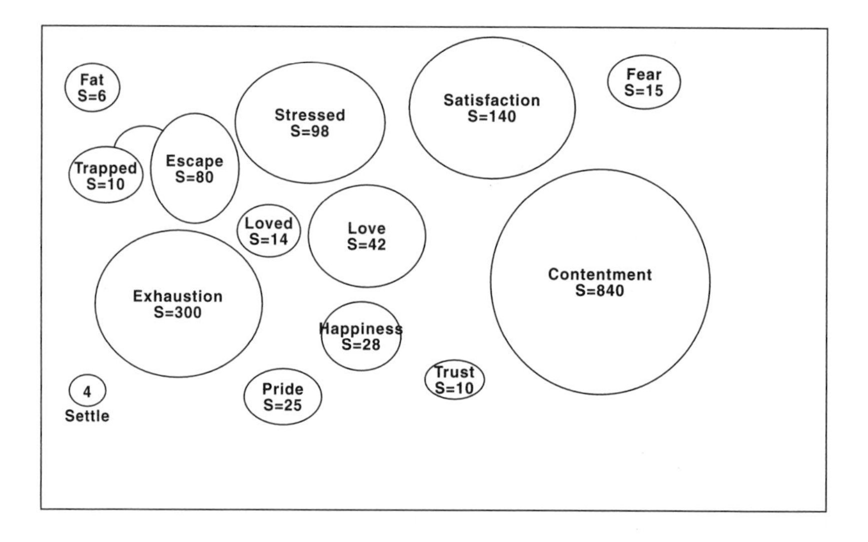

FIGURE 7.1 Frequency grouping of states/phases in design space.

* We have chosen to use the stability index from Table 7.5 to determine the relative differences in circle size.

have a greater number of paths leading to and from the attractor, namely, *approach trajectories* and *escape trajectories*. This is referred to a *trajectory density,* which is measured by adding together the number of approach and escape trajectories. If certain states/phases are chained together, they should be grouped closely together.

Two things will dictate the size of the circles. First, the more stable the state, the bigger the circle should be, so that a glance at design space will yield a clear picture of stability distribution. Second, the smallest circles must be big enough to represent state-bound behaviors inside of them if one intends to use the same diagram for this purpose.

For some interventions it is only necessary to represent the states/phases of the system, whereas for others it is necessary to represent the states, behaviors, trajectories, and stimuli, albeit for just a small region of design space. The stability index and/or probability amplitude also can be written inside the circles in order to keep all the information for the NeuroPrint on one page.

Step 8: Diagramming State-Bound Behaviors within States/Phases

Since behavior is a state-bound or state-dependent property of a system, a physiological state or emotion will necessarily specify a limited range of behavior available from within that state, thus limiting the degrees of freedom for a collective, coordinated action or macroscopic behavior. Therefore, when a neurocognitive system transitions into a particular biophysical state such as "anger," this phase transition to a new state severely narrows the range of collective, coordinated behavior possible. There are two obvious consequences resulting from this basic principle of system organization, since behavior is by necessity a subset of a biophysical state or phase of a system:

1. From within a state of "anger" or "rage," behaviors such as fighting and yelling are far more probable than kissing and hugging. For the behavior to change, the neurocognitive system must, by necessity of its organization, transition into a new biophysical state or phase which would then make kissing and hugging more probable state-bound behaviors. If the behaviors of kissing or hugging were attempted from within the state of "anger" or "rage," while this could in principle trigger a phase transition to the state of "loving" if the anger attractor is very shallow, it is highly unlikely. In such an instance it is more likely that the kissing and hugging behavior would be experienced by the recipient as a violent act.
2. One or more transition states or phases would likely separate the state of "anger" from the state of "love." Forcing the system to transition directly would be futile and possibly dangerous. If the state vector often gets trapped in this region of its phase path, the best intervention would be to build new transition states that provide the state vector with more escape paths.

We elicit state-bound behaviors by asking the subject questions such as, "What do you do when you feel (state/phase)?" or "Is there anything else that you do when you are in (state/phase)?"

The most important question we must ask when diagramming state-bound behaviors is whether the behaviors available in the given state *interrupt* the state/phase (shallow the attractor) or *stabilize* it (deepen the attractor). In some cases a successful intervention can be accomplished by adding new behaviors designed to interrupt the state, while allowing the state vector to escape.

While the reader may already have intervention tools available that can accomplish some of the interventions mentioned here, using a NeuroPrint as a guide helps one to plan exactly where and when to apply those tools, predict the effects of the intervention, and measure how close one is to the intended outcome.

Figure 7.2 shows three states of differential stability and their state-bound behaviors. This diagram of a section of a developing NeuroPrint makes the asymmetrical distribution of neurocognitive resources apparent. To determine the relative size of the circles representing each state/phase of the system, we have chosen to use the stability index. Contentment is shown here with a stability index of 840, relative to love with an index of 42 and frustration with an index of 30. When this subject is in the state of frustration there are only two state-bound behaviors available, yelling and retreating. While the subject may have thousands of behavioral choices that the system could potentially organize, when in the state of frustration the potential degrees of freedom are severely restricted.

One way to think about state-bound behavior is to recall the three states or phases of the atomic system, H_2O. While it is obviously a less complicated system, still it must obey the same laws and constraints; behavior is a property of a system's

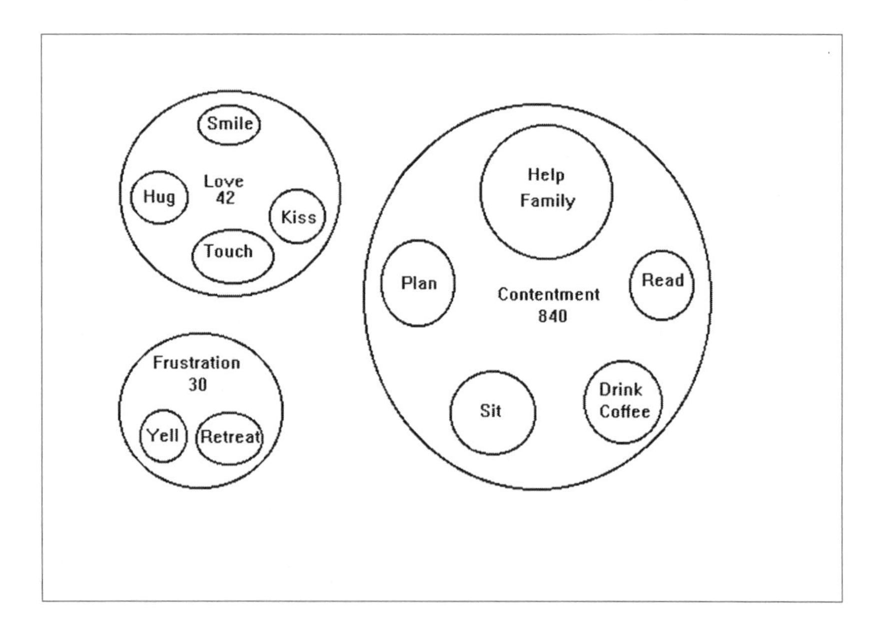

FIGURE 7.2 Diagramming state-bound behaviors within states/phases.

state or phase. Let us now imagine that the H_2O has been frozen inside of a large cube. The bottom of the cube has tiny pores that the H_2O can use to escape. However, in the solid phase (the block of ice), it cannot produce escape behavior. In order to fit through the tiny pores, it must necessarily transition from a state of solid to liquid. From within the state of liquid the escape behavior is quite easy. Now let us imagine what would happen if we turned the cube upside down. In this case the escape pores would be above the H_2O. In order to produce escape behavior from this set of circumstances, H_2O would necessarily have to be in the gaseous mode or state. Within the gaseous mode, it would have access to the behavior of rising or floating, a state-bound behavior unavailable to the system in the solid or liquid state. Once organized in a gaseous state, H_2O would easily be able to escape.

In order for the state vector of H_2O to be captured by the gaseous attractor, however, it must first pass through the transition state to liquid. The neurocognitive system has these same kinds of limitations. Frequently we shall find it necessary to construct a transition state in order for the system to gain access to necessary state-bound behaviors that are intrinsic properties of another state or phase.

Many people notice the phenomenon of state-bound behavior limiting their choices in daily life. The golf pro whose swing is off due to an argument, the student who knows that he knows the information but can't remember it at the time of the test, and the spouse who can't hug and kiss because he's angry are all examples of state-bound behaviors. In Figure 7.2, we notice that the subject must necessarily transition from frustration to contentment in order to feel like helping his family. Yet the intrinsic properties of contentment do not make hugging, kissing, and touching available. To produce these behaviors, the neurocognitive system must transition to the state of love.

State-bound behaviors are not limited to macroscopic, easily observable behaviors. Microscopic behaviors that support cognitive activity are also state-bound. Memory for information is no exception. One class of microscopic behaviors supporting thought, creativity, analysis, and other critical functions is called submodalities, which we shall say more about shortly.

State-bound behaviors are also asymmetrically distributed, which means that they use different amounts of spatiotemporal resources. The stability of state-bound behaviors is measured by the same algorithm that we use to measure the stability of states or phases (frequency, dwelling time, and phase velocity).

By measuring the stability for each behavior attractor, we may also predict the behavior that the neurocognitive system is most likely to select when in a particular state. In this way, we are able to also predict the most likely phase path the state vector will take from that point. This type of knowledge is vital when looking for high leverage points for therapeutic intervention.

The stability index for each behavior in both Figures 7.2 and 7.3 is represented by circle size in the same way that we represent states. The larger the circle, the more stable the behavior and the more likely that behavior will occur when the state vector is captured by that particular state or phase.

Step 9: Mapping Trajectories between Biophysical States/Phases (Defining the Basin of Attraction for Each of the System's States/Phases)

We now have a better idea of how states and state-bound behavior are organized in design space, resulting in an informational bio-architecture. The next step is to understand how the state vector travels through design space. In order to represent the neurocognitive system dynamically in four dimensions of space and time, we must now add its phase path.

The phase path is discovered one trajectory at a time. At this point we need to know how the system transitions from one state to another. To elicit this, we query the subjects with a question such as: "When you feel frustrated, what is the very next feeling or emotion that usually occurs?" We can also be more specific by asking, "When you feel frustrated and you retreat (state-bound behavior), what is the next thing that you do or feel?" This part of the elicitation is quite easy, since now the subjects are already scanning over their NeuroPrints, which contain all of their recurrent states and state-bound behaviors. If we find many escape trajectories leading away from an attractor, we can access the stability of a particular phase path by asking, "Which *transition* (e.g., read, sleep, walk) occurs most frequently?" The answer to this question provides a frequency index, the first measure of stability. If we want a more accurate assessment of the relative stability of several possible phase paths, we can then measure the phase velocity between competing trajectories.

In this regard, trajectories reveal the neurocognitive system *in motion*. We represent this by drawing connecting lines between states and behaviors and indicating the state vector's direction of movement with an arrow. The resulting picture tells us not only which states and state-bound behaviors are connected, but also the specific phase path the state vector must travel in order to be captured by a particular attractor.

In Figure 7.3, connecting lines are used to indicate the direction of transition or trajectories in the neurocognitive system. If the lines are followed in the direction of the arrows, a sense can be gained of what transition choices are available to the state vector for the region of the neurocognitive system represented.

In this neurocognitive system, in the state of frustration there are three possible paths that the state vector can take in order to be captured by the contentment attractor. In one case, the state of frustration itself becomes a stimulus to trigger the behavior of planning. When the behavior of planning has begun and dwelling time is sufficiently long, the state vector transitions completely into contentment, making available a whole new range of state-bound behaviors (sitting, drinking coffee, reading, helping family). In this particular example, a behavior (plan) is used by the system as a way to reach the trapping zone for the attractor "contentment." Therefore the transition from frustration to contentment is easily achieved. Since the frustration attractor is very shallow, only a small perturbation is necessary to initiate transition. Once the state vector is captured by the much deeper attractor of contentment, it is very unlikely to escape back to the frustration attractor. This aspect is indicated by the lack of *return trajectories*.

In the state of frustration, the system has two behaviors available to it, yelling and retreating. Both behaviors in this case sufficiently perturb the frustration

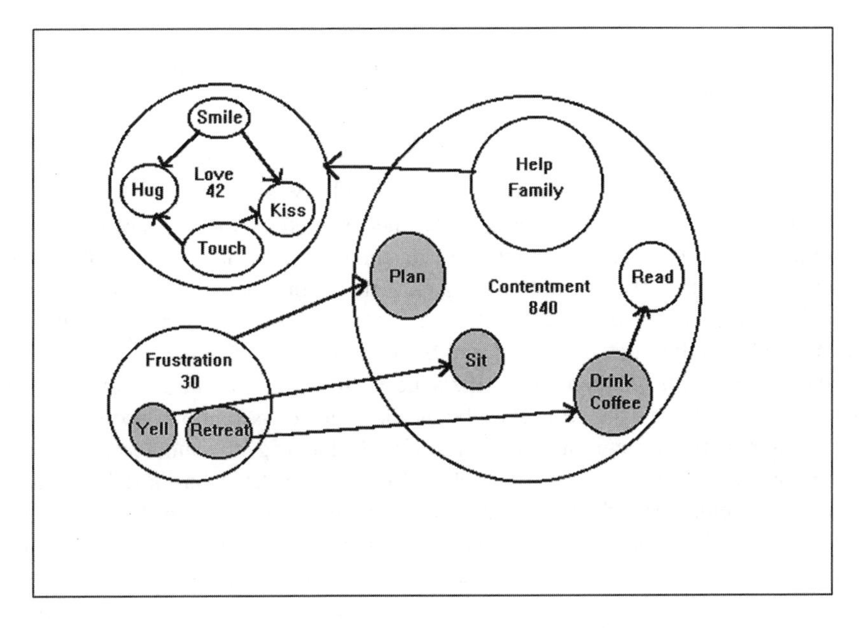

FIGURE 7.3 Mapping trajectories between biophysical states/phases.

attractor in order for transition to occur. As the NeuroPrint indicates, once perturbed, each behavior has a specific trajectory to a state-bound behavior located in the contentment attractor. That is, yelling transitions to sitting, and retreating transitions to drinking coffee, thus making an entire new range of state-bound behaviors available to the system. Notice also that the behavior of drinking coffee has a reliable trajectory that leads to the behavior of reading, which is also located within the contentment attractor.

The stability index for contentment indicates that dwelling time is quite high. This also makes it possible for the neurocognitive system to experience its full range of behavioral choices within that attractor. Once the state vector is captured by contentment, the most probable behavior (i.e., highest stability index) is "helping family." This NeuroPrint also indicates that the behavior of "helping family" initiates a phase transition to the state of love. It is unlikely for stable attractors, such as contentment, to give up the state vector to a shallow attractor like love, unless a particularly stable phase path exists. Here the phase path between the behavior of helping family and the state of love is quite stable. The stability of phase path or trajectory can also be determined by measuring the phase velocity of a particular trajectory. We shall come to this shortly.

MEMES AND THE CONSTRUCTION OF THE BRAIN'S INFORMATIONAL BIO-ARCHITECTURE

Phase paths (trajectories connecting stimuli, states, and behaviors) are generally wired together by memes (information patterns, thoughts) which internally influence state and behavior selection.

Once we have elicited the direction of a trajectory between connecting attractors (states or behaviors), we then want to know what information (memes) wired them together. This information helps us to determine the highest leverage point in the NeuroPrint for a therapeutic intervention. Since memes are self-replicating information patterns, we are looking for very stable thoughts that occur with a high frequency.

Referencing Figure 7.4, the subject was asked, "When you are excited about something, how do you become uncertain? Is there something that you say to yourself or remember, or is there something that you imagine will happen?" In this case, when the subject is excited about doing something, the meme "Look before you leap" is spontaneously triggered, so the state of excited about doing something is an internal stimulus that triggers this previously incorporated meme. The meme "Look before you leap" then causes the state vector to be captured by the state of uncertainty. From this state, the state-bound behavior of hesitation becomes available, which then becomes an internal stimulus that triggers another previously incorporated, conflicting meme, namely, "He who hesitates has lost." The two conflicting memes perturb both state attractors and within a relatively short time, the state vector is captured by the state of confusion. Within this state, another meme is recalled: "Successful people make quick decisions." The subject then goes ahead and resolves the confusion by making a quick decision and activating the excited attractor, which in turn is perturbed within a relatively short period of time transitioning the system to guilt and then back to confusion. This is a relatively messy problem, which expends enormous resources of the neurocognitive system.

The representation of this problem, shown in Figure 7.4, is extremely common. Similar structures and organizations result whenever conflicting memes have been incorporated. Such conflicting memes can rapidly cause *self-stabilizing* positive

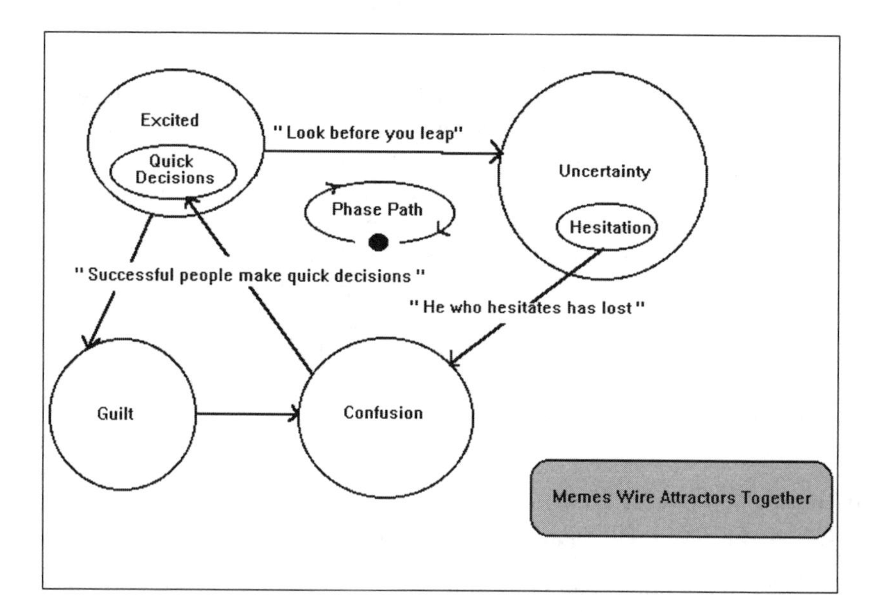

FIGURE 7.4 Memes assemble the phase path of a neurocognitive system.

feedback loops that force the neurocognitive system to continually reactivate each of these states, thus further stabilizing each attractor and the trajectories between them. The resulting phase path becomes so stable that it periodically whips the system into an inescapable Markov chain.

MEMES AND THE CONSTRUCTION OF DESIGN SPACE

Distinction memes can be stimuli, words, or names of states and behaviors.
Strategy memes specify what behaviors to do, how to do them, and when; they are state-bound information.
Association memes wire up phase path from external stimuli, thoughts (internal stimuli), biophysical states, and behaviors, telling us how to feel about connected bio-architectural structures and external stimuli.

A Note about Diagramming: Whenever a meme or internal thought is believed to be reliably influencing a state or behavior transition, it must be written directly above the trajectory indicating that transition (Figure 7.5).

The most important thing to remember here is that attractors (stimuli, behaviors, states) constantly get wired and rewired together by internally replicating and recombining information patterns called memes that in turn differentially stabilize and destabilize trajectories (choice points along a phase path). All of these resultant trajectories and choice points have a measurable stability and therefore yield a predictable phase path. It is this haphazard wiring that eventually leads to the breakdown of a healthy neurocognitive system and the myriad of bizarre human

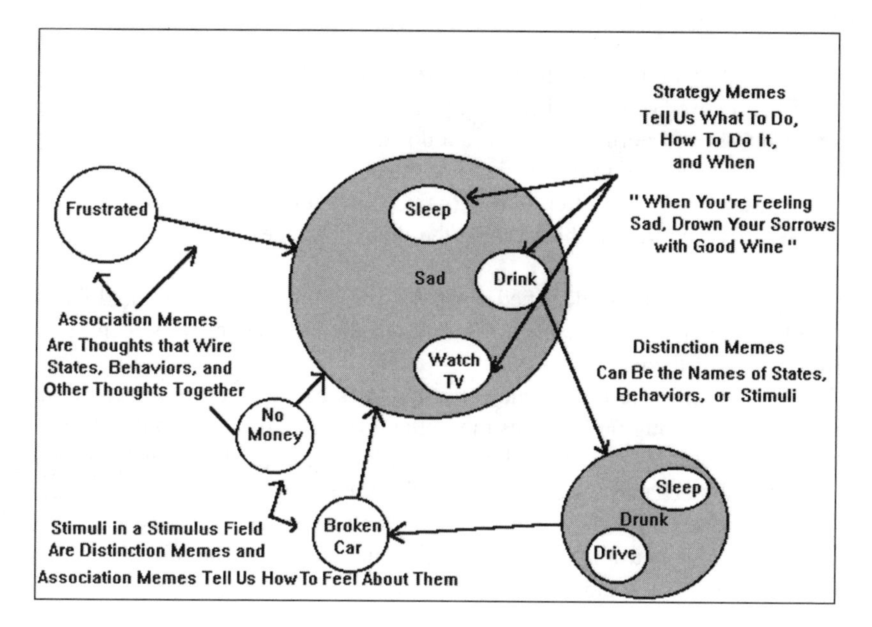

FIGURE 7.5 Memes influence the structure of design space long after incorporation.

maladies. Unlike external stimuli, memes are inclined to replicate and thus influence the structure of design space from the inside, long after they have been incorporated.

Step 10: Identifying Stimuli/Thoughts (Initial Conditions of Informed Matter and Energy) Influencing Transition into Each State/Phase

In this next step, we want to understand the initial conditions from which a neurocognitive system is released in order to start the state vector on its phase path through design space. The stimulus or information field is the source of the external initial conditions that assemble the neurocognitive system and drive the state vector through design space. As we saw in the last step, information from this field that is incorporated and frequently replicated (memes) continues to drive the assembly of attractors and the phase path (direction) of the state vector. However, while some stimuli become internally incorporated and replicated frequently, other stimuli from the information field simply release the system from its initial conditions and no longer have influence over the phase path once they are removed.

In many cases, it is much easier to remove a stimulus from the information field than it is to destabilize a meme that internally drives the system. For this reason, our next step is to identify the stimuli that lead the neurocognitive system into each state or state-bound behavior. This simple process can immediately reveal ways in which the information field can be reorganized in order to influence an internal restructuring of the neurocognitive system and a corresponding change in cognitive neurodynamics.

We elicit stimuli that influence cognitive neurodynamics by asking any or all of the following questions:

- *When* do you normally/most often feel (name the state)?
- What *causes* you to feel (name the state)?
- *When* do you normally/most often do (name the behavior)?
- What *causes* you to do (name the behavior)?
- *When* do you most often think* (name the thought/meme)?
- What *causes* you to think (name the thought/meme)?

In Figure 7.6 we have identified many stimuli in the information field that co-opt biochemical and biomechanical resources of the neurocognitive system. Let us examine a few of them.

Notice in Figure 7.6 how the anger attractor is activated by initial conditions. The stimulus of feeling fat becomes the initial condition to transition the system to one of two phases, anger or stressed. In this particular case, if the subject cries when she feels fat, the behavior of crying forces the state vector to the trapping zone of the anger attractor. The subject then enters the state of anger and has two additional state-bound behaviors available to her, fighting and driving. If she selects the behavior of driving, she is reminded that her car needs repairs, that she has no money for

* This question reveals both external stimuli and internal thought.

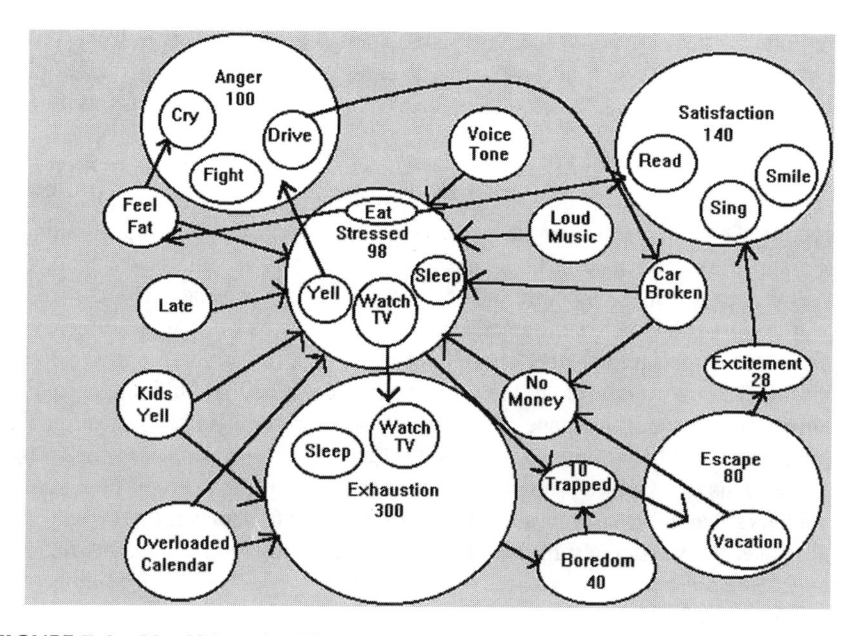

FIGURE 7.6 Identifying stimuli/thoughts (initial conditions influencing transition into each state/phase).

the repairs, which then causes the state vector to become captured by the stressed attractor. The subject can also transition directly to the stressed attractor just by getting into the broken car. It is not necessary for her to think about money.

If the subject does not cry when she feels fat, she then transitions directly to the stressed attractor. In either case, once the stressed attractor captures the state vector, she has four state-bound behaviors available to her: eating, sleeping, watching TV, and yelling. If she selects yelling, she immediately transitions back to anger, sending the system into a periodic cycle which results in the strengthening of both the anger and the stressed attractor. If this happens for a reasonable period of time, the stability index for both increases dramatically. If she chooses to eat, she may either transition to the state of satisfaction or may feel fat again, sending the system back through a periodic oscillation. If sleep is chosen, there is no direct, predictable transition. And if watching TV is chosen, the state vector is captured by the exhaustion attractor and held there for quite a while. The exhaustion attractor has a stability index of 300. It is the deepest attractor in the region of design space represented by this NeuroPrint.

By following the phase path that the state vector takes from attractor to attractor, we can see why stimuli from the information field are so valuable in predicting the direction in which a system will go. The NeuroPrint clearly reveals external information, which can be reorganized to prevent certain phase paths from being initiated. Notice that the stimulus of an overloaded calendar can release the state vector from a completely different part of the same region of design space. The overloaded calendar stimulus can lead to either the state of exhaustion or the state of stressed. The relative frequency and phase velocity of the two possible trajectories will determine the decision between phase paths. The system will most obviously choose the smoother of the two trajectories. In this case, we indicate the most likely phase path with the shorter line that connects the overloaded calendar stimulus to the state of exhaustion.

The state of exhaustion can make the behavior of sleep available to the system. But if sleep does not occur, then the only escape trajectory from exhaustion is boredom, which also has a single escape trajectory leading to the trapped attractor. The trapped attractor also has a single escape trajectory leading to the feeling for need for escape. Both boredom and trapped are transition states. These are states that are very weak (shallow attractors) that have a single escape path leading to more stable states with a larger degree of available behaviors. Unfortunately, if we continue to follow the state vector along the phase path, we notice that only one behavior is available in the state of escape. That is the behavior of planning a vacation. This behavior leads to the stimulus of no money, and then directly back to the stressed attractor. This is a system preparing for collapse. A rearrangement of phase path is absolutely essential to prevent this system from ultimate damage. Phase paths such as this one, with extremely limited degrees of freedom, are prime targets for intervention. The objective here is to free the state vector by adding alternative trajectories in this region of the phase path.

It should also be noted that at the time this NeuroPrint was developed for the subject, the stressed attractor had the largest number of approach trajectories. Two things are important here. First, the larger the number of approach trajectories for a particular attractor (its basin of attraction), the greater the potential number of times it will be activated (the state vector will be captured). Each activation of the attractor, which is undisturbed by perturbation, will result in an increase in stability of that attractor.

We can predict from this NeuroPrint that the stability index for "stressed" will increase dramatically over the next few weeks or months. Given any NeuroPrint that has a similarly high approach trajectory density, we can make the assumption that the attractor will further stabilize (deepen).

Second, the NeuroPrint clearly suggests key areas of intervention. We can eliminate some of the stimuli from the information field that trigger the "stressed" attractor; or we can add new state-bound behaviors that act to perturb that attractor and add additional, stable, escape trajectories to the stressed attractor that will predictably lead to a positive attractor landscape.

Also, it should be noted here that beliefs and values are closely related emergent properties of trajectories. Values occur mainly when stimuli from the information field get linked to particular emotion attractors. In Figure 7.6 the trajectory connecting the stressed attractor caused the subject to value TV. When asked what kind of TV she has, she indicated that it is the most expensive piece of furniture she has because it makes her feel better when she is stressed.

Beliefs are established mainly by cause and effect when trajectories connect stimuli to emotion attractors ("Too much TV can make you tense"), from behaviors to emotion attractors ("When people overeat it causes stress") or two state attractors ("Prolonged stress causes depression"). Beliefs and values that are culturally incorporated from the information field can also cause such trajectories to form and stabilize.

Unfortunately, when people try to alter the stimuli in the information field of their life unassisted by a map like NeuroPrint, they usually do more damage than repair, since they are unable to see the entire system at one time. Thus, in this regard, they are unable to predict the consequences of a particular rearrangement.

When people attempt to change their thoughts, feelings, and behaviors by rearranging the stimuli in their information field, they tend to haphazardly try the same things again and again. People might change their hair style monthly, rearrange their furniture, drink alcohol, overeat, watch TV, buy new clothing, change jobs, leave their spouses, redecorate, landscape their lawns, change the lighting at work, try aroma therapy, change their musical tastes, look at old pictures in their family album, vacation frequently, or even retreat from stimuli altogether and lock themselves in a quiet room and sleep for days. When they continue to do this, searching for the "cure" to their malady, they unwittingly release their neurocognitive system from several competing sets of initial conditions, resulting in greater conflict, periodicity, uncontrollable disorder, and damage. In some cases the pattern-entropy balance tips so far in one direction or the other that there is no return, as we saw in Section I. Rather, a thorough development of a NeuroPrint and a careful analysis can lead to more productive client participation in the therapeutic process, greater insight into the structure of a problem, and a quicker, more effective solution.

8 NeuroPrint Analysis: Understanding Cognitive Neurodynamics

CONTENTS

READING A NEUROPRINT: VISUALIZING COGNITIVE NEURODYNAMICS

In order to read a final NeuroPrint it must be set in motion. The neurocognitive system is a dynamic system, not a static one. The easiest way to do this is to make an internal mental movie of a ball or marble traveling along its phase path. If you have followed these steps in order to develop a NeuroPrint of your own neurocognitive activity, this will be quite easy.

Imagine the uneven terrain of a desert with its high sand dunes and low valleys breaking the symmetry of the flat terrain. The valleys represent the attractors: the more stable the attractor, the deeper the valley; the weaker the attractor, the more shallow the valley (Figure 8.1).

Phase velocity can be represented by the steepness of the approach to the valleys. If an attractor has a low phase velocity (with a high index such as 40 seconds to activate the state), imagine the ball slowly trying to climb a hill to get into a hole at the top (like a volcano). The lower the phase velocity (larger the index), the steeper the approach hill, the slower the ball travels. Attractors with very low phase velocities are sometimes referred to as *repellers*. A high phase velocity (with a low index such as 1 second) can be visualized as a ball traveling

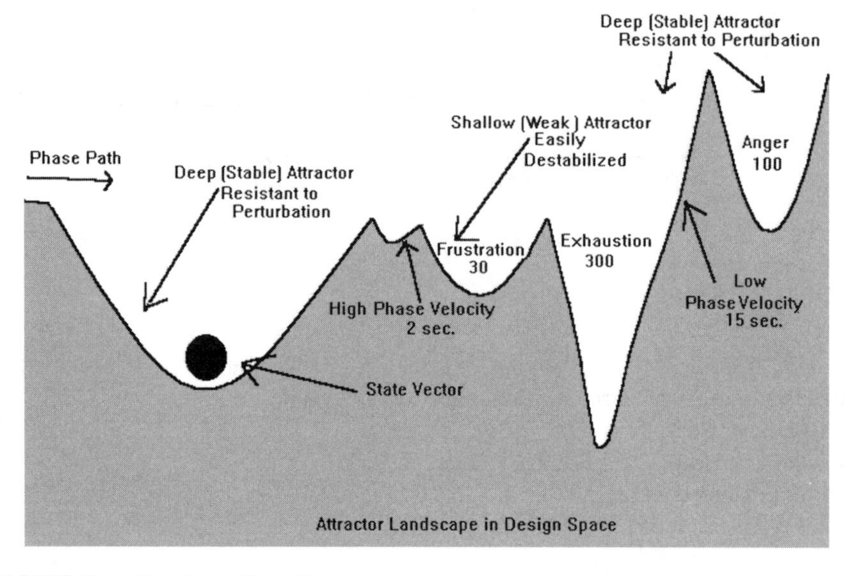

FIGURE 8.1 Reading a NeuroPrint.

very rapidly downhill to reach a hole. The higher the phase velocity, the steeper the downward slope and the faster it goes.

Now start from any point in design space on the NeuroPrint. A little imagination is recommended here. While looking at each attractor, imagine its depth. Remember that the deeper the attractor, the greater the perturbation needed to release the state vector. *Now imagine that you are the ball (state vector).* Once captured by the attractor you are limited to the state-bound behaviors available within that valley. If this is a negative attractor, ask yourself if you have any behaviors available from within that state to help you get out of it. If not, you will need a pretty big wind (perturbation) to move all that sand, and it could fall in on top of you or even get deeper. As you look at how many *escape* trajectories there are and how many ways there are to fall back in (*approach* trajectories), you may want to rearrange this. Can you create new *transition states* (dig new *steps* up the side of the slope to get out)?

Continue to travel along the phase path and when there are *decision points* (many ways to go), consider each of the phase velocities of the attractors before you — whichever is the highest (lowest index such as 2 or 3 seconds) is the one you get sucked into as if it were a vacuum.

As you travel through design space, notice if you get trapped in an isolated landscape that limits your availability of attractor choices (limits your degrees of freedom) and sends you traveling around in tight periodic circles. Notice the external stimuli that force you into different regions of design space. Ask yourself how easily this can be rearranged. Also notice the internal thoughts you have and the questions you ask yourself along each trajectory. Can this thought, question, belief, etc. be modified so as to lead to a different attractor?

When you reach deep attractors that lead to problems for the neurocognitive system, ask yourself what would happen if you filled it in with sand (shallow the attractor). Where would the approach trajectories go? Which of the attractors next on the phase path is stable enough to capture the state vector?

ANALYSIS OF THE NEUROPRINT — PRINCIPLES FOR ANALYZING COGNITIVE NEURODYNAMICS

Now that we have the steps to build a NeuroPrint, we must learn to read and analyze cognitive neurodynamics from the NeuroPrint itself before designing an intervention. Once we have learned to set the image in motion, we must know what to do when we get to decision points in a phase path. Decision points occur whenever the state vector is captured by an attractor that has more than one escape trajectory. In Figure 8.2, we have illustrated just such a case. Notice that the frustration attractor has one approach trajectory where the state vector escapes from the annoyed attractor, while having three escape trajectories (three possible directions for phase path). In this case, the state vector has three degrees of freedom for state selection from within the frustration attractor.

In order to properly read and analyze a NeuroPrint and design an effective intervention, we must be able to predict with reasonable accuracy the direction the state vector will take whenever it reaches a decision point in its phase path both before and after our intervention. There are three important clues that can be derived from the information we have so far:

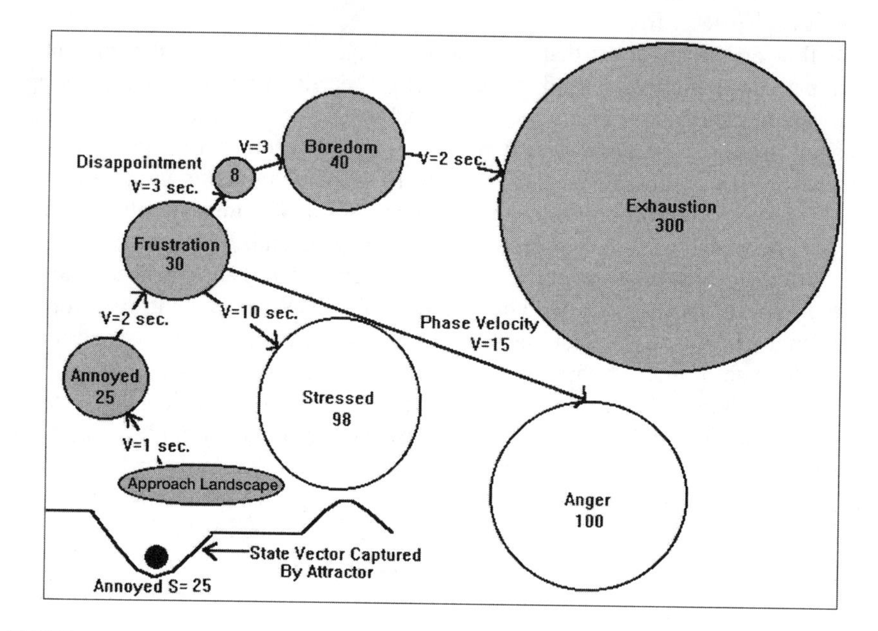

FIGURE 8.2 Decision points in a phase path.

- Frequency of occurrence
- Stimuli in the information field
- Phase velocity

Of the three predictors, phase velocity is the most reliable. However, if there is strong interference from stimuli in the information field that can simultaneously disturb the most likely phase path while specifying a particular attractor, it must be considered in conjunction with phase velocity. How do we do this?

The frequency of occurrence of any state/phase attractor is influenced by two factors: the stimuli in the information field that specify the attractor, and the number of approach trajectories for that attractor. Whenever phase velocity does not accurately predict the direction of the state vector, we then look at frequency mainly to help us uncover hidden (previously unreported) stimuli in the information field that may be influencing the phase path.

In Figure 8.2, once the state vector is captured by the frustration state attractor, its choices are "disappointment" with a phase velocity of 3 seconds, "anger" with a phase velocity of 15 seconds, and "stressed" with a phase velocity of 10 seconds. Providing that there is no strong interference from stimuli in the information field that are qualitatively different from the stimuli that assembled this phase path, the state vector will reliably select the "disappointment" attractor with a probability five times greater than "anger" and three times greater than "stressed." If the state vector is released from a set of initial conditions with several possible choice points in the phase path, the path with the greatest phase velocity will win, regardless of the depth of the attractors. Depth or stability of an attractor only tells us how well the attractor will hold onto the state vector once it captures it and how significant a perturbation is necessary to set it free.

Following the attractors that are shaded in Figure 8.2, observe that the entire phase path has a high velocity. The state vector is easily perturbed from the disappointment attractor, since that attractor only has a stability index of 8, being able to transition from disappointment to boredom in only 3 seconds. To transition from boredom to exhaustion, again only a relatively small perturbation is necessary to destabilize boredom, which has a stability index of 40. The transition time (phase velocity) between boredom and exhaustion is only 2 seconds. Once the exhaustion attractor captures the state vector, a highly significant perturbation is needed to set it free, since exhaustion has a stability index of 300. When reading a region of NeuroPrint where a state vector has multiple degrees of freedom for state selection, follow the path of highest phase velocity (lowest number of seconds).

If actual observations or reports of the subject differ from prediction, this dictates that we search for hidden stimuli from the external information field that may be disturbing the state vector's path. In other words, if after developing a NeuroPrint the subject reports becoming angry directly after being frustrated more often than stressed or bored, then we must search for unreported recent stimuli that may be triggering the anger attractor. Questions concerning what has recently caused anger in the subject may be revealing in this regard.

We look for a recent change in the information field, since if the stimuli had existed for a considerable length of time, the phase velocity between frustration and anger would have already significantly increased and the NeuroPrint alone would reliably indicate phase path. However, recent stimuli would not have had enough time to strengthen the phase path (increase the phase velocity). Once the stimuli that reliably trigger a transition to the anger attractor have been determined, simply add them to the NeuroPrint with connecting lines to the anger attractor, and predictions will again agree with observation. We can then also predict that if the recent stimuli remain in place, the anger attractor will deepen (stabilize) and its phase velocity will increase.

ISOLATED LANDSCAPES AND FEEDBACK LOOPS

As noted above, once a NeuroPrint can be read smoothly and effectively used for prediction, the first analysis that must be done is a global one. Viewing the entire NeuroPrint, we assess the flexibility of the neurocognitive system to move smoothly between its negative and positive states. There are three major indicators for this:

- Asymmetrical distribution of degrees of freedom between positive and negative states
- Asymmetrical distribution of stability between positive and negative states
- Isolation of positive and negative attractor landscapes

DEGREES OF FREEDOM

Let us suppose that the NeuroPrint shown in Figure 8.3 contains a complete representation of the state/phase attractors for this neurocognitive system. Ideally, a

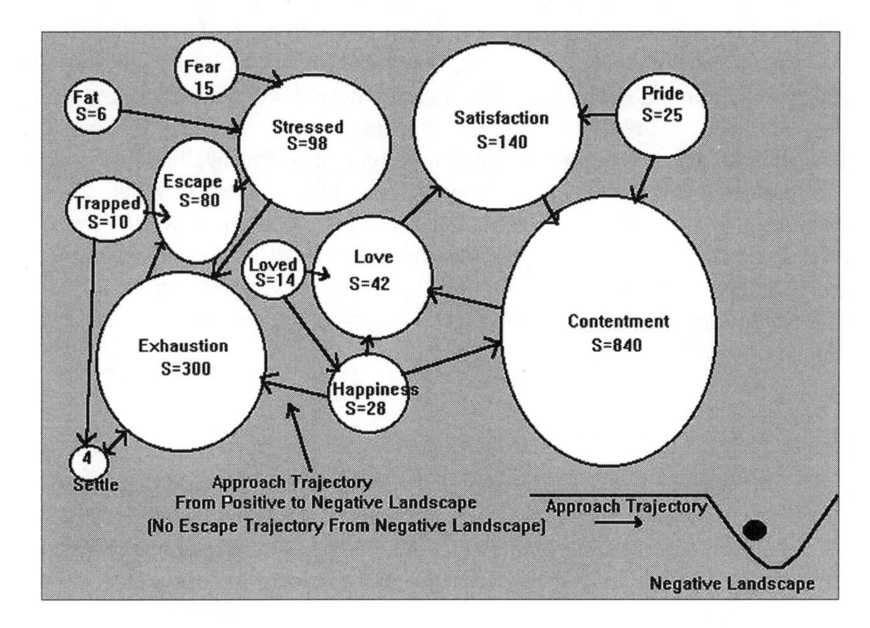

FIGURE 8.3 Isolated landscapes and feedback loops.

neurocognitive system would possess many more degrees of freedom for positive states than for negative ones. In actual practice, however, this is rarely the case. To check the positive/negative ratio of degrees of freedom, simply total the negative states and compare this to the total of positive state choices. In Figure 8.3, we can see that the positive landscape allows six degrees of freedom for the state vector (loved, happiness, love, satisfaction, pride, contentment), while the negative landscape allows seven degrees of freedom (fat, fear, stressed, escape, trapped, settle, exhaustion). This is a fairly symmetrical distribution and, considered by itself, a reasonably stable neurocognitive system.

Note that *escape* and *settle* were originally reported as positive states during the first six steps of the NeuroPrint (refer to Figures 7.2 through 7.6). It is only in Step 9 (mapping trajectories between biophysical states/phases) that we can truly assess whether a state attractor belongs to a positive or negative landscape. As it turns out, once the phase path of the state vector has been laid out by connecting trajectories (Figure 8.3), it is evident that *escape* and *settle* cause the state vector to become trapped in a negative landscape. It is also apparent that the *settle* attractor is used by the state vector as a transition state to *exhaustion.*

STABILITY DISTRIBUTION

The second global index to consider is the stability distribution of positive to negative state attractors. In the negative landscape, the only significant attractors are *escape* with a stability index of 80, *stressed* with a stability index of 98, and *exhaustion* with a stability index of 300. We compare that with the significant attractors in the positive landscape. *Love* has a stability index of 42, *satisfaction* has a stability index of 140, and *contentment* has a stability index of 840. Simply looking at the NeuroPrint gives us the sense that the state vector spends far more time in the positive landscape than in the negative one. We can accurately assess this simply by adding up the stability indexes for the negative and the positive landscapes separately. (Taking the frequency of occurrence and multiplying it by the dwelling time derives the stability index.) This gives us a fairly accurate relative measure of the amount of time spent in each attractor. It turns out that the stability index for the negative landscape is 513, while the stability index for the positive landscape is 1089. This is clearly an asymmetrical distribution of stability, and judging this ratio in isolation, we can safely predict that the neurocognitive system spends twice as much time organized into positive states.

ISOLATION OF LANDSCAPES

The third global analysis is done by tracing the phase path of the state vector through designed space, as in Figure 8.2. It is unnecessary to take into consideration the phase velocity at decision points when making this type of analysis. We can clearly see that the two escape trajectories departing from the stressed attractor both lead back into the negative landscape. The two escape trajectories for the loved attractor lead back to the positive landscape, as do the two escape trajectories for the pride attractor. The only escape trajectory from the positive landscape to the negative landscape departs from the happiness attractor, and there are none leading back to the positive landscape from the negative one. Even without taking into consideration phase velocities, this is an early predictor of possible trouble. Once the state vector leaves happiness and is captured by exhaustion, it becomes trapped in the negative landscape with no way back to the positive landscape, except for strong interference by stimuli in the information field or possibly a fundamental transition of the entire system to the state of sleep.

This attractor pattern is referred to as *isolated landscapes**. It signals potential danger for a few reasons. First, if the state vector is continually trapped by its phase path in the negative landscape, the stability of each of the connected attractors, along with their corresponding phase velocities, can be predicted to increase. Within a relatively short period of time, this will force the stability distribution of the neurocognitive system to change. The negative landscape will rapidly become far more stable than the positive one, eventually resulting in the formation of a Markov chain in the negative landscape, whereby the state vector will become periodic and continue its cyclical motion along the strongest phase path, unable to escape. We shall discuss Markov chains in more detail in our next example.

Second, people in this situation will likely try to solve this problem either by increasing the amount of time and the frequency of their sleep or by rearranging, adding, or subtracting the stimuli in the information field of their life. To do this haphazardly without a NeuroPrint is likely to cause a dramatic shift in the pattern-entropy balance of the system. Here we see an excellent, concrete example of what we discussed theoretically in Section I. The formation of the Markov chain in the negative landscape results in greater order (periodicity) in that landscape, and this increase in order is paid for by the rest of the neurocognitive system as it cycles into entropy (disorder and disorganization) in response to the introduction of haphazardly organized stimuli. The standard theory of pattern-entropy dynamics predicts that if this imbalance occurs, the neurocognitive system will necessarily seek ordered stimuli from the information field (environment). However, if it persists long enough or oscillates back and forth between order and disorder, the neurocognitive system will become less responsive over time to changing environmental information. This results in a lack of correspondence between the external environment and the neurocognitive system's internal model or map of that environment.

Lack of correspondence between map and territory always leads the system further into entropy with serious consequences. By using a NeuroPrint to represent and analyze neurocognitive activity, we can predict the early warning signs of such a system breakdown. NeuroPrint allows us to be able to create a successful intervention long before the problem gets out of hand. Shortly, we shall discuss in detail the type of intervention that can be used to protect the neurocognitive system against this.

---

* Severely isolate landscapes can lead to, and are characteristic of, several neurocognitive disorders including, but not limited to, conflicting "parts," feeling fractionated, dissociative identity disorder, clinical depression, obsessive compulsive disorder, and trauma-induced amnesia.

MARKOV CHAINS IN THE NEGATIVE ATTRACTOR LANDSCAPE (EXCESSIVE ORDER FORMATION)

We saw in our last example that Markov chains can form as a result of isolated landscapes in a normally healthy and flexible neurocognitive system. If they are identified early, simple interventions can be used to free the state vector and reestablish normal operation of the neurocognitive system.

In Figure 8.4 a Markov chain is shown that has formed in the negative landscape. Notice that the chain can be entered from either another negative approach landscape by transition to the *annoyed* attractor or through the positive landscape by exit from the *satisfaction* attractor and entrance to the *exhaustion* attractor.

Once the state vector is captured by either the *annoyed* attractor or the *exhaustion* attractor, there is no way out of the chain. The most important distinguishing feature of a Markov chain is that the state vector travels consistently in the same direction on the phase path. Each attractor in the path becomes a transition state for the next attractor.

Markov chaining occurs when the pattern-entropy balance of a neurocognitive system tips too far toward order (periodicity). It leads the system into disorder due to lack of correspondence with the rapidly chaining environment.

We can clearly predict that both the phase path, as well as each attractor in the phase path, will become more stable with each periodic cycle of the state vector. This occurs because there is no interference from *reverse direction* trajectories that could potentially perturb the phase path from inside the system, and there are no escape trajectories to other attractors in the positive landscape that may also perturb the phase path from within the system.

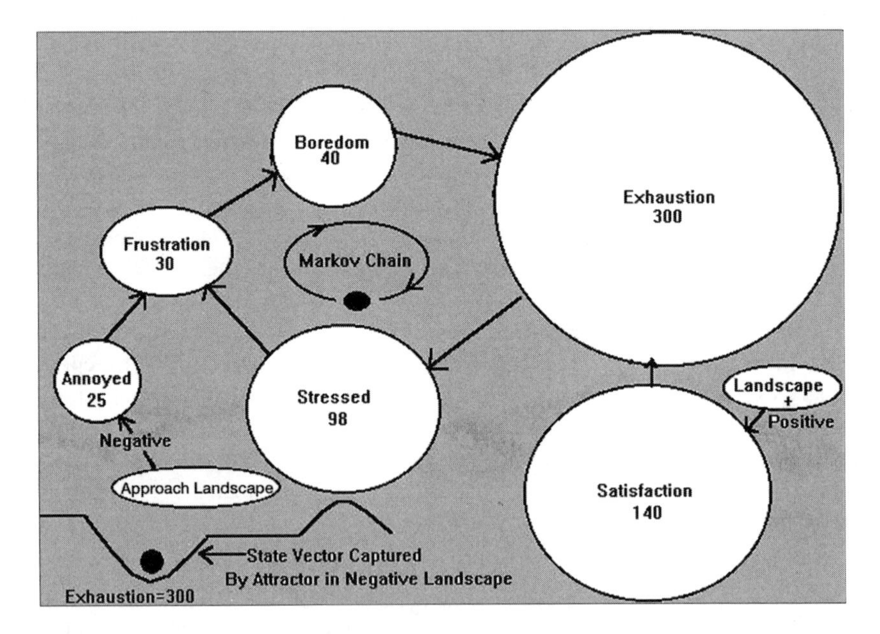

FIGURE 8.4 Markov chain in negative attractor landscape.

Since the neurocognitive system no longer possesses any internal options for perturbing the phase path, it must turn to the environment for stimuli from the information field of sufficient strength. This is why the neurocognitive system incorporates information not only to counterbalance entropy, but also to counterbalance too much pattern (order, periodicity).

For intervention purposes, it is best to identify potential Markov chains before they occur by looking for isolated landscapes. However, if this is not possible and Markov chains already exist, they can be broken easily by organizing and utilizing your intervention tools that best rearrange phase paths and destabilize attractors. Certain recently developed brief interventions are particularly useful for this purpose, namely, anchoring, Eye Movement Desensitization and Reprocessing (EMDR) (Shapiro, 1995), and a variety of energy psychology interventions such as Thought Field Therapy (TFT), Emotional Freedom Techniques (EFT), and Energy Diagnostic and Treatment Methods (EDxTM) (Gallo, 1998, 2000).

NEUROCOGNITIVE ENTROPY

The opposite of a Markov chain (order, periodicity) is *neurocognitive entropy*, both of which can be found to occur in just a single region of a neurocognitive system, or, in extreme cases, distributed throughout the entire system. In Figure 8.5, a NeuroPrint of design space is shown. The NeuroPrint topology indicates a dangerously high degree of neurocognitive entropy.

A system in neurocognitive entropy is characterized by a symmetrical distribution of phase intensity as measured by the SUD scale, each state varying only slightly in intensity and attractor stability (minimal stability asymmetry). The state/phase

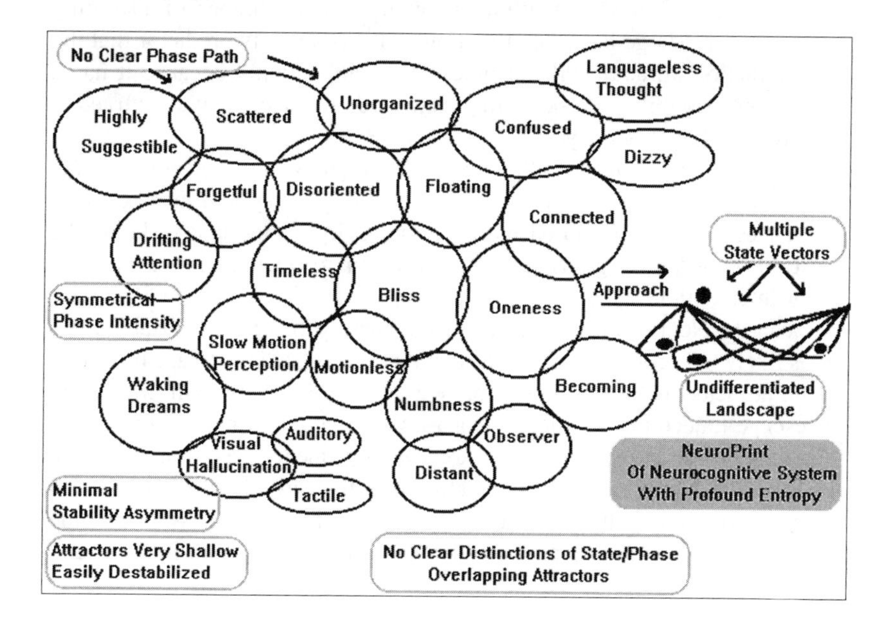

FIGURE 8.5 Systemic neurocognitive entropy represented by NeuroPrint.

attractors are very shallow and easily destabilized, overlapping with no clear differentiation between one state and another. The state vector has no clear phase path, and in many cases the subject has a sense of multiple state vectors (fractionated consciousness). It is difficult for the subject to concentrate for any period of time or to focus attention on a detailed task requiring critical, sequential, or ordered thought. A neurocognitive system characterized by high entropy may experience waking dreams, slow motion perception, disorientation, floating and numbness, confusion, languageless thought, "word salad" (jumbled speech), drifting attention, feeling of timelessness, etc. Neurocognitive systems characterized by a high degree of entropy normally respond very easily to suggestion. Occasionally, people in this condition will be found motionless for long periods of time (possibly cataleptic), appear distant, and report that they feel like an observer rather than a participant (dissociation and derealization). Logical, critical thought becomes nearly impossible. Design space organization such as the one shown in Figure 8.5 can result from many different causes such as severe trauma, cult conversion, information overload, prolonged sensory deprivation, stroke, etc.

UNDERSTANDING COGNITIVE NEURODYNAMICS: WHAT TO LOOK FOR IN THE NEUROPRINT

Before discussing intervention, let us add a few distinctions to assist in NeuroPrint analysis. In addition to the preceding, here are the seven most important things to look for:

1. Always scan the NeuroPrint for attractors with a high *approach trajectory density*. These are attractors with a large number of incoming trajectories (large basin of attraction). This is an early predictor that the destination attractor will become more stable simply because it has a greater possibility of becoming activated, relative to other attractors having a lower density.
2. Scan the NeuroPrint for attractors with a high *escape trajectory density*. These are regions of great flexibility, allowing the system many degrees of freedom once captured by an attractor. It is important to know the location of these regions and *how* and *where* to create them for purposes of intervention.
3. Scan the NeuroPrint for regions of the phase path where the state vector is likely to become trapped. This could be an isolated landscape or simply an attractor that has a high approach trajectory density with little or no escape trajectories. Regions of phase path that could potentially trap or severely restrict the state vector are always prime targets for intervention.
4. Scan the NeuroPrint for state attractors that are chained together in one direction with no reverse trajectories. These can easily become Markov chains if any portion of the chain links back to itself at any time in the future.

5. Scan the entire NeuroPrint for states that are predominantly activated by external stimuli, as well as those that are predominantly activated by internally incorporated stimuli/information (memes). Notice patterns in the types of memes that have been incorporated, the bio-architecture they created, and whether or not they seem to have a *common source* in the person's life (occupation, education, religious training, cultural traditions, family, peers, television, etc.).
6. Scan each state/phase attractor for the number and quality of state-bound behaviors available to the system (behavioral degrees of freedom). Notice if these behaviors are available to the system when needed to appropriately respond to external stimuli coming from the information field (environment).
7. Scan the NeuroPrint for state-bound behaviors that can potentially strengthen or destabilize their own or other state/phase attractors.

9 Intervention and Change

CONTENTS

COGNITIVE NEUROPHYSICS — INTERVENTION AND CHANGE

The most important reason for developing a NeuroPrint is to uncover the hidden structure (pattern) of a problem in order to guide intervention. Since the pattern-entropy principle of matter and energy dynamics reveals that structure equals function, a change in structure will necessarily produce a change in function. In order to gain deeper insight into what it means to conduct therapeutic intervention or to create lasting change, we will be defining intervention along several lines, from the general to the very specific.

The broadest and most fundamental way to view intervention is from the structure/function principle. Since a change in structure necessarily produces a change in function, the aim of therapeutic intervention is to make a change in the structure of the neurocognitive system's design space. To accomplish this, it follows that we must have the ability to see the current structure, which NeuroPrint affords.

As we learn from NeuroPrint, the structure or pattern of the neurocognitive system can be geometrized and hence viewed in four dimensions. A complete NeuroPrint makes it possible for us to see clearly the neurocognitive system's asymmetrical distribution of matter and energy resources, which is comprised of attractors (state, behavior, etc.) and trajectories of differential stability which make up its overall pattern in design space. Bearing this in mind, we can expand our definition of intervention as follows: The very essence of intervention is the modification of the structure and therefore the function of a neurocognitive system. To accomplish this, we must either modify the system's stability distribution (attractors, trajectories) or rearrange its phase path (the state vector's transition path through design space). Every successful intervention we have had the opportunity to study accomplishes one or both of these objectives.

CLUES TO LOCATING LEVERAGE POINTS IN DESIGN SPACE

If the development, reading, and analysis of the NeuroPrint is done thoroughly, then the job of designing an intervention is a relatively simple one. Most of our work is already done, although there are two important questions that will help us form a bridge from analysis to intervention.

1. **What is holding this problem together?**
   Is it the *stimuli* in the external information field?
   Is it the direction of the *trajectories* forming the phase path?
   Is it the *memes* (internally incorporated stimuli from the information field)?
   Is it the *state/phase* attractor stability?
   Is it the *behavior* attractor stability?
   Is it the lack of state-bound behaviors that can interrupt the state/phase?
   Is it the *limited degrees of freedom* for state/phase, behavior, phase path, or stimuli?
2. **Where is the keystone? What is the most elegant, efficient way to modify this structure so that it results in the intended function?**

SIX CLUES TO DESIGNING A NEUROCOGNITIVE INTERVENTION

The first five clues to designing successful neurocognitive interventions come from the five behaviors of pattern discussed in Section I. It is these five fundamental behaviors of pattern, interacting in highly complex ways, that are responsible for creating the structural topology that is visible in the Neuro-Print:

- **Incorporation:** absorption of one element, pattern, or aggregate by another
- **Replication:** duplication, reactivation, or copying of an element aggregate or process
- **Cleaving:** separating or dismembering of an aggregate into two or more parts/elements
- **Recombination:** aggregation and rearrangement of elements or larger aggregates
- **Transmitting:** isomorphic propagation and exchange of pattern/information through and between elements, aggregates, or coordinate space

Since it is from combinations of these fundamental behaviors of pattern that all neurocognitive systems are created, we must look for clues to design intervention from the same building blocks. When used in complex combination, the five behaviors of pattern give us as many options for changing design space as there are for creating it in the first place.

The sixth clue for designing a neurocognitive intervention comes from the principle of pattern-entropy balance. If we refer to our definition of intervention and expand upon it, it becomes evident that the modification of design space, whether by changing attractor stability or rearranging phase path, necessitates that we differentially stabilize parts of the neurocognitive pattern, while at the same time destabilizing other parts. The keystone to modifying stability of any portion of design space represented by the NeuroPrint is the creation of intervention tools that tip the pattern-entropy balance in the intended direction.

Destabilizing Energy Barriers = Increasing Entropy

Whenever destabilizing a region of design space, however large or small, we must always increase entropy in order to overcome the energy barriers that hold the existing pattern, and hence structure and function, in place.

Stabilizing Energy Barriers = Decreasing Entropy and Increasing Pattern/Order

Whenever stabilizing a region of design space, however large or small, we must always increase pattern/order to fortify existing energy barriers or create new ones in precise locations of design space.

Recall that these energy barriers exist at every level of structural organization in the brain, from atomic-level electron clouds and the molecular-level covalent bonds to the neural-net plasticity subserved by the modification of dendrite branches and synaptic effectiveness. All of these are affected when the pattern-entropy balance is tipped one way or the other.

The most important decision we must make as change agents is how narrow or broad the target area for intervention will be, so that we may precisely adjust the focus of our pattern-entropy beam. This is why we topologically represent neurocognitive activity and ask all of the questions cited in the preparatory stages of development, analysis, and intervention. We must clearly decide on the size of our target area and accordingly design our intervention tools to seek the specific target chosen. It is easier to do this when the structure and function of a neurocognitive system is topologically represented as it is in a NeuroPrint. All successful tools and methods of neurocognitive intervention either increase entropy (disorder) or increase pattern (order) in some region of design space by use of some combination of the five behaviors of pattern. Understanding this is the first key to organizing the intervention tools and methods available in one's repertoire and custom designing an unlimited variety of new and more effective tools.

Let us take a moment to illustrate the principles discussed. In nearly every intervention that seeks to rearrange phase path or modify differential stability of portions of that path or its connecting attractors, we must employ the pattern behavior referred to as cleaving (sever). When we cleave an approach trajectory to an attractor, we not only rearrange the phase path of the state vector, but we also prevent the activation of that attractor by this approach trajectory. In turn, this weakens the attractor stability (shallows the attractor). Thus, by cleaving an approach trajectory the phase path of the state vector is altered and eventually the stability of the connecting attractor is weakened as a result of the drop in frequency of activation. Notice in Figure 9.1 that timing is the critical element that determines whether a destabilizing force cleaves a trajectory or directly weakens an attractor.

In order to cleave an approach trajectory, we must interrupt the state vector at the *trapping zone* of the attractor just prior to being captured. In this way, we initially destabilize only the approach trajectory. Later, the attractor stability will weaken in direct proportion to the trajectories we have removed vs. the ones that remain.

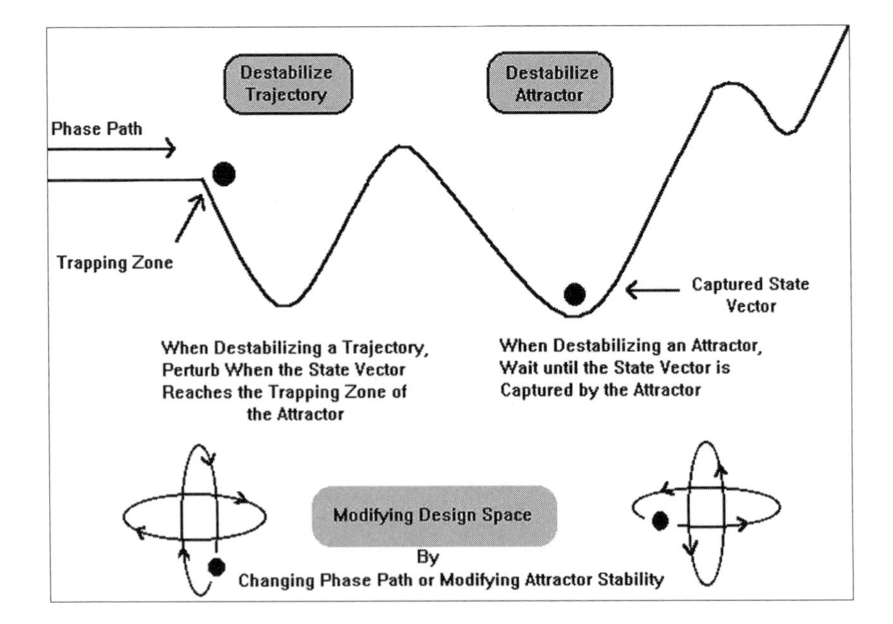

FIGURE 9.1 When, where and how to interrupt the phase path.

If our target of intervention is to directly weaken (destabilize) an attractor (state, behavior, etc.), we must wait until the state vector is captured by the attractor. It is impossible to destabilize a single attractor, and only that attractor, unless it is being activated by the neurocognitive system. Systemic, entropy-increasing tools, such as hypnotic trance and electroconvulsive therapy, are broad-focus interventions capable of temporary or even permanent disruption of the brain's energy barriers. Although these two intervention tools vary greatly in scope and intensity, neither is precise enough to cleave a selected trajectory or destabilize a single attractor without dramatically increasing systemic entropy at the same time. Prolonged or pernicious use of either may result in systemic neurocognitive entropy, similar to the NeuroPrint in Chapter 8, Figure 8.5.

Over the last three decades, numerous schools of brief therapy have arisen which make use of a much more tightly focused arsenal of entropy-increasing tools. Some of the more successful of these are NLP, EMDR, and various energy psychology approaches (Gallo, 1998, 2000; Gallo and Vincenzi, 2000). The field of NLP makes use of anchoring, strategies, and submodalities, which are extremely tightly focused tools. Energy psychology methods are also tightly focused with respect to targeting a variety of contextualized negative affects and variously promoting pattern and entropy. These tools can be very easily adopted to make precisely targeted changes in design space.

Before moving on, we should also note that we can precisely measure the degree of destabilization that has occurred at any point in our intervention simply by taking a new measure of phase velocity or dwelling time with a stopwatch. When a trajectory or an attractor begins to weaken, the decrease in phase velocity is referred to as *critical slowing*. The principle of critical slowing states that the longer it takes for the state vector to be captured (the attractor to be activated), the more shallow (unstable) it has become. Thus, critical slowing is a valid and reliable method of measuring stability and, by implication, the degree of change attained after intervention. We do this simply by comparing the initial values for phase velocity against the current ones.

The most reliable methods of measuring the stability of an attractor are frequency and dwelling time. While dwelling time can be measured immediately, the client tracks frequency of occurrence over the course of the following week or weeks. Both dwelling time and frequency of occurrence should be taken into account when measuring the success of any intervention.

In Figure 9.2, we show a simple intervention designed to cleave a Markov chain (see Chapter 8, Figure 8.4 for an example of a Markov chain) in order to free the state vector from this region of design space and, in turn, reestablish more healthy, flexible functioning of the neurocognitive system.

INTERVENTION AND CHANGE

Step 1: Predetermining the New Direction for the State Vector

It is obvious from the structure of the Markov chain that we must cleave some part of it (trajectory) in order to free the state vector and rearrange the phase path.

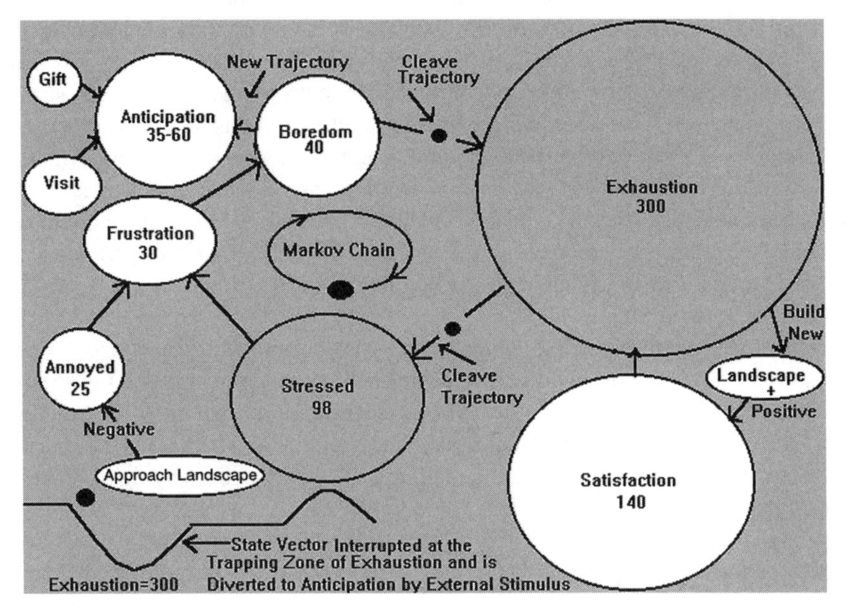

FIGURE 9.2 Cleaving a Markov chain and diverting the phase path to a positive attractor landscape.

If we do not cleave a trajectory, each attractor and trajectory in the periodic cycle will increase in relative stability over the rest of the regions of design space. However, before arbitrarily cleaving any trajectory, we must determine which direction the state vector should take (i.e., predetermine how the pattern is to be recombined). If we do not do this first, an attractor with a relatively high phase velocity and high stability index will arbitrarily capture the state vector. This can lead to unpredictable and undesirable abreaction. Many NLP and EMDR practitioners saw this type of problem in the early stages of their work with trauma and phobia. While these techniques are extremely effective at destabilizing memories, parts of memories, their somatosensory (kinesthetic) components, and reactions to stimuli, they unfortunately led to serious abreaction in many subjects. The source of the abreaction can easily be attributed to the fact that an earlier-formed attractor that produces a high-intensity response arbitrarily captures the now-freed state vector, having no prespecified place to go.

In Figure 9.2, we determined that the best place to interrupt the phase path was between boredom and exhaustion, so as to prevent capture of the state vector by the enormously deep exhaustion attractor, and cause a new trajectory to form from boredom to anticipation. When analyzing the entire NeuroPrint, it seemed that "anticipation" was the most effective place to enter the positive landscape. The only problem we saw is that the anticipation attractor (at an initial stability index of 35) was slightly shallower than the boredom attraction (with a stability index of 40).

Step 2: Modifying the Stability of the Destination Attractor (*Replication*)

Considering the problem of relative stability between anticipation and boredom, it is clear that the second step in our intervention is to strengthen (deepen) the anticipation attractor. After this is achieved, it is much more likely to hold onto the state vector once it is captured, rather than giving it back to the boredom attractor. There are many intervention tools available to strengthen an attractor. In this case, we had the subject vividly reexperience (replicate) the feeling of anticipation by remembering several occasions that triggered the emotion and reliving each experience intensely. Once this was done, a new measure of dwelling time alone changed the stability index for anticipation from 35 to 60. Consequently, the phase velocity also increased, making it much easier to transition into that state from boredom. By having the subject continually reenter the state of boredom in-between reexperiencing each instance of anticipation, we built a stable new trajectory between the boredom attractor and the anticipation attractor. Each time the stability of the new trajectory was measured by its phase velocity, it increased as expected.

When evaluating any two attractors in a phase path, there is a second way of measuring their relative stability and the stability of the new trajectory erected between them. We refer to this measure as *switching time*. The switching time from a shallow attractor to a deeper attractor will always be shorter than the switching time going in the opposite direction. So, to be sure which is the deeper attractor, we can have the subject transition between boredom and anticipation and then from anticipation back to boredom, measuring and comparing the phase velocities in both directions. The state vector is ready to be freed when the transition from boredom to anticipation occurs significantly faster than the opposite (two times faster is ideal).

Step 3: Cleaving the Trajectory and Freeing the State Vector (*Cleaving*)

We have now successfully differentially stabilized the portions of the negative and positive landscapes for our intervention. The next step is to cleave the trajectory between the boredom attractor and the exhaustion attractor (destabilize the energy barrier). Any internal or external influence that continually interrupts the state vector at this transition point will do. Brief therapies are brimming with tools that can accomplish this. Many of them are classified under the name pattern interruption or pattern disruption. Pattern interruption occurs when a competing pattern-integrity (state of motion) dis-integrates an existing self-stabilizing pattern-integrity.

Step 4: Establishing the New Phase Path for the State Vector (*Recombination*)

Immediately following the disturbance of the boredom-exhaustion trajectory, the anticipation attractor must be specified to guarantee that the state vector is captured there. We simply accomplish this by the presentation of one or both of the stimuli that reliably activate that attractor. In this case, receiving a surprise gift or visit, or even imagining these contextual cues, guaranteed the capture of the state vector by the anticipation attractor. This step should be initiated nearly simultaneously with Step 3 (interrupt the path and then specify a new path). Timing is critical. The procedure of intentionally adding, subtracting, introducing, or rearranging stimuli that reliably trigger an attractor is referred to as classical, respondent, or Pavlovian conditioning (stimulus–response conditioning). In NLP it is referred to as anchoring.

Step 5: Measuring the Degree of Change

To determine if the change is complete, or how close we are to the intended change, we simply measure all of the necessary phase velocities and stability indexes of the attractors involved. Ideally, we want boredom to become a transition state for anticipation, so in this case we would like to see "anticipation" continue to stabilize relative to "boredom." We would also like to decrease the stability of the exhaustion attractor that begins to happen simply by cleaving approach trajectories, as we have done. We can also do this more directly by simply activating (replicate) the state attractor for exhaustion and using a destabilizing intervention (entropy increasing).

Notice that in Figure 9.2 we also cleave the escape trajectory from exhaustion to stressed with the intention of shallowing the stressed attractor as well. We then use the exhaustion attractor as a transition state to another area of the positive attractor landscape by building a new trajectory in the exact same way we did between boredom and anticipation. This entire intervention can be planned and completed in less than 40 minutes once a NeuroPrint has been developed.

The Markov chain has now effectively been disrupted, resulting in the state vector being freed and diverted down the newly specified phase path that leads back to the positive landscape at several points. In this brief intervention example we have both modified the differential stability between selected regions of design space and rearranged the phase path of the state vector. Numerous tools that can be tailored for such an intervention are covered in Section III.

NEUROCOGNITIVE MODELING: THE REPLICATION, TRANSFER, AND INCORPORATION OF A REGION OF DESIGN SPACE

In the above Markov chain intervention, we made use of both entropy-increasing and pattern/order-increasing tools. The pattern behaviors of the tools used were replicating, cleaving, and recombining. In this next example, we illustrate another type of intervention that makes use of the remaining two pattern behaviors of transmission (transfer) and incorporation. This method is much more detailed and comprehensive in its scope, utilizing all of the principles of NeuroPrint. We refer to this type of intervention as neurocognitive modeling, which involves the transfer and incorporation of a region of design space from one person to another. In other words, if someone has spent many years developing a particular neurocognitive

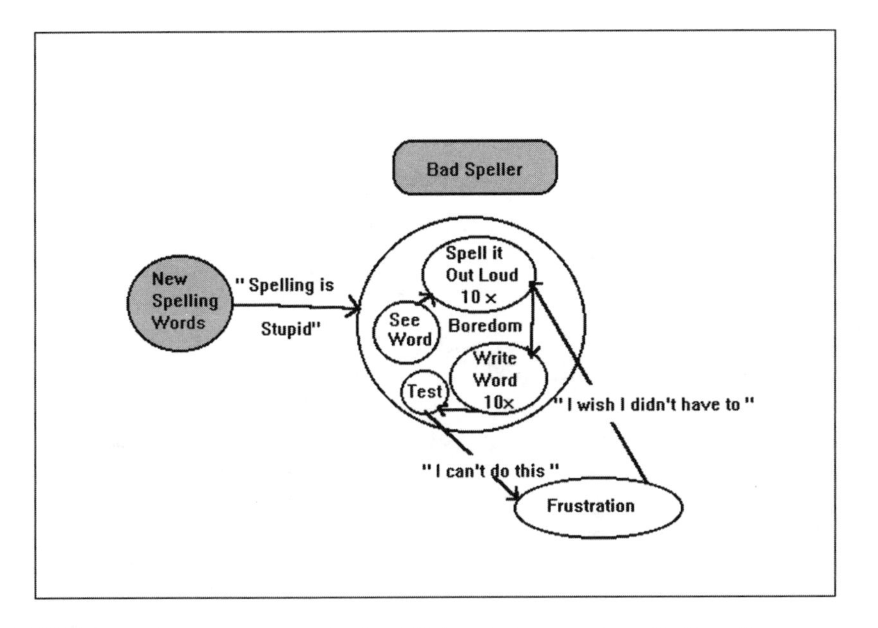

FIGURE 9.3 Representing models for skill/knowledge transfer with NeuroPrint — poor spelling strategy.

skill, that skill can be represented by a NeuroPrint (replicated) and transferred (transmitted) to another human being by a complete incorporation of the same structure. We do this when a redesign of an existing area of phase space would require too many steps. Instead, we may just bypass that region design space that we want to fall into disuse (destabilize) by incorporating a "complete" model of a skill and creating a trajectory that triggers that particular landscape instead of the old one. In Figure 9.3, we have a NeuroPrint of a child who has difficulties in spelling.

In this case, the attractor landscape is triggered (unfolded) by the stimulus (new spelling words). The meme that forms the trajectory leading to this landscape is "spelling is stupid." The meme triggers the state of boredom from which four state-bound behaviors are available. The child sees the word, spells it out loud ten times, writes the word down on paper ten times, tests himself, and every time he gets one wrong he says, "I can't do this." This activates the new state/phase of frustration, which leads back to spelling the word out loud ten times again, reactivating and deepening the attractor for boredom. This continues until either all the words are finished or the child gives up out of frustration and boredom.

Modifying this landscape and the tightly chained behaviors (strategy memes) previously incorporated would be too time consuming. Instead we bypass this region altogether and use the stimulus "new spelling words" to build a new trajectory that leads to a completely different landscape.

We develop that new landscape by first eliciting the strategy that an excellent speller uses and then representing it in a NeuroPrint. The NeuroPrint resulting from this process can be seen in Figure 9.4. Here we can see immediately that the phase

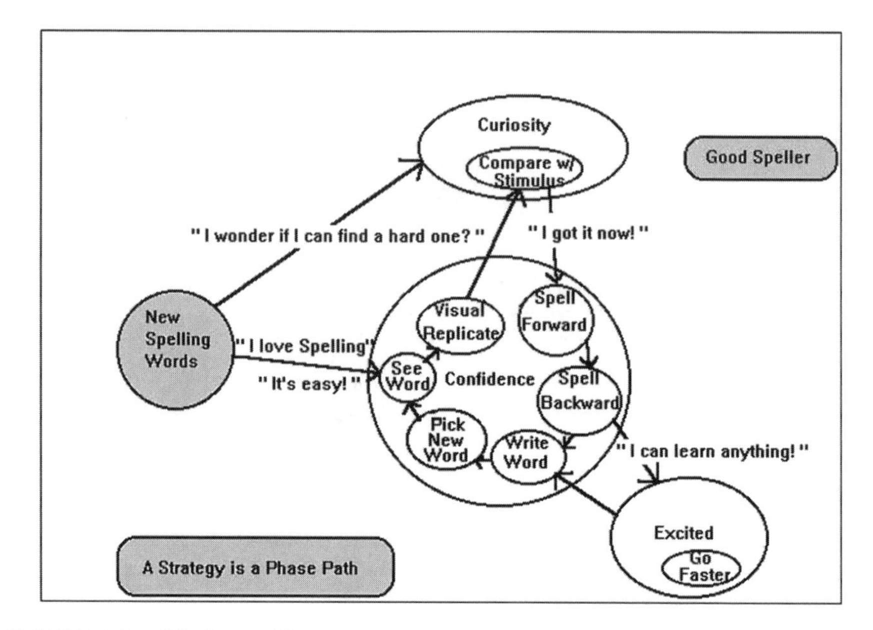

FIGURE 9.4 Effective spelling strategy.

path of the state vector (its state of motion) is qualitatively quite different. The state vector, which is released from the initial conditions "new spelling words," maintains a phase path that stays well within the positive attractor landscape. The state of confidence is triggered by the meme "I love spelling. It's easy." From within the state of confidence, six state-bound behaviors become available. Seeing the word and visually replicating it are the first two. The behavior of visual replication has a trajectory that leads to the curiosity attractor. This causes the child to compare the visually imagined word with the original stimulus (the spelling word to be incorporated). When they match, the confidence attractor is then triggered by the meme "I got it now." The child immediately continues along the chain of state-bound behaviors for confidence, spelling the word forward, then backward, then writing the word down, and then picking a new word. At this point, the process starts all over again.

From time to time, the state vector will depart from confidence and trigger excitement with the meme "I can learn anything." The excited attractor contains the state-bound behavior of increasing the child's speed through the strategy. This continues on until all the spelling words are learned. The quality of the strategy is judged by its result, which is comprised of the number of words that are spelled correctly, the time it takes to complete the task, and the feeling that the student has about the task upon entrance and exit from the landscape. Many of the tools necessary to transfer this model between students have already been discussed. However, modeling is discussed in much greater detail in Section III. The potential of modeling as a method of intervention and change is only limited by the creativity of the practitioner.

Looking closely at Figures 9.3 and 9.4, notice that they are both, in a sense, Markov chains. However, they are qualitatively different if judged by the states, behaviors, and memes that are self-stabilized by the use of each of these strategies. Clearly, the resulting effect on the neurocognitive system will be very different.

Many useful neurocognitive strategies are highly complex and well hidden in microscopic internal behavior of the neocortex. But models of these skills are well worth eliciting, representing, and transferring as they may dramatically shorten the amount of time necessary to learn a complex skill. In order to capture and represent the activity of the state vector for such skills, we need to represent microscopic-level states and behaviors. The microscopic-level states, or submodalities, may entail internal visualization, auditory imaging, etc., and the microscopic-level behaviors are the behaviors of pattern available within each of those states.

REPRESENTING MODE-BOUND BEHAVIORS WITH NEUROPRINT (SUBMODALITIES)

Submodalities are an important class of complex behaviors performed by the neurocognitive system. The major difference between these behaviors and the behaviors previously discussed is that they are much more difficult to observe than the collective, coordinated, macroscopic behaviors performed by our musculoskeletal system (e.g., expressions, vocalization, movement, etc.). Part of the difficulty is that only a small portion of the behavior is visible from the surface. The greater portion of the behavior is visible to the performer as soon as he/she is trained to pay attention to it. The other difficulty is that many of these behaviors can occur simultaneously. (See Figure 9.5.)

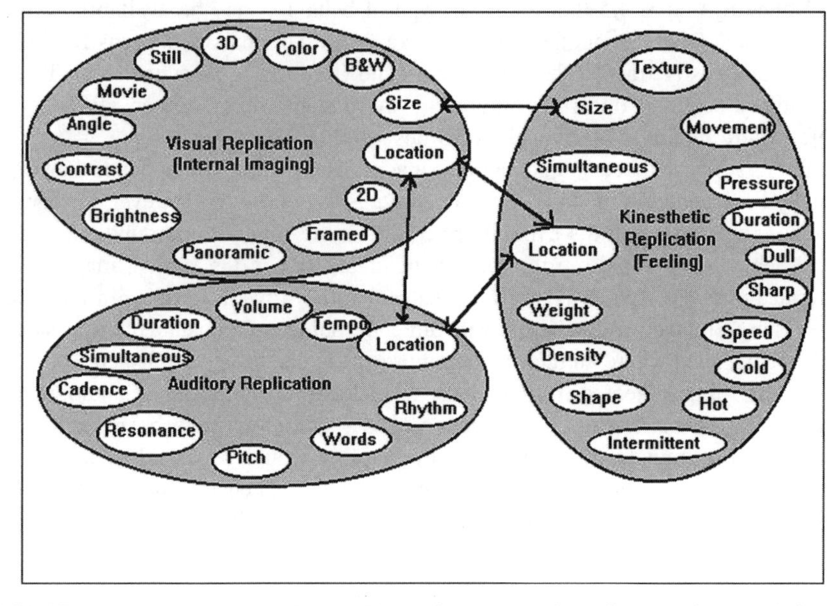

FIGURE 9.5 Representing submodalities with NeuroPrint: diagramming mode-bound behaviors.

Submodality is a short way of saying *sub*modal behavior, meaning that these behaviors occur within a *mode* (state or phase), just like macroscopic behaviors do. So when represented by NeuroPrint, they are represented the same way — within a state/phase or mode (visual replication, auditory replication, somatosensory replication, etc.).

In NLP these modes are referred to as visual internal representation, auditory internal representation, etc. NLP practitioners prefer to use the word *re-present* rather than replicate. However, this term can be misleading as it presupposes that there is some tiny internal observer in the brain that we are representing to (a homunculus).

These more complicated submodal behaviors arise from the primary behaviors of pattern: incorporation, replication, cleaving, recombining, and transmitting. Many of them occur during the state of perception (incorporation) as well as during the state of internal representation (replication) of environmental information/pattern. For the purpose of knowledge/skill transfer, both types are important.

Submodalities are connected to other submodal behaviors and modes/states by trajectories, and they are organized into phase paths referred to as strategies by behavioral modelers. Submodalities are discussed in more detail in Section III*. For the time being, it is only important to understand how to represent their activity with NeuroPrint. These microscopic state and behavior transitions are influenced by information just like their macroscopic counterparts and they are the keystone to understanding the structure of human excellence.

The elicitation, representation, and transfer of this internal structure to others are referred to as *human performance modeling*. When the represented models are transferred to computers, we call the result an *expert system*. Expert system modeling is a branch of Artificial Intelligence (AI).

Notice in Figure 9.5 that the same type of behavior may be available in more than one state/phase or mode (location, size, etc.). This makes patterns replicated in one state easily transferable to another state (referred to as *synethesia*). This is particularly useful if one state attractor is more stable than another, allowing the subject more facility with the submodalities within.

Let us suppose that a subject has a headache. The pattern of that headache would initially be detected in the kinesthetic or *somatosensory mode*. However, the subject may not be practiced in the explicit use of the microscopic, state-bound, neurocognitive behaviors available in the kinesthetic state. Rather, the subject may be able to replicate images (patterns) for far greater duration (dwelling time) in the visual state. Therefore, we would alternatively direct the subject to translate the pain pattern to the visual state where an intervention more easily can be accomplished. From the visual state the subject can change the size, location, duration, orientation, and distance of the pattern until the somatosensory attractors have been sufficiently perturbed to eliminate the pain. Next the subject would be asked to attempt to replicate the pain pattern and record the phase velocity, intensity, and dwelling time, if any. If the subject is unable to replicate the pain pattern within a reasonable period, the intervention is considered to be complete.

---

* A comprehensive listing of submodal behaviors can be found in Appendix 2 of this section.

ORGANIZING INTERVENTION TOOLS

As mentioned previously, the very essence of intervention is the modification of the function (state of motion) of a neurocognitive system. The function of the neurocognitive system is totally dependent upon its pattern (informational bio-architecture). Therefore, in order to modify the function of the neurocognitive system we must necessarily do one or more of the following things to change that pattern:

1. Modify the system's attractor/trajectory stability distribution by stabilizing and destabilizing attractors.
2. Rearrange its phase path through design space to free trapped state vectors.

It is important to understand that it is not necessary to discard any intervention tools previously learned. In order to use previously learned tools effectively, they simply must be organized and classified so that their effects on design space and the neurocognitive system are well understood. There are three rules that easily can guide the classification and organization of existing intervention tools and methods:

- Pattern/entropy effect
- Scope of effect of the tool or method
- Behavior of the tool

Let us take a moment to describe each of these in detail.

PATTERN/ENTROPY EFFECT

It would be helpful for the practitioner to start by listing all of the tools and methods commonly used. Once this is accomplished, develop another list divided into two sections: *entropy-increasing* and *pattern/order/structure-increasing*. Then for each tool or method listed, determine if it increases or decreases entropy, classifying each intervention tool or method under one of these two categories. The pattern-increasing tools are applied any time that we wish to stabilize a region of design space or add new structures to design space. The entropy-increasing tools will be used any time we wish to destabilize or eliminate regions of design space (destabilize energy barriers).

Increasing order can be achieved in many ways. We may add or stabilize states, behaviors (macroscopic and microscopic), stimuli, memes, or trajectories. Anything that increases the asymmetrical distribution of matter and energy resources will necessarily increase pattern, order, and structure.

When we increase entropy, we are breaking energy barriers and thus, either temporarily or permanently, reducing the asymmetry of design space and, by implication, the neurocognitive system.

SCOPE OF EFFECT (OF THE TOOL OR METHOD)

Once we have determined which tools are entropy-increasing and which tools are pattern-increasing, we next want to consider each tool or method by the scope of its effect. That is, is the tool or method broad or narrow in its effect?

- Does it affect the entire neurocognitive system?
- Does it affect a specific landscape (positive or negative)?
- Does it affect a specific region of a landscape?
- Does it affect a trajectory in the phase path?
- Does it affect a meme or thought?
- Does it affect a macroscopic state/phase attractor?
- Does it affect a microscopic state/phase attractor (mode)?
- Does it affect a macroscopic behavior attractor?
- Does it affect a microscopic behavior attractor (submodality)?

Each of the tools should then be subclassified in this manner.

Behavior of Tool or Method

Once the tools and methods are subclassified by the scope of their effect, to more deeply understand their flexibility and appropriate use, they can be further subclassified by the pattern behavior of each of the tools. That is, which of the following does the tool or method accomplish?

- Incorporate a pattern
- Replicate a pattern
- Cleave a pattern
- Recombine a pattern
- Transmit a pattern

Once all of the intervention tools have been classified and organized by these three methods, we have a much greater understanding of their applicability and utility for neurocognitive intervention. Whenever the practitioner is in a quandary about how to approach an intervention, he/she can start by looking at the NeuroPrint and the variously classified intervention tools, and then consider the following options:

- Modify the stability of a structure (deepen, shallow, or eliminate a target area) such as a meme, state, behavior, trajectory, landscape, etc.
- Increase or decrease state vector choice for selection of state or behavior.
- Rearrange a phase path.
- Add attractors (macrobehavior, submodal behavior, macrostate, microstate, meme, etc.).
- Add trajectories.
- Change existing trajectories.
- Chain (recombine) the order of states or behaviors in a phase path.
- Add, subtract, or rearrange stimuli.
- Add disrupting or stabilizing state-bound behaviors (macroscopic or submodal).

By reviewing these few options and available intervention tools, as well as reviewing the development, analysis, and intervention portions of this section, the practitioner should be able to generate numerous solutions with his/her existing tools. If available tools and methods are not sufficient to accomplish the desired outcome, then it is time to begin to design new tools.

DESIGNING NEW INTERVENTION TOOLS AND METHODS

There are two circumstances that propel the development of new intervention tools and methods. The first occurs when we find that after classifying and organizing our intervention tools, many of the categories mentioned are empty. The second occurs when the NeuroPrint reveals insight about a problem in the neurocognitive system, but we find that we do not have a tool or method appropriate for the scope of the problem, desired entropy effect, or entailing the necessary combination of pattern behavior. In such cases, the guidelines in this section, along with the principles in Section I, provide a sufficient foundation to custom design tools and methods appropriate for the intended intervention. Section III will also be helpful, as it contains an in-depth analysis of several important intervention tools.

NEUROPRINT SHORTCUTS

- Measuring velocity (V) and dwelling time (D) at the same time
- Measuring stability by F, D, or V alone when one measure of stability will suffice
- Using only the NeuroPrint development steps you need (1–10)
- Selecting the NeuroPrint view most appropriate: frequency distribution, dwelling time, phase velocity, stability index, probability amplitude, comparing a few states only, seeing entire matrix, etc.
- NeuroPrint diagramming of only the portion of topology that you need: states only, states and behaviors, region, landscape, etc.
- Modeling only the "problem space" with NeuroPrint

In certain circumstances due to time constraints, scope of problem, or when attempting to rapidly gain rapport with a new client, it may be advisable to skip the complete inventory of neurocognitive states and state-bound behaviors in design space, and model only the design space for the presenting problem. In Figure 9.6, we provide an example of how to NeuroPrint only the regions of design space immediately involved, temporarily ignoring systemic effects. As far as necessary, we can continue to map out the *problem space* in order to design an effective intervention. Later, a systemic NeuroPrint can be started by the client between sessions and completed over a series of sessions designed to guide the client through each step of the process, allowing as much independence as possible.

Figure 9.6 shows a NeuroPrint of the problem space only. Numerous interventions can be designed without going beyond these boundaries. However, in the case of choosing to divert the state vector to another landscape, we can diagram only as much of that landscape as is needed. This shortcut is not recommended until the

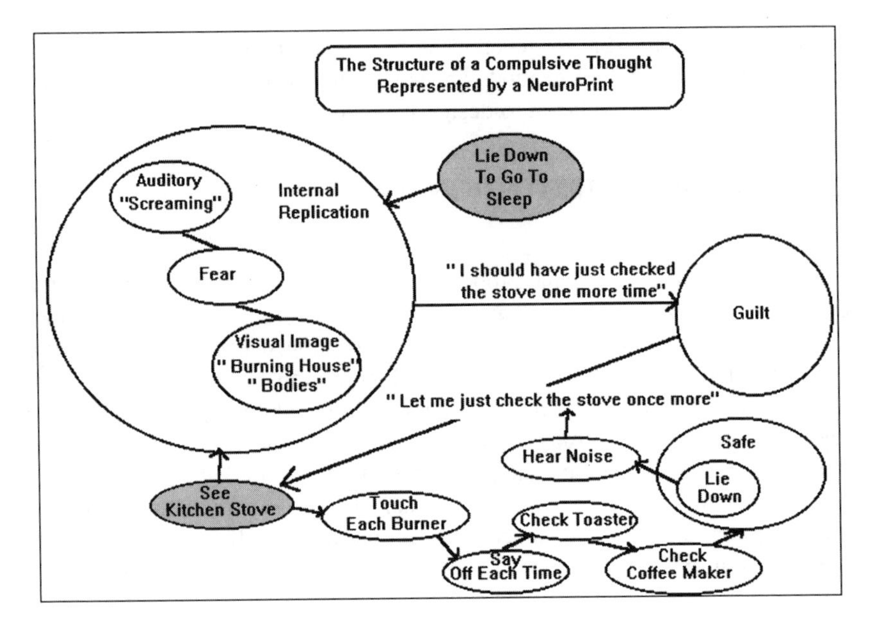

FIGURE 9.6 Modeling problem space with NeuroPrint.

practitioner has previously completed several NeuroPrints in order to be able to anticipate the systemic effects of interventions.

NEUROPRINT — ALTERNATE VIEWS AND APPLICATIONS

In addition to NeuroPrint being a content-neutral algorithm, it is also systemically neutral. The same steps can be used to topologically model larger systems than an individual. NeuroPrint is easily adaptable to couples and family counseling, corporate team building, performance modeling, academic education, etc. These and several other topics will be the focus of future texts. This section on NeuroPrint was designed to be a brief overview of the essential fundamentals.

SUGGESTIONS FOR COMPUTER SIMULATION OF NEUROPRINT

NeuroPrint can be represented dynamically (in four dimensions) by constructing a landscape with all the attractors of a particular neurocognitive system represented. The dynamic computer image can be set in motion by releasing a ball (the state vector) from any selected set of initial conditions to see what the system will do. All of the values that direct the state vector along its phase path (frequency, dwelling time, and phase velocity) can be entered in a spreadsheet or database from which the dynamic graphic image is constructed.

THE UNFOLDING NEUROCOGNITIVE SYSTEM

In this section we have presented a method whereby information, thought, states, and their effects on behavior can be perceived. In the systematic unfolding of our cognitive neurodynamics layer by layer, we acquire deep insight into the complex informational bio-architecture of a human life.

NeuroPrint is possible because the human brain is a finite state system; it repeats itself. If it did not, the emergent properties that we cherish and that make us human — our memories, emotions, behaviors, beliefs, values, language, traditions — would not exist. At the very least, all of these things require the incorporation and replication of pattern. To accomplish this, the brain must erect energy barriers in order to make the incorporated information readily available to the neurocognitive system.

Slowly but surely these energy barriers stabilize that information into a dynamic informational bio-architecture, constructing an internal model of the external environment from the "informed" matter and energy we incorporate. At the same time, these energy barriers become waveguides that direct the flow of new incoming information, and systematically diminish the total degrees of freedom available to the neurocognitive system. As our external environment continues to evolve away from our internal models, some of these patterns continue to serve us, while others do not. Conflicting information patterns vie for neurodynamic control. Our cognitive neurodynamics become enslaved by the incessant tug of pattern and entropy and the unbridled internal interplay between the five behaviors of pattern. New and sometimes inappropriate behaviors spontaneously arise in attempts to control the flow of information across our semipermeable information barriers — filtering, deleting, and distorting information essential to our continued correspondence with the external environment.

Conscious awareness provides us with the opportunity to *see* these patterns if we know how to look and the ability to change them if we know how to change. As human beings we are endowed with the unprecedented opportunity to influence our own design.

Appendix 1

Understanding Cognitive-Neurodynamic States, Experiences, and Cognitive Phenomena

CONTENTS

EXAMPLES OF STATES, EXPERIENCES, AND PHENOMENA CHARACTERIZED BY A FUNDAMENTAL SYSTEM OR SUBSYSTEM PHASE TRANSITION FROM PATTERN (ORDER) TO ENTROPY (DISORDER)

Cleaving: Creativity, dissociative identity disorder
Confusion, disorientation, dizziness, going "blank," forgetting
Vague speech
Relaxation, flow, fluidity, flexibility
Hypnotic trance, meditation, hallucination
Feeling fractionated, conflicting parts, indecisive, hesitant
REM stage sleep, dreaming
Interruption, illness
Macroscopic stillness
Temperature equilibrium between bodily regions
Clinical depression, psychotic behavior
Electroconvulsive therapy

Sensory deprivation, sensory overload, conflicting information patterns
Catatonia, catalepsy
Death

EXAMPLES OF STATES, EXPERIENCES, AND PHENOMENA CHARACTERIZED BY A FUNDAMENTAL SYSTEM OR SUBSYSTEM PHASE TRANSITION FROM ENTROPY (DISORDER) TO PATTERN (ORDER)

Incorporation: Reading, television, and learning
Replication: Memory, remembering
Recombining: Mathematics, painting, sculpting, understanding, building, designing, decorating, and developing
Transmitting: Talking, teaching, singing, and making conversation
Distinction memes, strategy memes, association memes
Behavioral predictability, event predictability
Playing video games
Stress, inflexibility, stuck, rut
Rhythm, poetry, music, dance (incorporating, replicating, or transmitting)
Sports, exercise, hobbies
Macroscopic sequential movement
Identity, personality, habits, beliefs, traditions
Asymmetrical body temperature
Obsessive-compulsive thought or behavior, neurotic behavior
Feeling integrated, whole
Healing

Appendix 2

Submodal Behaviors (Submodalities)

CONTENTS

VISUAL MODE

1. Remembered/constructed
2. Movie/still
3. Whole/part
4. Detailed/contextual
5. Content digital or analog
6. Solid/transparent/vapor
7. Color intense or muted
8. Right/left/center
9. Up/middle/down
10. Bright/dim/dark
11. Degree of contrast
12. Life-size/bigger/smaller
13. Proportion of image
14. Proximity
15. Fast/medium/slow
16. Specific focus
17. Self in/out of picture
18. Frame/panorama

19. 3D/2D
20. Particular color
21. Viewpoint/angle viewed from
22. Number of images/pictures
23. Simultaneous/sequential
24. Number of picture shifts

AUDITORY MODE

1. Remembered or constructed
2. Associated/disassociated
3. Self/others
4. Content
5. How it's said
6. Volume
7. Voice tonality
8. Voice quality/timbre
9. Tempo
10. Location
11. Distance or proximity
12. Sound move around (spatial)
13. Simultaneous/sequential
14. Relation of sound to image
15. Resonance (size/power of sound)
16. Harmony/dissonance
17. Rhythm regular/irregular
18. Cadence (interruptions, groupings)
19. Pitch high or low
20. Inflection
21. Duration

KINESTHETIC MODE

1. Associated/disassociated
2. Internal/external
3. Simultaneous/sequential
4. Temperature change
5. Texture change
6. Rigid/flexible
7. Vibration
8. Pressure
9. Location of feeling
10. Tension/relaxation
11. Movement/direction/speed
12. Breathing (hot, cold, easy, strained, fast, slow, rhythmic, arrhythmic)

13. Weight sensations: heavy or light
14. Steady/intermittent (throbbing)
15. Size: large or small, radiate/localized
16. Shape change (shape of the sensations)
17. Direction (were feelings coming into body or going out?)
18. Density
19. Intensity
20. Duration

SOMATOSENSORY SUBMODALITIES FOR PAIN

1. Tingling
2. Hot/cold
3. Muscle tension
4. Sharp/dull
5. Pressure
6. Duration
7. Intermittent (such as throbbing)
8. Location

GUSTATORY MODE (TASTE)

1. Sweet/sour/bitter/salty
2. Texture (smooth, chewy, crunchy, slimy)
3. Strong or mild

OLFACTORY MODE (SMELL)

1. Odors pleasant/unpleasant
2. Strong/faint
3. Type of odor (perfume, cooking, animal, nature, trash, cleaning, sterile)

FUNDAMENTAL MODE–NEUTRAL BEHAVIORS FOR (VISUAL, AUDITORY, KINESTHETIC, OLFACTORY, GUSTATORY)

Incorporate
Replicate
Cleave
Recombine
Transmit*

* Transmission is the displacement of a pattern-integrity through a substrate, medium, aggregate, or coordinate space by way of enfolding, unfolding, rotation, inversion, solition wave (translation), vibration, expansion/contraction (spatio-temporally, symmetrically, asymmetrically), or parity (mirror image, inside-outing, retrogradation), whereby all information is conserved. Transmission is an isomorphic transformation.

Appendix 3

Classification and Organization of Intervention Tools and Methods: NLP, Eye Movement Methods, and Energy Psychology

CONTENTS

TOOL/METHOD: ANCHORING

Entropy Effect: Pattern increasing (stabilizing)
Scope: Attractor (behavior/state/mode/submodality), trajectory
Behavior: Replication, recombination
Description: The intentional adding, subtracting, or rearranging of stimuli in the information field with the intention of specifying a specific attractor or trajectory; can be used to recombine stimuli (internal/external) with specified state attractors; can also be used to rearrange phase path for behaviors, (submodal or macroscopic) memes, thoughts, and states/phases.

TOOL/METHOD: COLLAPSING ANCHORS

Entropy Effect: Entropy increasing (destabilizing)
Scope: System, landscape, attractors (state, mode, submodalities), trajectory
Behavior: Replication, cleaving
Description: Releases the state vector from two or more initial conditions simultaneously causing destabilization (shallowing of both attractors); produces a light, temporary, trance state.

TOOL/METHOD: CHAINING ANCHORS

Entropy Effect: Pattern increasing/ordering (self-stabilizing)
Scope: Attractors (state, behavior, mode, submodal behaviors, memes, thoughts), trajectories
Behavior: Replicating, cleaving, and recombining of phase path
Description: Sequential ordering of attractors into a phase path with the intention of forming a narrow waveguide for the state vector to follow; useful for redirecting a freed state vector to a positive landscape or forming pathways having limited degrees of freedom.

TOOL/METHOD: BEHAVIOR GENERATOR (STATE-BOUND BEHAVIOR CHAINING)

Entropy Effect: Pattern/order increasing (self-stabilizing)
Scope: Attractors (state-bound and mode-bound behaviors), trajectories (behavior to behavior)

Behavior: Replication, recombination

Description: Increases state-bound or mode-bound behaviors available when state/phase attractor is activated by stimuli from the information field; forms trajectories between those new behaviors to combine them sequentially, limiting degrees of freedom for behavior within that state. The intention of this method is to make more effective state-bound behavioral strategies available to the neurocognitive system when external stimuli dictate state/phase selection.

TOOL/METHOD: THE SWISH PATTERN

Entropy Effect: Pattern/order increasing (stabilizing)

Scope: Submodal behavior

Behavior: Replication, recombination, cleaving, and transmission/translation

Description: The replication and recombination of mode-bound *representations* of macroscopic system behaviors usually in visual mode. The intention is to chain alternative behaviors, memes, and thoughts to unwanted ones and then transfer to macroscopic-level behavior.

TOOL/METHOD: PERFORMANCE MODELING/KNOWLEDGE ENGINEERING

Entropy Effect: Pattern/order increasing (self-stabilizing)

Scope: Attractor landscape, system

Behavior: Incorporation, replication, cleaving, recombining, and transmitting

Description: The elicitation, representation, and transfer of a region of design space to another neurocognitive system. The intention is to cut training and education necessary to acquire new skills or knowledge. Sometimes this is called knowledge engineering or performance engineering.

TOOL/METHOD: SWITCHING PERCEPTUAL POSITIONS

Entropy Effect: Pattern increasing (self-stabilizing)

Scope: Submodal behaviors, state attractors

Behavior: Incorporation, replication, and recombination

Description: The incorporation of viewpoints from two other people, one inside the communication loop with you and one outside. These viewpoints (visual mode, emotional state, and thoughts) are incorporated, then replicated internally, and finally chained together in a phase path. The intention is to help an individual gain deeper insight into another communicator's model of the world.

TOOL/METHOD: TIME LINE THERAPY/CHANGE HISTORY

Entropy Effect: Entropy increasing → to pattern increasing
Scope: Visual submodal behavior/internal stimuli that specify attractors
Behavior: Incorporation, replication, cleaving, and recombining
Description: Memory management — the incorporation, replicating, cleaving, and reordering (recombining) of memories and constructed images of the future with intention of effecting macroscopic state and behavior. In time line therapy we are mainly manipulating the *internal stimuli* that specify particular state/phase attractors and, by implication, their state-bound behaviors.

TOOL/METHOD: SUBMODALITY PATTERN DISRUPTION

Entropy Effect: Entropy increasing (destabilizing)
Scope: Submodal behaviors, state attractors
Behavior: Replication, cleaving, recombination, and transmit/transfer
Description: Numerous manipulations by submodal behaviors with the intention of destabilizing mode and state attractors; the introduction of a pattern-integrity (a state of motion) that dis-integrates an existing self-stabilizing pattern-integrity.

TOOL/METHOD: VISUAL/KINESTHETIC DISSOCIATION (V/KD)

Entropy Effect: Entropy increasing (destabilizing)
Scope: Visual mode — mode-bound behavior/state attractors
Behavior: Replicating, cleaving, recombination, and transmitting
Description: The internal replication of an external or internal stimulus from the information field, transmission or translation of pattern to different regions of visual working memory (visual field), and destabilization of internal replication of stimulus with the intention of cleaving the phase path prior to activation of unwanted state/phase attractors and their state-bound behaviors.

TOOL/METHOD: EYE MOVEMENT DESENSITIZATION AND REPROCESSING (EMDR) AND EYE MOVEMENT INTEGRATION (EMI)

Entropy Effect: Entropy increasing (destabilizing)
Scope: Cleaves, phase path, destabilizes attractors of submodal behaviors, modes, states (narrow focus)

Behavior: Cleaving

Description: Rapid eye movements led by a finger or wand during the time a state vector is at the trapping zone, in the attractor well, or on the phase path; interrupts neurocognitive processing by cleaving phase path at a chosen point and shallowing selected attractors. A similar technique used by NLP practitioners is called pattern interruption, and can be targeted to any mode (visual, auditory, somatosensory, etc.).

TOOL/METHOD: THOUGHT FIELD THERAPY (TFT) ALGORITHMS

Entropy Effect: Entropy increasing (destabilizing)

Scope: System, landscape, attractors (state, mode, submodalities), trajectory

Behavior: Replication, cleaving

Description: These procedures, described in detail elsewhere (Gallo, 1998, 2000), release the state vector from one or more initial conditions simultaneously causing destabilization (shallowing of attractors). In TFT theory the algorithm is hypothesized to collapse "perturbations" in the thought field. From the standpoint of pattern-entropy dynamics, tapping on the defined acupuncture meridian points effectively perturbs the stability of the attractors, thereby shallowing them, freeing the state vector to form new "healthier" trajectories. These algorithms often produce a light, temporary, trance state.

TOOL/METHOD: NEGATIVE AFFECT ERASING METHOD (NAEM)

Entropy Effect: Entropy increasing (destabilizing)

Scope: System, landscape, attractors (state, mode, submodalities), trajectory

Behavior: Replication, cleaving, recombining

Description: Globally releases the state vector from one or more initial conditions simultaneously causing destabilization (shallowing of attractors). Tapping on the specified acupoints, chakras, and thymus location effectively perturbs the stability of the attractors, thereby shallowing them and freeing the state vector to form new "healthier" trajectories. When NAEM is combined with the OPP (see below), which is often the case, pattern/order increasing also occurs. NAEM often produces a light, temporary, trance state. This procedure is described in detail elsewhere (Gallo, 1998, 2000) and is also referred to as the *Midline Energy Treatment* (MET) (Gallo and Vincenzi, 2000).

TOOL/METHOD: EMOTIONAL FREEDOM TECHNIQUES (EFT)

Entropy Effect: Entropy increasing (destabilizing)
Scope: System, landscape, attractors (state, mode, submodalities), trajectory
Behavior: Replication, cleaving
Description: Globally releases the state vector from one or more initial conditions simultaneously causing destabilization (shallowing of attractors). EFT involves tapping on a significant number of TFT acupoints, thus effectively perturbing the stability of the attractors, thereby shallowing them and freeing the state vector to form new "healthier" trajectories. Often produces a light, temporary, trance state. This procedure is described in detail elsewhere (Craig and Fowlie, 1995; Gallo, 1998).

TOOL/METHOD: OUTCOME PROJECTION PROCEDURE (OPP) (STATE-BOUND BEHAVIOR CHAINING)

Entropy Effect: Pattern/order increasing (self-stabilizing)
Scope: Attractors (state-bound and mode-bound behaviors), trajectories (behavior to behavior)
Behavior: Replication, recombination, and transmission
Description: This procedure increases state-bound or mode-bound behaviors available when the state/phase attractor is activated by stimuli from the information field. OPP forms trajectories between those new behaviors to combine them sequentially, limiting degrees of freedom for behavior within that state. The intention of this method is to make more effective state-bound behavioral strategies available to the neurocognitive system when external stimuli dictate state/phase selection. This procedure is described in detail elsewhere (Gallo, 2000).

TOOL/METHOD: ED×TM SINGLE-POINT AND MULTIPOINT PROTOCOLS

Entropy Effect: Entropy increasing (destabilizing)
Scope: System, landscape, attractors (state, mode, submodalities), trajectory
Behavior: Replication, cleaving, recombining, transmission
Description: In a highly focused manner, these protocols release the state vector from one or more initial conditions simultaneously causing destabilization (shallowing of attractors). Tapping on the specified acupoints effectively perturbs the stability of the attractors, thereby shallowing them and freeing the state vector to form new "healthier" trajectories. These protocols also are combined with the OPP, thus resulting in pattern/order-increasing features. These procedures are described in detail elsewhere (Gallo, 2000).

TOOL/METHOD: COLLARBONE BREATHING (CBB) EXERCISES

Entropy Effect: Pattern/order increasing (stabilizing)

Scope: Attractors (state-bound and mode-bound behaviors), trajectories (behavior to behavior)

Behavior: Replication, recombination

Description: CBB is employed to alleviate what is referred to as *switching* or *neurologic disorganization*, a condition that interferes with the accuracy of manual muscle testing and treatment efficiency (Gallo, 1998). When this procedure is needed, the neurocognitive and energy system as a whole is in a state of destabilization. Therefore interventions designed to destabilize that which is already destabilized cannot work. CBB serves to increase order, pattern, and rhythmicity in a disorganized system. This and related procedures are described in detail elsewhere (Gallo, 1998, 2000).

TOOL/METHOD: OVERENERGY CORRECTION

Entropy Effect: Pattern/order increasing (stabilizing)

Scope: Attractors (state-bound and mode-bound behaviors), trajectories (behavior to behavior)

Behavior: Replication, recombination

Description: When this procedure is needed, the neurocognitive and energy system as a whole is destabilized. Therefore interventions designed to destabilize that which is already destabilized cannot work. The overenergy correction serves to increase order, pattern, and rhythmicity in a disorganized system. This procedure is described in detail elsewhere (Gallo, 1998, 2000; Gallo and Vincenzi, 2000).

TOOL/METHOD: BASIC (THREE POLARITIES) UNSWITCHING PROCEDURE

Entropy Effect: Pattern/order increasing

Scope: Attractors (state-bound and mode-bound behaviors), trajectories (behavior to behavior)

Behavior: Replication, recombination

Description: When this procedure is needed, the neurocognitive and energy system as a whole is destabilized. Therefore interventions designed to destabilize that which is already destabilized cannot work. The basic unswitching procedure serves to increase order, pattern, and rhythmicity in a disorganized system. This procedure is described in detail elsewhere, and is also referred to as the Three Polarities Unswitching Procedure (Gallo, 1998, 2000).

TOOL/METHOD: FLOOR-TO-CEILING EYE ROLL AND ELABORATED EYE ROLL

Entropy Effect: Entropy increasing (destabilizing)
Scope: System, landscape, attractors (state, mode, submodalities), trajectory
Behavior: Replication, cleaving
Description: The client/patient is directed to replicate a troubling cognition while slowly moving his/her eyes in a vertical direction while tapping at a specified acupoint on the Triple Energizer meridian. The procedures release the state vector from one or more initial conditions simultaneously causing destabilization (further shallowing of attractors). Tapping on the specified acupoint while vertically elevating the eyes effectively perturbs the stability of the attractors, thereby shallowing them and freeing the state vector to form new "healthier" trajectories. The eye movements destabilize visual mode representations via ocular-motor perturbation and tapping provides both tactile and motor perturbations to the associated attractors. Often produces a light, temporary, trance state. These procedures are described in detail elsewhere (Gallo, 1998, 2000).

TOOL/METHOD: NINE GAMUT TREATMENTS AND BRAIN BALANCING PROCEDURE

Entropy Effect: Entropy increasing
Scope: Visual, auditory, and somatosensory mode attractors and trajectories
Behavior: Replicate, cleave
Description: The client/patient is directed to open, close, and move his/her eyes in specific directions, hum, and count while tapping at specified acupoints on the Triple Energizer meridian, while simultaneously replicating a troubling cognition. The technique simultaneously destabilizes trajectories and attractors in visual, auditory, and somatosensory modes. The eye movements destabilize visual mode representations via ocular-motor perturbation, humming destabilizes auditory dialog, and tapping provides both tactile and motor perturbations to the associated attractors. These procedures are described in detail elsewhere (Gallo, 1998, 2000).

Appendix 4

NeuroPrint Templates for Data Collection and Organization

CONTENTS

The following templates have been provided to assist the scientist, practitioner, and student in organizing data elicited from the client that will be essential for constructing a NeuroPrint.

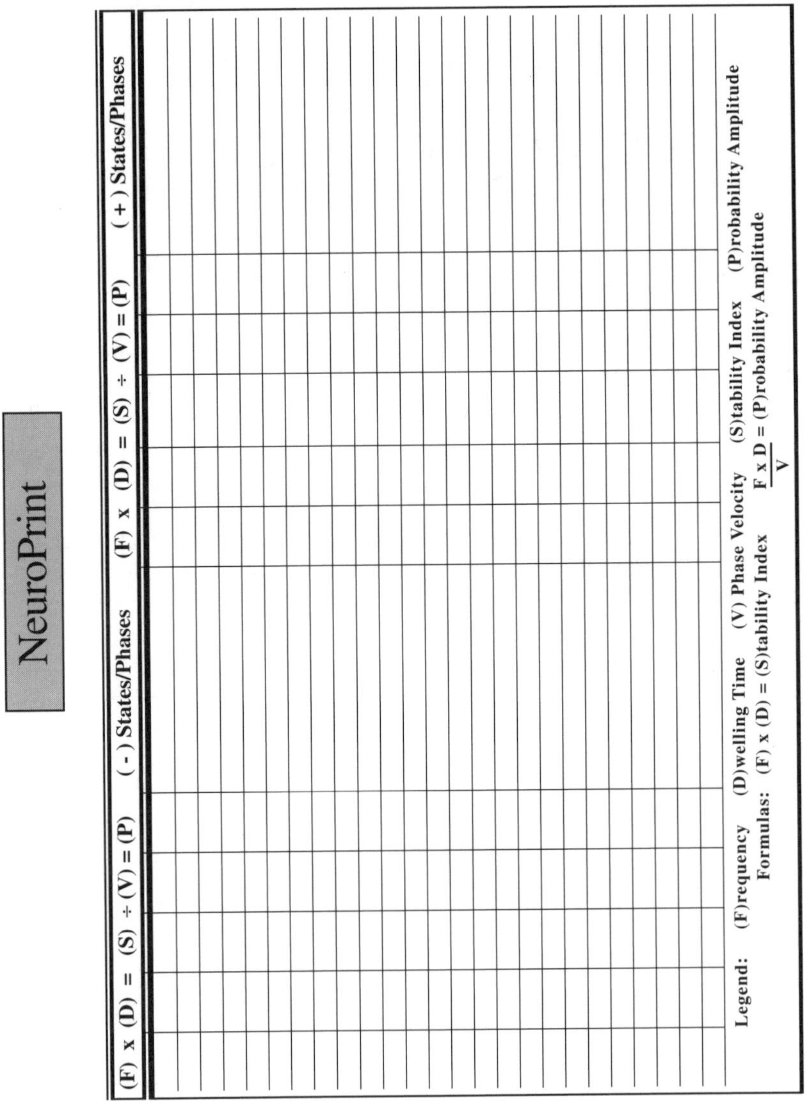

NeuroPrint

(F) x (D) = (S) ÷ (V) = (P)	(-) States/Phases	(F) x (D) = (S) ÷ (V) = (P)	(+) States/Phases

Legend: (F)requency (D)welling Time (V) Phase Velocity (S)tability Index (P)robability Amplitude

Formulas: (F) x (D) = (S)tability Index $\frac{F \times D}{V}$ = (P)robability Amplitude

FIGURE A4.1 Template 1 — States/Phases.

NeuroPrint

(F) x (D) = (S) ÷ (V) = (P)	(-) Behaviors	(F) x (D) = (S) ÷ (V) = (P)	(+) Behaviors

Legend: (F)requency (D)welling Time (V) Phase Velocity (S)tability Index (P)robability Amplitude

Formulas: (F) x (D) = (S)tability Index $\frac{F \times D}{V}$ = (P)robability Amplitude

FIGURE A4.2 Template 2 — Behaviors.

NeuroPrint

(F) x (D) = (S) ÷ (V) = (P)	(-) Memes	(F) x (D) = (S) ÷ (V) = (P)	(+) Memes

Legend: (F)requency (D)welling Time (V) Phase Velocity (S)tability Index (P)robability Amplitude

Formulas: (F) x (D) = (S)tability Index $\frac{F \times D}{V}$ = (P)robability Amplitude

FIGURE A4.3 Template 3 — Memes.

REFERENCES

Craig, G. and Fowlie, A. (1995). *Emotional Freedom Techniques: The Manual*. The Sea Ranch, CA.

Furman, M. (1995). Pattern Interruption: Redesigning the Pathways of Thought in the Brain. Unpublished paper submitted to Florida State University during "Active Ingredient" study.

Furman, M. (1996a). Submodalities through the eyes of a neuroscientist. *Anchor Point*, May, 19, 10–15, 19.

Furman, M. (1996b). Foundation of neurocognitive modeling: eye movement — a window to the brain. *Anchor Point*, December, 14, 10–12.

Furman, M. (1997a). NeuroPrint — Human Performance Modeling and Engineering. Unpublished lecture on video. Orlando, FL.

Furman, M. (1997b). NeuroPrint — Nonlinear Systems Approach to Change. Unpublished lecture. Spain.

Furman, M. (1998a). Memetics: the behavior of information. *Anchor Point*, July, 27, 12–17.

Furman, M. (1998b). Intelligent learning systems. *Anchor Point*, December, 12(12), 21–29.

Gallo, F. (1998). *Energy Psychology: Explorations at the Interface of Energy, Cognition, Behavior, and Health*. Boca Raton, FL: CRC Press.

Gallo, F. (2000). *Energy Diagnostic and Treatment Methods*. New York: W. W. Norton & Company.

Gallo, F. and Vincenzi, H. (2000). *Energy Tapping: How to Rapidly Eliminate Anxiety, Depression, Cravings, and More Using Energy Psychology*. Berkley, CA: New Harbinger.

Shapiro, F. (1995). *Eye Movement Desensitization and Reprocessing: Basic Principles, Protocols, and Procedures*. New York: Guilford Press.

Section III

Tools for Intervention: Modeling, Influencing, and Changing Cognitive Neurodynamics

As we enter the dawn of a new millennium and reflect on the technological quantum leaps of the last century, the critical imbalance between the evolution of man and machine becomes painfully apparent. It is now time to reorient our focus and expend maximum effort in developing tools and technologies to accelerate human cognitive development and rebalance a system headed for trouble. As decentralized, parallel-processing computer technology rapidly replaces physical and mental activities once performed by humans, and as nanotechnology learns to replace worn out circuits of the human brain with silicone neural networks, we must pool our resources and position ourselves strategically to redefine our meaning and purpose in the world. We must continue to evolve and learn to do things that cannot immediately be replaced by decentralized automatons.

This section contains an in-depth look at the tools and methods that can be implemented in order to more precisely influence cognitive neurodynamics and lasting change. This section provides the clinical and educational practitioner with many creative options for altering neurocognitive design space. In some cases we shall want to make direct changes to attractor landscapes that appear to be creating information-processing problems for the neurocognitive system. In other cases, we shall want to bypass existing landscapes and provide the state vector with access to a newly incorporated landscape. We refer to this process as *human performance modeling*. The tools and methods presented here, used in conjunction with NeuroPrint, will allow the practitioner to begin to capture, encode, and transfer neurocognitive models of human excellence and expertise.

Our greatest resource for accelerating human development and improving our quality of life is the untapped potential trapped within the human brain, comprising a lifetime of learning and adaptations unable to be represented by language, and thus nontransferable with present technology. It has long been a human dilemma that a skill is only worth transferring once perfected. Yet in order to perfect, it must be translated into a compressed, energy-efficient phase path that resides in part or in totality, within the subcortical information processing pathways of the brain. When incorporated information is finally transferred to these more stable encoded mediums, it is just beyond reach of our conscious awareness and we are no longer able to accurately translate these patterns within linguistic representational regions of the brain (Baddeley, 1986; Benson, 1994; Baron et al., 1990; Chance, 1994; Kissin, 1986; Partridge and Partridge, 1993). Without such ability, we cannot expect accurate or complete transfer of knowledge or skill to another human being. Without modeling tools and a physical theory to guide their application, this wealth of "informed" matter and energy remains inaccessible.

The main intention of this section is to take an in-depth look at the neurocognitive effect of a wide variety of intervention tools, which can be used to alter neurocognitive dynamics. Although only a paucity of scientific studies have been conducted toward understanding the common elements that all successful intervention methods share, scientific literature abounds with information that has proven to be an invaluable resource. The scientific literature extant has done well to guide us in accurately predicting the effects that a given intervention tool or method will have on cognitive neurodynamics.

Practitioners of Neurolinguistic Programming will find this section especially relevant as it contains explanatory models that unveil many of the mysteries surrounding the effectiveness of NLP. This section is also relevant to many other "rapid-change" therapeutic interventions, including eye movement approaches such as Eye Movement Desensitization and Reprocessing (Shapiro, 1995), and energy psychology approaches such as Thought Field Therapy and Energy Diagnostic and Treatment Methods (EM×TM) (Gallo, 1998, 2000; Gallo and Vincenzi, 2000).

In short, this section reveals many hidden forces in nature and culture influencing our daily thoughts, emotions, and our sometimes perplexing behaviors, while providing tools and principles necessary to take immediate control of our mental and physical life by influencing our own design. It is our hope that the tools of information/pattern engineering contained here will provide the basis by which humans can begin to use their unique gift of conscious self-perception to become aware of, gain control over, and intentionally organize the biophysical, cultural, and social influences that so often rob human life of meaning, purpose, and happiness.

To accomplish this, we must first come to realize that the journey is an internal rather than an external one. The remainder of this section explores some of the most important tools and principles necessary for the precise reorganization of design space in an intended direction and the informational bio-architecture that we call a human life.

USING THIS SECTION

This section can be thought of as an intervention handbook to be used in conjunction with NeuroPrint and the Standard Theory of Pattern-Entropy Dynamics. It is divided into five parts:

Part I: Preliminary Considerations for Modeling Human Expertise is an introduction to the modeling process, reviewing the basics of organizing modeling intentions and tasks, elicitation of an expert model, and the model selection process. These steps are completed prior to the actual neurocognitive modeling of attractor landscapes encoding such expertise to be transferred.

Part II: Performance Modeling — Tools of Calibration and Intervention provides some fundamental neurophysiological calibration tools necessary to peel back the superficial levels of expertise and capture the essence of human performance. In this part, we introduce the Cortical Field Activation Cue Model (CFAC) for calibrating neurocognitive mode and submodal behavior.

Part III: Information Processing, Replication, Organization, and Destabilization in the Human Brain further explores how external and internal sensory information is processed, replicated, organized, and destabilized in the brain. These chapters provide a deeper understanding of how the brain prepares for change and how information influences thoughts, emotions, and behaviors. In this part we elucidate the fundamental biomechanics behind the effectiveness of broad-scope pattern-disrupting (entropy-increasing) interventions, as well as those useful for rearranging phase path.

Part IV: Performance Engineering — Tools and Principles of Neurocognitive Pattern Modification describes the fundamental tools and principles necessary for narrow scope, neurocognitive performance engineering. In these chapters we continue to explore the tools and principles most appropriate for the modification and transfer of beliefs, values, cognitive strategies, submodalities, enabling emotions, and other keystone information patterns.

Part V: Encyclopedic Glossary provides an interdisciplinary text of interest to practitioners in several disparate fields who are unfamiliar with some of the terminology used. Many of the terms, while defined initially in the text, can also be found in this glossary. Those who are new to the terminology of nonlinear physics will find the encyclopedic glossary particularly useful. We suggest that the reader browse through it first and revisit it often, in order to develop a richer understanding and greater facility with the conceptual tools discussed throughout this work.

Here, the reader can also find an extensive resource guide; an exhaustive reference section listing source materials and recommended readings from nearly 500 books and journals that have been influential in the development of this work. Some of the information contained in Section III was originally published in 42 countries between 1995 and 1997, by Anchor Point, a practical journal for NLP. The international response received from the articles and lectures compelled us to reorganize the individual monographs into a coherent order that would prove more useful to the therapeutic intervention and education communities.

Section III — Part 1

Preliminary Considerations for Modeling Human Expertise

10 The Science and Practice of Human Performance Modeling and Engineering

CONTENTS

Modeling, once the great mecca of NLP visionaries, is now, 25 years later, in danger of extinction. Why has this happened? Too often many of us are unwilling or unable to make highly refined distinctions in a given area. We become complacent about the mechanisms that make something work and are happy to simply flip the switch and get a result. In time, if enough people begin to think and act this way, the vital secrets die with the founders. It certainly would not be the first time in human history that such vital tools for human evolution have been lost, buried with the mind that first created them. Those intrigued by the story of Nikola Tesla will certainly agree that humankind was robbed of many decades of technological advancement. This is the risk we take when neglecting the curiosity and tenacity to uncover what makes things work.

In the 1930s Alfred Korzybski coined the term "time binders" (1933, 1994). While most scientists were content simply to conclude that our ability to create language sets us apart from other life forms, Korzybski revealed that the truly significant difference is that the complexity of language allows us to carry into the future knowledge that would otherwise be "interred with our bones." That is, the complexity of human language allows us to "bind time," and the binding of time is what has allowed us to evolve rapidly to the point where we, as a species, are literally co-creators of the world in which we live. Through the development of symbolic languages, we have gained the ability to pass down a significant portion of our cumulative informational bio-architecture, a Lamarckian-like transfer of informationally inherited biophysical characteristics, beyond the capability of genes.

Over the last few years, we have had the pleasure of conversing with some great thinkers in the NLP world, as well as those with the determination to continue to seek the answers to unanswered questions. What has surprised us most is the realization that many who were attracted to NLP during its early stages were inspired most by the promise of modeling — the hope that one day, after mastering the basic skills, human excellence could be captured, encoded, and transferred. Where are all the models of human excellence that were elicited in the last 25 years? As far as we are aware, there have only been three books printed in the NLP literature providing an ancillary treatment of modeling: *NLP Volume I* (Dilts et al., 1980), *The Emprint Method* (Cameron-Bandler et al., 1985), and *Modeling with NLP* (Dilts, 1998). These books mainly propose basic elicitation and encoding structures. Additionally, Wyatt Woodsmall published two monographs entitled *The Science of Advanced Behavioral Modeling* (1988) and *Strategies* (1988). Five years later, he and Marilyne Woodsmall published a taped lecture entitled *Introduction to Modeling and Performance Enhancement* (1993). Much of Woodsmall's work in the behavioral modeling field has proven to be consistent with scientific observation and experiment, containing tools, principles, and methods that can be effectively applied to a NeuroPrint. We explore this more thoroughly in the present chapter.

Before proceeding, however, let us dispel one myth. The incorporation of an "expert" model cannot be accomplished by simply reading the linguistically encoded translation of that model from the pages of a book alone. As discussed in the previous sections, the energy barriers engendering our informational bio-architecture necessarily cause deletion and distortion of such passively incorporated information. The active process of elicitation, however, disturbs the stability of that bio-architecture and readies the neurocognitive system for incorporation of a qualitatively different attractor landscape resulting in change. No "statically encoded" expert model by itself could possibly alter cognitive neurodynamics in the intended direction.

LACK OF SPECIFIC APPLICATION

There are many factors that have led to the present condition of the discipline of behavioral and neurocognitive modeling, the first of which is the lack of specific application. If we could build models of human excellence in any area, what model would be worth building? In what context would it be applied? What are the benefits of this type of technology? How does it differ from traditional training methodologies? Without the ability to differentiate modeling from other forms of skill acquisition, there would be no reason for its application. Without the ability to clearly illustrate its effect, there would be no way to encourage practitioners to select this methodology over another.

LACK OF NEUROSCIENTIFIC KNOWLEDGE

The second factor that has thus far prevented modeling from becoming the remedy that it is capable of being is lack of knowledge of the supporting multidisciplinary fields of neuroscience. These fields have shed light on the relationship between observable behavior, both verbal and nonverbal, and its relationship to brain function. Without an understanding of the basic principles of neurophysics, neurophysiology, neurobiology, and nonlinear dynamical systems, building a transferable model of how the neurocognitive system accomplishes what is unavailable to conscious awareness is futile. Until recently, NLP has concentrated on the detection of correlative patterns between behavior (verbal and nonverbal) and mind strategies, submodalities, etc. The next correlation that must be well understood is that of behavior, cognition, and biophysics, and the effect of information on these properties.

CALIBRATION SKILL

The third factor that must be addressed is calibration skill. Again, without the knowledge of some basic principles of neuroscience and nonlinear physics, calibration models such as eye-accessing cues, strategies, representational systems, submodalities, and language patterns remain too general to be useful in capturing the heuristics of information processing we refer to as unconscious competence. Practitioners of neurocognitive intervention must learn to see much more and make increasingly more highly refined distinctions with their sensory organs. Even calibration, the most fundamental practitioner skill, has not been modeled sufficiently to be incorporated by a novice of the intervention sciences.

INSTALLATION: HUMAN PERFORMANCE ENGINEERING

The fourth factor that accounts for the lack of widespread use of modeling is an inadequate facility with the techniques collectively referred to as installation (pattern transmission) or performance engineering. The skills of installation have never been delineated and taught precisely enough so that a complex model of neurocognitive function can be transferred from one person to another. What little that has been written and taught has never adequately approached the issue of integrating an elicited, formalized model with the complex cognitive neurodynamics of another human being. Without transferability, there is no purpose for eliciting an expert model. Altering informational bio-architecture and preparing the neurocognitive system for the incorporation of a formalized model is the domain of dynamical systems physics and NeuroPrint. Although the vast architecture of such a science cannot be completely expressed within the scope of this book, we intend this to serve as a foundation for one of the most valuable future directions available to humanity.

BASIC PRELIMINARY STEPS TO HUMAN PERFORMANCE MODELING AND ENGINEERING

There are eight preliminary steps that should precede the elicitation and formalization of a model by NeuroPrint.

STEP 1: DEFINE A PURPOSE FOR THE MODEL

The first and most important step to complete before beginning to model is to define a purpose. Here are some questions to consider:

1. What will be the purpose of eliciting and formalizing the model?
2. Will the modeler be incorporating it within himself or transferring it to others?
3. Will the modeler be formalizing it for the purpose of modifying the phase path or stability distribution of the neurocognitive system from which it was elicited?
4. Do we want it to be permanent or temporary?
5. Should the heuristics generalize or be discriminatory (activated by specific context)?
6. Do we want to improve something that is already good or copy an unconscious competence exactly?
7. Will we be creating a training program for many people or just one?
8. Will we be communicating this information in written form, such as a book?
9. Do we want to know how somebody does what he or she does or do we want to know just what he or she does? What will be the scope of our inquiry?

STEP 2: SELECTION OF AN EXPERT TO BE MODELED

The next step is to select the model of excellence.

1. Who will be the model?
2. From how many models will information be elicited?
3. What criteria will be used to select a model?
4. How will we convince the expert to work with us?
5. How will the expert benefit by allowing such elicitation?
6. In what context will the model be elicited?
7. Does a model of excellence even exist for the tasks that we want to accomplish?

Step 3: Select the Type of Model Appropriate to the Outcome

The next step is to decide on the type of model that we wish to construct. The type of model will be interdependent on both our purpose and the expert(s) to be modeled.

1. Will we be building an exact copy of what and how someone does something?
2. Will we be eliciting different models from different people with the intention of forming a single composite?
3. Will we be eliciting the unconscious competence of two or more experts and exchanging those models between them with the intention of improving the performance of all the experts involved?
4. If a model of excellence is not available, do we know enough about the physics and physiology of neurocognitive function to create a theoretical model?

Step 4: Decomposition of the Process

The next step is to begin to divide the skill into the smallest possible outcomes definable. In essence, at this point we decompose the skill or unconscious competence into the smallest possible aggregates that define an outcome.

1. What are the microlevel behaviors and the fixed action programs that the expert has been able to chain together on the phase path to form the unconscious competence?
2. How is each of these fixed action programs stabilized and adjusted by its environment? A NeuroPrint will reveal more extensive initial conditions later on.
3. Which of these were explicitly learned through training and which are heuristics spontaneously occurring due to changes in contextual cues and a necessity to adapt to those contextual cues?
4. Which contextual cues are providing the necessary feedback to continually build spontaneously occurring strategies to novel situations?
5. Which contextual cues thought to be vital upon initial training are no longer given attention or determine the behavior of fixed action programs in the expert?

By asking these few preliminary questions, we may now have a greater appreciation for the complexity of neurocognitive modeling and the need for a highly refined degree of anchoring skill and calibration, necessary for both detection and installation of an expert model.

Step 5: Selection of Appropriate Calibration Tools

Out of the thousands of tools available to us that extend our senses, giving us greater precision to calibrate the invisible, we must decide what tools are necessary and capable of capturing and calibrating the behavior we want to model. In part the selection of tools will be dependent upon the scope of the model.

1. Are we modeling motor movements or glandular excretions?
2. Are we modeling brain waves or temperature?
3. Are we modeling regional cerebral blood flow or biochemistry?
4. Are we modeling T-cells of the immune system or the molecular behavior of hemoglobin?
5. Will we be modeling neural networks or the expression of DNA in the nerve cell of those networks?
6. Will we use EEGs, ERPs, or PET scans?
7. Will we need a pulse meter, GSR, or a SQUID?

The tools chosen for initial calibration will also be used as a guidance system to extend one's senses during installation (performance engineering).

If we wanted to train an athlete's heart to beat between 68 and 72 beats per minute in order to support superior long distance running, we may select a metronome as a signal anchor and a clicker as a selection anchor (see Chapter 18). Of course, the clicker would have to be transformed into a conditioned reinforcer before using it to select successive approximations of the desired heart rate, and we would continue to fire the selection anchor until the heart rate pattern formed a distinct attractor basin. Once the heartbeat has been brought into range, we may again want to employ the clicker as a selection anchor to engineer a supporting breathing pattern.

Only after building this general framework, can we begin extensive elicitation of the model, understanding that the results of that elicitation will be substantially different based on the context that we choose to elicit in, as well as the physiological state of the subject and our own state.

The rest of the journey becomes a labyrinth of neural network complexity, every factor being totally interrelated and cybernetically influenced by the dynamical interplay of perceptual filters, behavioral determinants, and decision heuristics. NLP provides many convenience models for the elicitation of behavioral determinants and perceptual filters, yet it provides virtually no models to conceptualize the dynamical interplay of decision heuristics as the human brain interacts with its complex, sensory-rich internal and external environment. For such a model, we must rely on NeuroPrint.

Step 6: Identifying Perceptual Filters

In order to build a model of unconscious competence, we must know what sensory systems are being used for the collection and representation of available information and how the infinite array of available information is filtered and assigned to useful or not useful categories according to fluctuations in physiology. NeuroPrint provides the most comprehensive system for recording the incorporation and replication of information through the sensory systems. It can also be used, as we have shown in Section II, to model internal stimuli (interoceptive), which influences our behavior and unconscious competence.

Sensory systems, for which we are unaware, form the matrix of unconscious competence that seems so elusive. The somatosensory system carries information about pain and pleasure, our relationship to gravity, balance, limb position in space, vibration, pressure, fluctuations in blood volume, familiar and unfamiliar positions and postures, neurochemical releases in the brain, immune system, gastrointestinal tract, autonomic nervous system, changes in blood gases, etc. (Kandell et al., 1991, 1995; Hobson, 1994; Roland, 1993). NLP currently generalizes and classifies this enormous flow of incoming information under the kinesthetic representational system. The current level of distinction this classification provides is not useful, since many of the determinants of human behavior and the rapidly changing architecture of external perceptual filters (exteroceptive) are dependent in part upon this flow of information (Miller et al., 1986; Neelakanta and De Groff, 1994; Pavlov, 1927, 1928, 1941; Pribram, 1991; Skinner, 1938, 1953, 1968; Watson, 1919). This information cybernetically affects what representational systems (visual, auditory, kinesthetic, etc.) are selected for specific tasks, how they are used, submodality encoding, selection of available strategies, and the destruction and creation of new representational system sequences (trajectories).

It is also important to note that no two representational systems or sensory organs in different brains perform exactly the same level of highly refined discrimination. Unconscious competence and the transference of it can vary in complexity by the ability of the expert or recipient to make a distinction in a particular representational system. If it is the case that the recipient's representational systems are incapable presently of making that level of distinction, either because of lack of training or defect, a decision must be made concerning time and effectiveness as to whether it must be bypassed and isomorphically assembled in another representational system (transmission) (Woodsmall1988; Woodsmall and Woodsmall, 1993).

Step 7: Defining and Identifying Behavioral Determinants

Behavioral determinants are factors which, in part, determine or predict the way in which our representational systems will spontaneously develop heuristics capable of adapting to novel situations beyond the boundary conditions of initial training. They are a product of, and reciprocally produce changes in, interior and exterior sensory systems. The field of NLP has defined a few of these behavioral determinants, including values, beliefs, meta programs, criteria, and anchors, all of which are context and content dependent (state-bound properties) in the world of modeling. In addition, there are laws of neurophysiology, such as that of perceptual contrast, which spontaneously influence everything from submodalities to beliefs and even decision heuristics without conscious awareness (Kandel et al., 1991, 1995).

A simple example of this can be illustrated with three basins of water — one cold, one hot, and one room temperature. If you were to put your left hand in the cold basin and your right hand in the hot one for a few minutes, and then simultaneously place both hands in the room temperature water, your beliefs about the temperature of the water and your kinesthetic submodalities would vary in the relative accuracy of the information, based on which hand you chose to process data to determine the temperature of the water. If detecting a change in the temperature of the water were a decision point that variably accessed different fixed action programs in a strategy, the heuristics would obviously be influenced, resulting in a grossly different outcome respective to the hand whose thermoreceptors processed the incoming data.

Step 8: Defining Decision Heuristics

The elicitation and encoding of decision heuristics is an extensive field unto itself. Impressively large pieces of the puzzle have been put together by two branches of artificial intelligence. *Expert systems* and *knowledge engineering* are fields that explore the way in which an expert, confronted with a myriad of possible responses, begins to narrow these choices in favor of one decision. What NLP delineates as nothing more than a subclass of the strategies model (the "decision" strategy), expert systems devote countless volumes (Meyer and Booker, 1991; Miller et al., 1986; Neelakanta and De Groff, 1994; Van Someren et al., 1994). Decision heuristics are decisively influenced by nonlinear dynamics, regional cerebral blood flow, neurosynaptic chemistry, volume transmission by neuropeptides, synaptic effectiveness, etc. (Davidson and Hugdahl, 1995; Haken, 1988, 1996; Hobson, 1994; Houk et al., 1995; Kelso, 1995; Kissin, 1986; Pribram, 1991; Roland, 1993; Rugg and Coles, 1995).

Is it worth the time and effort to master a methodology that would allow humankind to share unconscious competence and carry it to future generations? Absolutely! If we could have mastery of these skills now, what would we want to model for ourselves, our loved ones, or our business? And whom would we identify as the possessor of that human excellence? Performance modeling is a human level example of the most fundamental behaviors of pattern: incorporation, replication, and transmission. This leads us to neurocognitive modeling.

REFERENCES

Baddeley, A. (1986). *Working Memory: Oxford Psychology Series — II*. New York: Oxford University Press.

Baldwin, J. D. and Baldwin, J. I. (1986). *Behavior Principles in Everyday Life (2nd Edition)*. Englewood Cliffs, NJ: Prentice-Hall.

Baron, S., Kruser, D. S., and Huey, B. M. (1990). *Quantitative Modeling of Human Performance in Complex, Dynamic Systems*. Washington, D.C.: National Academy Press, National Research Council.

Bellack, A. S., Hersen, M., and Kazdin, A. E. (1990). *International Handbook of Behavior Modification and Therapy (2nd Edition)*. New York: Plenum Press.

Benson, D. F. (1994). *The Neurology of Thinking*. New York: Oxford University Press.

Cameron-Bandler, L., Gordon, D., and Lebeau, M. (1985). *The Emprint Method: A Guide to Reproducing Competence*. Moab, UT: Real People Press.

Chance, P. (1994). *Learning and Behavior (3rd Edition)*. Pacific Grove, CA: Brooks/Cole Publishing Company.

Checkland, P. (1993). *Systems Thinking, Systems Practice*. New York: John Wiley & Sons.

Davidson, R. J. and Hugdahl, K. (1995). *Brain Asymmetry*. Cambridge, MS: MIT Press.

Dilts, R., Grinder, J., Bandler, R., and DeLozier, J. (1980). *Neuro-Linguistic Programming: Volume I: The Study of the Structure of Subjective Experience*. Cupertino, CA: Meta Publications.

Dilts, R. (1998). *Modeling with NLP.* Capitola, CA: Meta Publications.

Druckman, D. and Bjork, R. A. (1991). In *The Mind's Eye: Enhancing Human Performance*. Washington, D.C.: National Academy Press, National Research Council.

Druckman, D. and Bjork, R. A. (1994). *Learning, Remembering, Believing: Enhancing Human Performance*. Washington, D.C.: National Academy Press, National Research Council.

Druckman, D. and Swets, J. A. (1988). *Enhancing Human Performance: Issues, Theories and Techniques*. Washington, D.C.: National Academy Press, National Research Council.

Flood, R. L. and Jackson, M. C. (1991). *Creative Problem Solving: Total Systems Intervention*. New York: John Wiley & Sons.

Furman, M. E. (1996). The science and practice of human performance modeling and engineering. *Anchor Point*, September, 31–43.

Fuster, J. M. (1995). *Memory in the Cerebral Cortex: An Empirical Approach to Neural Networks in the Human and Nonhuman Primate*. Cambridge, MA: MIT Press.

Haken, H. (1988). *Information and Self-Organization: A Macroscopic Approach to Complex Systems*. New York: Springer-Verlag.

Haken, H. (1996). *Principles of Brain Functioning: A Synergetic Approach to Brain Activity, Behavior and Cognition*. New York: Springer-Verlag.

Hannon, B. and Ruth, M. (1994). *Dynamic Modeling*. New York: Springer-Verlag.

Hobson, J. A. (1994). *The Chemistry of Conscious States: How the Brain Changes Its Mind*. New York: Little, Brown & Company.

Houk, J. C., Davis, J. L., and Beiser, D. G. (1995). *Models of Information Processing in the Basal Ganglia*. Cambridge, MA: MIT Press.

Kandel, E. R. and Hawkins, R. D. (Eds.). (1992). The biological basis of learning and individuality. *Scientific American*, September.

Kandel, E. R., Schwartz, J. H., and Jessell, T. M. (Eds.). (1991). *Principles of Neural Science*, 3rd ed. Norwalk, CT: Appleton & Lange.

Kandel, E. R., Schwartz, J. H., and Jessell, T. M. (Eds.). (1995). *Essentials of Neural Science and Behavior*. Norwalk, CT: Appleton & Lange.

Kazdin, A. E. (1994). *Behavior Modification in Applied Settings*. Pacific Grove, CA: Brooks/Cole.

Kelso, J. A. S. (1995). *Dynamic Patterns*. Cambridge, MS: MIT Press.

Kissin, B. (1986). *Psychobiology of Human Behavior (Volume I): Conscious and Unconscious Programs in the Brain*. New York: Plenum Press.

Klir, G. J. (1985). *Architecture of Systems Problem Solving*. New York: Plenum Press.

Korzybski, A. (1933). *Science and Sanity.* Englewood, NJ: Institute of General Semantics.

Korzybski, A. (1994). *Science and Sanity: An Introduction to Non-Aristotelian Systems and General Semantics (5th Edition)*. Englewood, NJ: Institute of General Semantics.

Kosslyn, S. M. (1994). *Image and Brain*. Cambridge, MS: MIT Press.

Kruse, P. and Stadler, M. (1995). *Ambiguity in Mind and Nature*. New York: Springer-Verlag.

Martin, G. and Pear, J. (1996). *Behavior Modification: What it is and How to Do It (5th Edition)*. Upper Saddle River, NJ: Prentice-Hall.

Maslow, A. H. (1987). *Motivation and Personality (Third Edition)*. New York: Harper & Row.

McClelland, D. C. (1987). *Human Motivation*. Cambridge, MS: Cambridge University Press.

Meichenbaum, D. (1977). *Cognitive-Behavior Modification: An Integrative Approach*. New York: Plenum Press.

Meyer, M. A. and Booker, J. M. (1991). *Eliciting and Analyzing Expert Judgement: A Practical Guide*. New York: Academic Press.

Miller, G. A., Galanter, E., and Pribram, K. H. (1986). *Plans and the Structure of Behavior.* New York: Adams-Bannister-Cox.

Neelakanta, P. S. and De Groff, D. F. (1994). *Neural Network Modeling: Statistical Mechanics and Cybernetic Perspectives.* Boca Raton, FL: CRC Press.

Nunez, P. L. (1995). *Neocortical Dynamics and Human EEG Rhythms.* New York: Oxford University Press.

Paillard, J. (1991). *Brain and Space.* New York: Oxford University Press.

Partridge, L. D. and Partridge, L. D. (1993). *The Nervous System: Its Function and Its Interaction with the World.* Cambridge, MS: MIT Press.

Pavlov, I. P. (1927). *Conditioned Reflexes.* New York: Oxford University Press.

Pavlov, I. P. (1928). *Lectures on Conditioned Reflexes (Vol. I).* New York: International.

Pavlov, I. P. (1941). *Lectures on Conditioned Reflexes (Vol. II): Conditioned Reflexes and Psychiatry.* New York: International.

Pribram, K. H. (1991). *Brain and Perception: Holonomy and Structure in Figural Processing.* Hillsdale, NJ: Lawrence Erlbaum Associates.

Pryor, K. (1995). *On Behavior: Essays & Research.* North Bend, WA: Sunshine Books.

Roland, P. E. (1993). *Brain Activation.* New York: Wiley-Liss.

Rugg, M. D. and Coles, M. G. (1995). *Electrophysiology of Mind: Event-Related Brain Potentials and Cognition.* New York: Oxford University Press.

Skinner, B. F. (1938). *The Behavior of Organisms: An Experimental Analysis.* New York: Appleton-Century-Crofts.

Skinner, B. F. (1953). *Science and Human Behavior.* New York: The Free Press.

Skinner, B. F. (1968). *The Technology of Teaching.* New York: Appleton-Century-Crofts.

Van Gigch, J. P. (1991). *System Design Modeling and Metamodeling.* New York: Plenum Press.

Van Someren, M. W., Barnard, Y. F., and Sandberg, J. A. C. (1994). *The Think Aloud Method: A Practical Guide to Modelling Cognitive Processes.* New York: Academic Press.

Watson, J. B. (1919). *Psychology from the Standpoint of a Behaviorist.* Philadelphia, PA: J.B. Lippincott.

Weinberg, G. M. (1975). *An Introduction to General Systems Thinking.* New York: John Wiley & Sons.

Woodsmall, W. (1988). *Strategies.* Arlington, VA: Advanced Behavioral Modeling, Inc.

Woodsmall, W. (1988). *The Science of Advanced Behavioral Modeling.* Arlington, VA: Advanced Behavioral Modeling, Inc.

Woodsmall, W. and Woodsmall, M. (1993). *Introduction to Modeling and Performance Enhancement.* Arlington, VA: Creative Core (taped lecture).

Section III — Part 2

Performance Modeling and Tools of Intervention

11 Neurocognitive Modeling: The Art and Science of Capturing the Invisible

CONTENTS

Have you ever wondered what causes our eyes to move when we think? Why do some people have more trouble visualizing than others do? For readers familiar with the NLP eye-accessing cue model, did it ever occur to you that there are three eye positions for the auditory modality and only one for kinesthetic? Where are the eye-accessing cues for motor, tactile, and emotional representations? We have a visual construct, an auditory construct, but where is the kinesthetic construct? Have you ever noticed people visually remembering in the place where they are supposed to visually construct? (That is, eyes upper right instead of upper left.) Why can people still visualize when their eyes are in kinesthetic position? Why does it seem to be universal that visual is always up and kinesthetic is always down? Did you know that if you can see yourself (i.e., dissociated position) in a visual memory, it's not a memory, but a visual construct? What do patterns of eye movements actually reveal about our neurocognitive processes? If questions like these have caused conflict for you in the past or the inability to answer them has actually stopped you from using eye patterns as a calibration modality, you are not alone.

Thousands of NLP practitioners around the world have seen the logic behind the use of eye-accessing patterns, yet the myriad of unexplainable inconsistencies and counterexamples has led to massive confusion. What was the seed of one of the most valuable indicators of cognitive activity available to us, withered and died before the taproot ever found water. The eyes, the only part of the brain visible to the outside world, are the windows to the brain and its activity. Although the field of NLP initiated the notion of tracking these observable patterns, the original model has not been revised or updated with new distinctions for almost 20 years. The future of human performance modeling on a neurocognitive level depends on precise and predictable calibration models that reveal otherwise invisible cognitive and biophysical function. The advancement of any model, such as the eye-accessing cue model, depends upon people who sort by "difference," that is, people who make highly refined distinctions in the face of blatant counterexamples, rather than those who willingly accept the sweeping generalizations of a static model.

BUILDING THE RIGHT FOUNDATION

As discussed in Section I, the field of NLP teaches practitioners to pay attention to patterns. These patterns fall mainly under two broad categories. The first are those patterns that reoccur in "time." The second are correlative patterns between language and behavior, such as eye movements, gestures, and voice tones. The problem of understanding eye-accessing cues and their relation to neurocognitive function is that the correlative patterns necessary to reveal that connection were unable to be tracked 20 years ago. So we relied on the relationship between language and eye movements. Anyone who has closely tracked these patterns may have noticed one basic problem: the speed at which eye scanning patterns occur are on a much faster time scale than the speed at which our language describes our inner mental representations. As a result of this gross difference in time scale, it is not possible for the current eye-accessing cue model to give us an accurate representation of cognitive activity. However, the last 10 years of brain imaging studies, along with a 100-year-old foundation in cytoarchitectonics, gives us a solid-enough foundation to more deeply understand why our eyes move when we think and to substantially revise previous thinking about eye scanning patterns.

In short, the solution to the problem is to track not only the correlation between language, gestures, and eye movements, but also the correlation between those eye movements and the localized brain activation that subserves them. It is in studying this localized brain activation that clues are revealed as to what areas of the brain are being used (Baddeley, 1986; Davidson and Hugdahl, 1995). So let us begin to answer some questions.

RESOLVING THE MYSTERIES OF EYE MOVEMENT

Why do many people have trouble visualizing and some even believe that they do not visualize at all? Visualizing requires an immense amount of neurophysical and neurochemical energy. The human brain, weighing in at only 3 pounds, utilizes between 20 and 25% of the body's entire blood supply to efficiently nourish our 100 billion plus brain cells with oxygen and nutrients (Roland, 1993). Of all the cortical fields subserving collection and representation of sensory information, the visual cortex appears to require more oxygen and blood flow than any other (Cornoldi et al., 1996; Finke, 1989; Roland, 1993). While it takes only 16 milliseconds for an auditory stimulus to reach the brain stem, it takes twice that for a tactile stimulus (32 milliseconds), and as much as 65 to 100 milliseconds to assemble a visual image (Davidson and Hugdahl, 1995; Stein and Meredith, 1993). Any visual image that we assemble fades and must be refreshed every 250 milliseconds. This allows our eyes to inspect an external visual scene without having an after image every time we move our eyes (Kosslyn, 1994). This makes "seeing" very efficient but visualizing very difficult, since we use mainly all of the same cortical pathways, and both the inside world and the outside world compete for territorial control of the association cortices. Quite simply, visualizing, the act of making and maintaining visual representations in the visual and association cortices, is hard work and takes much practice. As many readers realize, most of the effective, narrow-scope NLP change techniques require a highly refined ability to visualize in order to be effective.

WHAT CAUSES OUR EYES TO MOVE WHEN WE THINK?

This is a highly complicated area that will be simplified for the purposes of laying a solid foundation upon which a reliable model of eye scanning patterns can be constructed. Our eye movements are part of a functional synergy; where our eyes move, our head follows. Therefore, eye movements by themselves only tell part of the story of internal cognitive functioning. We must also understand what happens physiologically in our brain when our head moves.

The coordination of eye, head, and hand movements is under the control of the *superior colliculus*, an area of the brain stem. This structure possesses a spatial map of visual, auditory, kinesthetic (somatosensory), and motor activity so that all the senses can be coordinated together in space (Stein and Meredith, 1993). This is the part of the brain without which you could not swat a mosquito that has just taken a bite out of your leg. This spatial tracking system allows your hand to move toward the mosquito, whether its presence was captured by auditory (the buzzing of its wings), visual, or tactile sensory input. This and other related brain functions are also responsible for the reconstruction of the external world as an internal representation. It allows us to reproduce that world in an accurate, body-centered, spatial representation. As this internal representation (replication) reconstructs the external world, our eye movements both help activate the correct cortical area via the vestibular system as well as maintain spatial location of the representation via the visuoparietal-prefrontal cortices (Baddeley, 1986). The prefrontal cortex acts as a movie screen allowing the stored visual image to be projected from the visual cortex (transmission). The parietal cortex holds a spatial representation (replication) of the entire image so that the eyes and head can move around that image, as it remains stationary in body-centered coordinates (Stein and Meredith, 1993). This process happens only after we have accessed the image with the help of a quick saccadic eye movement. This eye movement activates the vestibular system to move the head in certain optimum positions so that increased blood flow and oxygen can be maintained to the part of the brain being activated (Roland, 1993).

This means that when one looks up in order to visualize, optimum blood flow increases in the visual cortex toward the back of the head due to gravity, thus maintaining a higher level of regional blood flow and oxygen in the visual cortex. Gravity is a source of mechanical disequilibrium needed for pattern formation (a decrease in entropy), and it creates a sustained asymmetrical distribution of matter and energy in the brain, which is necessary for the replication of previously incorporated patterns. In this instance the replication is referred to as internal visualization.

This same relationship can be found for all other sensory areas as well. The understanding of this relationship helps us not only to answer the questions posed in the beginning of this chapter, but it can also assist us in constructing a highly predictable model of calibration with which to track neurocognitive information processing.

In the next chapter we will continue to unravel the mysteries of eye movements and begin to discuss how we can use this knowledge to model the elusive obvious through the cognitive window of the brain.

REFERENCES

Baddeley, A. (1986). *Working Memory: Oxford Psychology Series — II*. New York: Oxford University Press.

Cornoldi, C., Logie, R. H., Brandimonte, M. A., Kaufmann, G., and Reisberg, D. (1996). *Stretching the Imagination: Representation and Transformation in Mental Imagery.* New York: Oxford University Press.

Davidson, R. J. and Hugdahl, K. (1995). *Brain Asymmetry.* Cambridge, MS: MIT Press.

Finke, R. A. (1989). *Principles of Mental Imagery.* Cambridge, MS: MIT Press.

Furman, M. E. (1996). Neurocognitive modeling: the art and science of capturing the invisible. *Anchor Point*, November, 40–42.

Furman, M. E. (1997). Modeling at the speed of sight. *Anchor Point*, January, 29–33.

Kandel, E. R., Schwartz, J. H., and Jessell, T. M. (Eds.). (1991). *Principles of Neural Science (3rd Edition).* Norwalk, CT: Appleton & Lange.

Kosslyn, S. M. (1994). *Image and Brain.* Cambridge, MS: MIT Press.

Luer, G., Lass, U., and Shallo-Hoffmann, J. (1988). *Eye Movement Research: Physiological and Psychological Aspects.* Lewiston, NY: C.J. Hogrefe.

Roland, P. E. (1993). *Brain Activation.* New York: Wiley-Liss.

Stein, B. E. and Meredith, M. A. (1993). *The Merging of the Senses.* Cambridge, MS: MIT Press.

12 Foundations of Neurocognitive Modeling: Eye Movement, A Window to the Brain

CONTENTS

THE WINDOW TO NEUROCOGNITIVE ACTIVITY

Since the inception of NLP, its founders believed that it was possible to model virtually any form of human excellence, even if the essence of that excellence was trapped invisibly in cognitive space. Although it is relatively easy to model collective, coordinated, macroscopically observable behavior subserved by the sensory motor system, and it is equally possible to model some cognitive activity from language, the largest missing piece which makes it all possible is the neurocognitive activity that puts the whole process in motion.

The first two types of modeling had very reliable external pattern indicators. However, the third type relied on eye scanning patterns to provide a window into cognitive activity. To model elusive cognitive activity more accurately, it is necessary to learn how to use eye scanning patterns more accurately, for these patterns hold the key to the competence for which we are completely unconscious and incapable of linguistically accessing. Understanding eye movement with greater precision gives the modeler access to unconscious information that previously might not have been recognized.

BUILDING NEW DISTINCTIONS THROUGH COUNTEREXAMPLES

In order to start assembling a more accurate model of eye-accessing cues, we can certainly begin where NLP left off 20 years ago and build new distinctions through the analysis of counterexamples. With a little understanding of neurophysics and the old NLP eye-accessing model in hand, we are ready to begin.

Previously, we discussed why some people have more trouble visualizing than others, and we also established a neurophysiological and neurophysical basis for why our eyes have to move while we think. Neurophysiologically, it is nearly impossible to internally replicate certain mode-dependent information without the appropriate eye and head movement. As noted, much of this movement is controlled via the brain stem.

COGNITIVE ILLUSIONS AND VERTICAL EYE MOVEMENTS

Have you ever noticed why it seems to be so universal that accessing visual information requires the eyes to be up and accessing kinesthetic information requires the eyes to be down? In actuality this is just an illusion. First of all, diligent calibration of eye-accessing cues reveals that people can in fact visualize with their eyes down in the kinesthetic position. This is the first counterexample to analyze. If it is possible to visualize when the eyes are in the kinesthetic position, then the NLP eye-accessing cue model does not give us an accurate representation of sensory system activation. Since sensory system activation is vital for modeling neurocognitive processes, such counterexamples have to be resolved.

The first distinction that we must make is that eye positions indicate initial activation, maintenance, and transmission of an image. When someone is asked a question that requires visual accessing, his or her eyes dart up to the left, right, or center for a short period of time. This is an accessing (initial replication) cue. However, when the person holds or maintains that image in mind, he or she can move (transmit) it to virtually any part of his/her visual field, paradoxically overlapping with auditory and kinesthetic cues. Consistent with the principles of inertia, the initial activation of an image always requires greater neurophysical activity than the maintenance of that image. Therefore, upon activation we may observe that the eyes move up and that the head moves back to allow gravity-assisted regional cerebral blood flow to increase to the visual cortex. This increase in blood volume temporarily boosts the energy level of visual neural networks needed to assemble the image (replication). Once the image is assembled, it can be moved (transmitted) to the prefrontal cortex (just behind the forehead) and projected to almost any area in the visual field. The nerve cells that hold (encode) this image are the *pyramidal cells*. Through a property of neurophysics called *degeneracy*, any of the pyramidal cells in this area can be selected by the brain to represent this image. This means that an image initially generated in the upper portion of the visual field can be expanded or contracted and moved to virtually any location (Finke, 1989; Kosslyn, 1994). Without this flexibility, thinking as we know it would not be possible.

As a previously replicated image is now transmitted (a gliding solition wave) from one portion of the visual field to another, different aspects of sensory representation are enhanced (Stein and Meredith, 1993). Physiologically, information from the sensory systems overlaps. This is why in some portions of the visual field, one can just see a picture very well; while in other portions of the visual field, the visual and auditory portions of the experience are replicated equally well. At still other locations, one can simultaneously and richly see an image, hear the sounds, and feel the feelings that were felt from that experience. This enhancement of visual, auditory, and kinesthetic signals is made possible by nerve cells in the brain, *bimodal* and *trimodal neurons*, which combine and integrate sensory information from more than one system. These neurons reside in the association cortices, and the prefrontal cortex is one of those association cortices. Bimodal neurons carry information from two sensory systems such as visual and auditory, auditory and kinesthetic, and visual and kinesthetic. Trimodal neurons carry information from all three sensory systems (Baddeley, 1986; Gazzaniga, 1995; Stein and Meredith, 1993).

As one slides an image over specific receptor fields of bimodal or trimodal neurons (pattern transmission by solition wave), consequently different combinations of auditory, visual, and kinesthetic information are enhanced. So one of the most valuable calibrations that we can make depends on correlative patterns between eye position and type of association neurons being used. If we were to divide the visual field into three horizontal sections (upper, middle, and lower), when the eyes are in the uppermost section, visual information patterns will overlap with auditory information patterns that originate from eye level and above and also will overlap with kinesthetic information patterns originating from the upper head area. Therefore, it is possible to "tune in" kinesthetic information while looking straight up, although the kinesthetic information available would only be that which originates from the upper head area, since the nerve cells that overlap with that area of the visual association cortex are only nerve cells that carry kinesthetic information from the upper head area. Thus it is easily possible to "tune in" or enhance kinesthetic information about a previously experienced headache or skull injury by looking up.

As the eyes move into the middle area of the horizontally divided visual field, we are now able to enhance auditory sensory information that originated from ear level down to approximately the level of the sternum, and kinesthetic information becomes enhanced from ear level to just below the shoulders and upper chest. This means that if we wanted to "tune in" kinesthetic information from ear level to upper chest, it would be best to move our visual pictures (transmit the visual pattern) to the center horizontal division of the visual field. Also, since the ears have their focal point of sensory information collection at this level, it is best to enhance auditory processing by maintaining the eyes at midline level.

As the eyes are moved down to the lowest portion of the visual field, we are able to clearly "tune in" or enhance auditory and kinesthetic information from the upper chest down. This is why we so commonly notice kinesthetic accessing to be down, as most of our feelings — emotional, tactile, and motor (somatosensory and proprioceptive systems) — will be generated between the upper chest and our feet (Gazzaniga, 1995; Stein and Meredith, 1993). Obviously there is more body below the upper chest than above it.

With this understanding, it is easier to see how a person remembering the sound of a cat's meow from the ground would need to place his or her eyes in what we now call kinesthetic position. In actuality, they are just overlapping the visual image they have of the cat with the sound of its meow that originated from below their chest. Since the meow sound came from ground level, the neurons that represent it would be in the lower portion of the visual field (prefrontal cortex).

To notice this, a fine degree of sensory acuity is needed. It is best for modelers to begin by noticing this in themselves before trying to notice it in others. Becoming aware of this distinction will allow an observer to fairly accurately predict the mode being activated during thinking or remembering, as well as many of the submodal behaviors available to the subject at that moment (Davidson and Hugdahl, 1995; Kosslyn, 1994; Luer et al., 1988; Stein and Meredith, 1993).

Obviously bimodal and trimodal neurons make it organically impossible to separate visual, auditory, and kinesthetic processing into the clean divisions that the original NLP eye-accessing cues map suggested. Without this additional physiological information, revealed by the brain science of *cytoarchitectonics*, we would certainly experience some difficulty when trying to model internal neurocognitive behaviors from the vantage point of only the tip of the iceberg that protrudes from the surface.

This neurophysiological information gives us an essential part of the foundation for neurocognitive modeling. Since all functions of mind require the activation of biological tissue in the brain (Baddeley, 1986; Benson, 1994; Fuster, 1995; Roland, 1993), modeling technologies that hope to capture the invisible essence of mind must take into account the neural substrate of a given neurocognitive behavior.

UNCOVERING THE MYSTERY OF HORIZONTAL EYE SCANNING PATTERNS

Thus far, this information clarifies why our eyes move vertically during thinking. However, this is only half of the story, because as we have experienced, our eyes also move horizontally during thinking. Just as the head follows eye movement in vertical patterns, it does so similarly in horizontal patterns. Eye movement is the lead system for vestibular functioning. Our eyes help us maintain head position and balance, and wherever our eyes move, our head and body follow. This, of course, also helps to direct regional cerebral blood flow to the sensory circuits being most heavily activated (Kandel et al., 1991; Roland, 1993).

So why do our eyes move both left and right when we think? It has been known for quite some time that the brain is divided into right and left hemispheres. Each hemisphere has a different cellular structure allowing for different types of function. Our eyes will move left and right depending upon the cortical function (submodal behavior) we need to perform. For example, to see motion we may have to look left, and to move or transmit portions of an image, we may need to look to the right depending upon which hemisphere controls high resolution/frequency activity vs. low resolution/frequency activity.

Let us take the visual cortex first. Why does it seem that many right-handed people look up to the left when retrieving (replicating) a visual image and up to the right when constructing (spatial cleaving and recombining) a visual image or thinking of a future event? Again, much of this is an illusion, since the same people can retrieve memories (replicate) looking up to the right (opposite side) and they can visually construct (sequential, temporal recombination) or think of the future by looking up to the left. There is more going on here than "meets the eye." Whenever there is a counterexample, there is a new distinction to be learned. To better understand the functional areas of left visual field (LVF) and right visual field (RVF), refer to Figure 12.1, which shows how a house will be represented on visual receptor fields from each hemisphere. These distinctions will be consistent with right-hand dominance.

CORTICAL ASYMMETRY AND THE VISUAL FIELD

The right hemisphere subserves the left visual field. Brain cells in the right visual hemisphere have very broad overlapping receptor fields (see Figure 12.1). Because these receptor fields are so broad, fewer are needed to encode an image than in the

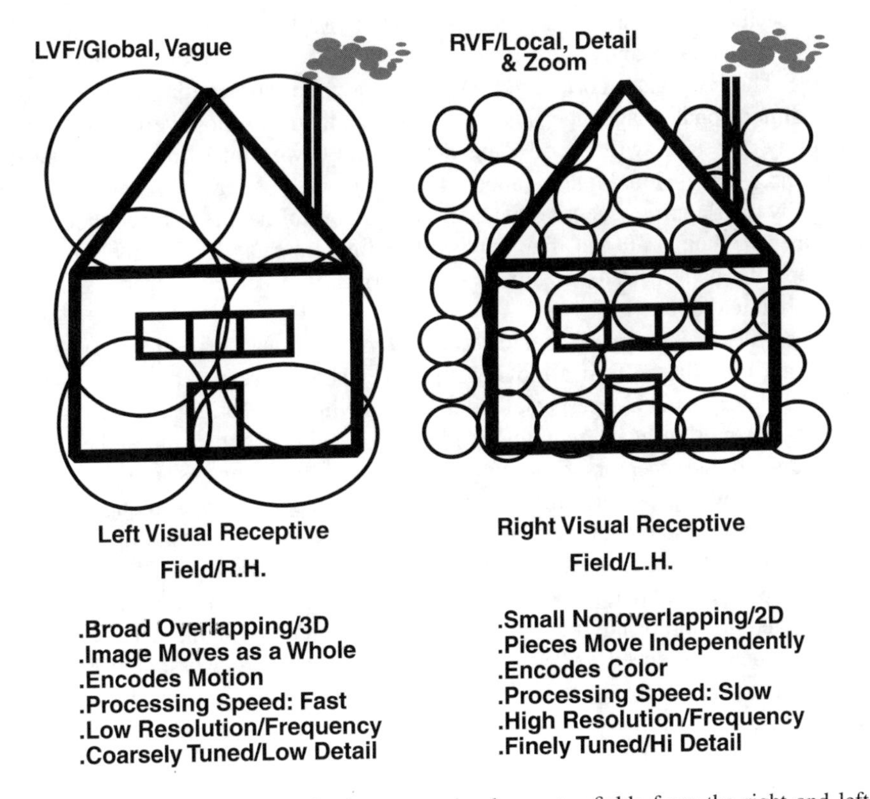

FIGURE 12.1 Representation of a house on visual receptor fields from the right and left hemispheres.

RVF. In the RVF, as can be seen from the diagram, the receptor fields of brain cells in the visual cortex are much smaller and nonoverlapping. The first property to understand is that the processing speed of the LVF will be faster than the RVF (Davidson and Hugdahl, 1995; Finke, 1989; Gazzaniga, 1995). This is because less neurons are involved in the encoding and decoding process (incorporation and replication). This one physiological fact sheds light on a large number of neurocognitive activities and submodal behaviors. Since the speed of image replication is faster in the LVF than the RVF, any retrieved image will assemble quicker, with less energy expenditure, in the LVF. Since speed of retrieval is consistently reinforced in our culture, this operant becomes instrumentally conditioned. This is why we notice memory being frequently accessed upper left. As the image is being assembled and the correct cytoarchitectonic structure is being selected to represent (replicate) it, the eyes are drawn immediately to that field by attention networks in the parietal cortex, cingulate gyrus, and reticular formation. In cases of memory retrieval, the eyes will rarely be drawn up to the right unless functions (pattern behaviors such as cleaving) of the left visual cortex are necessary for retrieval. Recent PET scans have also verified that regional blood volume increases in the right prefrontal cortex during retrieval (replication) of a memory, and the left prefrontal cortex during encoding (incorporation) of sensory information (Kosslyn, 1994; Posner and Raichle, 1994; Roland, 1993).

Since the right visual cortex is made up of overlapping receptor fields, it is possible to encode and decode (incorporate and replicate) three-dimensional images. The left visual cortex and RVF are not adept at encoding and decoding three-dimensional images, because the receptive fields are nonoverlapping. Consequently, the RVF will encode and decode only two-dimensional flat images. Since three-dimensional images appear more real and were originally encoded and stored by overlapping receptive fields, it is common for people to place their "past timeline" in their LVF and their "future timeline" in their RVF. This eliminates the need for pattern transmission (translation) and conserves neurophysical and neurochemical energy.

The LVF also incorporates and replicates moving images (solition wave) more efficiently than the RVF. If a moving image needs to be retrieved, it will be most likely retrieved in the LVF, especially if it were moving when originally incorporated. Replicating motion again requires overlapping receptive fields. Since the LVF is made up of broad receptors, fewer in number than the RVF, images are very coarsely tuned with low degree of detail (low resolution). As a person is asked questions that require greater visual detail, he or she will move the image (transmitting it) to the opposite visual field. This movement assists retrieval of a more detailed image with a two-step process. First, the area of the image to be zoomed in on is marked. Next, the marked portion of the image is refreshed in the RVF, utilizing nerve cells with a slow processing speed and very high resolution or detail (Kosslyn, 1994). This process is made possible by the principle of degeneracy, a process by which different neural networks can represent the same information (Edelman, 1987).

The cerebral hemispheres also differ in their ability to process color. The encoding and decoding of color information is done by parvocellular pathways more densely populating the left visual cortex (projecting information to the RVF).

The encoding and decoding of black, white, and gray scales are best done by magnocellular pathways located in the right visual cortex (LVF) (Kandel et al., 1991). Therefore, if a neurocognitive task requires color information, an image originally retrieved in the LVF will have to be enhanced via degeneracy by sliding (transmitting) it to the RVF.

Another interesting cognitive function that differs by hemisphere is the ability to transmit, cleave, and recombine pieces of an image independently of one another. This function is not possible with overlapping receptive fields that bind the image into one coherent, uncleavable piece. Consequently, it is much more difficult to separate and recombine pieces of an image with the LVF, which is subserved mostly by overlapping receptor fields.

Since the RVF has nonoverlapping receptors, pieces are easily cleaved, separated, and recombined independently. This physiological information explains why right-handed people move images to the right when performing most "visual construct" operations requiring spatial cleaving and recombining, and to the left when performing sequential, temporal cleaving, and recombining of a spatially coherent image. It is important to note that although the fundamental pattern behaviors of incorporation, replication, cleaving, recombining, and transmitting are not mode-bound, there are cytoarchitectonic regions within each system that subserve a particular mode, better equipped to perform each of these pattern behaviors.

By adding a little knowledge of neurophysiology and cytoarchitectonics to earlier observations of eye scanning patterns, we can make many times the number of distinctions originally possible with greater accuracy. This significantly increases the efficiency of the elicitation of unconscious cognitive processes which appear to underlie skills and abilities as elusive as photographic memory, problem solving, decision making, creative innovation, genius, and the like.

In order to properly calibrate unconscious cognitive activity, it must also be understood that eye movements alone do not tell the entire story. Proper calibration requires an understanding of how head movements, both vertical and horizontal, act in functional synergy along with eye scanning patterns to produce what is referred to as the *Cortical Field Activation Cues* (CFAC) *Calibration Model* (Furman, 1996). This model allows us to elicit the behavior of information at the neurocognitive level by comparing functional synergies of the eyes and head with the corresponding cortical areas of the brain being activated during information encoding and decoding. With this knowledge, it is possible to build models of neurocognitive function far below the existing level of submodality distinctions extant, and far more precisely. It should also be noted that all submodal behaviors are to some degree under the control of the neurocognitive system's cytoarchitecture (asymmetrical distribution of neurocellular morphology).

HORIZONTAL EYE SCANNING PATTERNS OF THE AUDITORY SYSTEM

Now that a few of the functions of the right and left visual fields and their corresponding cortical field activation cues have been introduced, let us briefly discuss some of the functions of the right and left auditory fields. In NLP it was previously thought that the LVF indicated auditory-remembered information and the RVF indicated auditory-constructed information. This is also a sweeping generalization similar to those made about visual remembered and visual construct. Upon closer examination, we notice that the left auditory/visual field indicates phonetic, sequential, and rhythmic processing, while the right auditory/visual field processes the analogical components of language and sound referred to as *prosody*. Analogical changes in elements such as pitch and inflection are decoded (replicated) when the eyes move toward the right auditory/visual field (located at the right, middle region of the visual field). These are just a few of the distinctions that are responsible for the gross counterexamples to the original NLP eye-accessing cue model.

When asking a right-handed person to retrieve a memory of a conversation, he/she will look to the left when hearing the exact words that were said and to the right when attending to the tone of voice being used or specific inflection that certain words were given.

Now we come to the elusive "third" auditory position, which purportedly resides down and to the left (for a right-handed person). To unravel this confusion, it would be valuable to understand a little more about cytoarchitectonics of the auditory cortex. The functions previously discussed are carried out in a part of the brain called *Wernicke's area*. These regions exist approximately just above each ear and are correlated with the horizontal auditory eye movements right and left.

In the frontal cortex, a part of the brain called *Broca's area* controls *articulatory motor movements*. As the right-handed person looks down to the left and tilts his/her head in the same direction, the change in head orientation facilitates a gravity-assisted, regional cerebral blood flow increase within Broca's area, making it easier to construct speech (articulatory motor patterns) (Gazzaniga, 1995; Kosslyn, 1994; Posner and Raichle, 1994; Roland, 1993). Consequently, rather than the distinction of *auditory digital*, it now becomes apparent that when the eyes move down to the left we are accessing (replicating and recombining) articulatory motor patterns which help to translate (transmit) an internal auditory image of "dialog" into jaw, tongue, and larynx movements. This can be thought of as the kinesthetic/motor counterpart to auditory internal dialog. At last, the elusive "kinesthetic construct" is revealed. Many who have closely studied eye movements and their correlative patterns notice that when a person's eyes are positioned down to the left, his lips and larynx may exhibit unintentional leakage of the actual words he is in the process of thinking about. Given the neurophysiology, this phenomenon becomes understandable.

Finally, we are prepared to discuss the kinesthetic position, which is classically located down and to the right for most right-handed individuals. In actuality this eye movement helps us to enhance information coming from the dominant somatosensory cortex which is located in the right hemisphere (Davidson and Hugdahl, 1995; Roland, 1993). It is no accident that when a person's eyes move down and to the right, his or her head also tilts down in that direction thanks to the vestibular system. This head reorientation with respect to gravity assists a temporary regional cerebral increasing in blood volume to the *somatosensory cortex*. This cortical area is responsible for collecting all transmitted motor, tactile, vestibular, and emotional (biochemical) information from inside the body and relaying (transmitting and translating) it back into other relevant cortical areas for cognitive representation (replication) (Kandel et al., 1991). As the eyes move down and to the right, information carried by these pathways from the upper chest and below becomes enhanced.

This has been a brief introduction to the myriad of neurocognitive functions that lie just beneath the surface of eye scanning patterns. A significant amount of previously hidden information about cognitive processing can be readily discerned through this valuable window to the brain by using the CFAC calibration model.

This information provides the neurocognitive interventionist with two important tools. The first is a precise way of discovering activated modes and submodal behaviors that subserve neurocognitive abilities or perpetuate neurocognitive problems. NeuroPrint can easily represent this information. Once this information is accurately represented in a NeuroPrint model, it can be used for skill transfer applications or for intervention. When using this information for intervention it is handled in the same way as we do for collective macroscopic states and behaviors. We can significantly modify cognitive neurodynamics by either modifying the stability distribution of modes, submodal behaviors, and their connecting trajectories or we can rearrange the phase path taken by the state vector through design space on the modality level. In Part III, we discuss how information is processed, represented, organized, and destabilized in the brain.

REFERENCES

Baddeley, A. (1986). *Working Memory, Oxford Psychology Series — II*. New York: Oxford University Press.

Benson, D. F. (1994). The *Neurology of Thinking*. New York: Oxford University Press.

Cornoldi, C., Logie, R. H., Brandimonte, M. A., Kaufmann, G., and Reisberg, D. (1996). *Stretching the Imagination: Representation and Transformation in Mental Imagery*. New York: Oxford University Press.

Davidson, R. J. and Hugdahl, K. (1995). *Brain Asymmetry*. Cambridge, MA: MIT Press.

Edelman, G. M. (1987). *Neural Darwinsim: The Theory of Neuronal Group Selection*. New York: Basic Books.

Finke, R. A. (1989). *Principles of Mental Imagery*. Cambridge, MA: MIT Press.

Furman, M. E. (1996). Foundations of neurocognitive modeling: eye movement — a window to the brain. *Anchor Point*, December, 14–22.

Furman, M. E. (1997). Modeling at the speed of sight. *Anchor Point*, January, 29–33.

Fuster, J. M. (1995). *Memory in the Cerebral Cortex: An Empirical Approach to Neural Networks in the Human and Nonhuman Primate*. Cambridge, MA: MIT Press.

Gazzaniga, M. S. (1995). *The Cognitive Neurosciences*. Cambridge, MA: MIT Press.

Kandel, E. R., Schwartz, J. H., and Jessell, T. M. (Eds.). (1991). *Principles of Neural Science (3rd Edition)*. Norwalk, CT: Appleton & Lange.

Kosslyn, S. M. (1994). *Image and Brain*. Cambridge, MA: MIT Press.

Luer, G., Lass, U., and Shallo-Hoffmann, J. (1988). *Eye Movement Research: Physiological and Psychological Aspects*. Lewiston, NY: C.J. Hogrefe.

Posner, M. and Raichle, M. (1994). *Images of Mind*. New York: Scientific American Library.

Roland, P. E. (1993). *Brain Activation*. New York: Wiley-Liss.

Stein, B. E. and Meredith, M. A. (1993). *The Merging of the Senses*. Cambridge, MA: MIT Press.

Section III — Part 3

Information Processing, Representation, Organization, and Destabilization in the Human Brain

13 Neocortical Dynamics of Persuasion and Influence

CONTENTS

The human brain is the most complex organization of matter agglomerated in a 3-pound package in the known universe. Containing over 100 billion neurons richly interconnecting with between 1000 and 200,000 others, the brain forms a complex nonlinear, far from equilibrium, dynamical system. The potential number of emergent states and behaviors is quite large, and the brain's vast neurochemical and neurophysical systems never return to exactly (molecule for molecule, atom for atom) the same state twice (Kelso, 1995).

Why, then, do we approach the influencing of this dynamic system by training a change agent or salesperson with an arsenal of static "techniques" and "lines"? With this classic style of persuasion and influence training, we leave much of the results up to chance, leaving therapists, educators, salespeople, negotiators, and managers to the inevitable fate of no better than casino averages. There are few programs to be found that have been built from a solid foundation of what has been uncovered in the last 100 years of neuroscientific research.

Persuasion and influence are both a science and an art form. Successful training and implementation of such skills require a basic understanding of the principles governing brain function and exquisite understanding of the art of *utilization*. The fields of investigation useful for the integration of persuasion and neocortical dynamics range from neurobiology to hypnosis. While the scope of this chapter will not allow for an exhaustive treatment of this subject matter, the purpose is to acquaint the reader with the potential application of this important area of knowledge for intervention and change.

THE BRAIN AS A NONLINEAR DYNAMIC SYSTEM

Imagine an aerial view of a life-size maze. Your objective is to enter one side and, after much torturous trial and error, to emerge successfully on the other side. The maze is a square with walls that are 8 feet high, and inside the square the walls form intricate maze-like convolutions, all but one pathway leading to dead ends. If this were a linear maze, we could enter it always expecting to find the entrance in the same place. Once inside, we could turn to the right and, if that led to a dead end, we could turn back and try to the left leaving the maze unchanged. Eventually we would exit successfully, always expecting to find that opening in the same place. It would be nice if the neurocognitive system we are influencing were a linear system, but it is not.

Now imagine an aerial view of a similar maze. Each time we take a closer look, however, the opening is not where we expect it to be. In fact, each time we intend to engage in maze walking, the entrance is never found in the same place. We circle the perimeter looking for the entrance. As we finally enter, we have a choice of going left, right, or forward. As we turn right we realize that we have reached a dead end, so we return back to the first decision point to find that there is no longer a left. Instead, we find three new pathways. Suddenly we realize that each step taken causes the walls of the maze to rearrange and self-organize into a new maze pattern. It is impossible for us to make any decision without irreversibly altering the maze forever. We suddenly realize there is no way to start over from the same place. There is no way to memorize the landscape or the topology of this maze. Our only hope to get out is by distilling the principles by which every action causes it to self-organize. Welcome. You have just entered the nonlinear world of the human brain.

DESTROYING THE MYTHS

Much of what is taught to change agents is idiosyncratic behavior, for which the teachers attribute their own personal successes, as well as others' failures to succeed. Unfortunately, when an expert transfers an unconscious skill, he or she rarely if ever transfers the keystone. Rather than a set of guiding principles, they tend to teach static sets of behaviors.

Novice salespeople are taught to extend a hand, have a firm handshake, make eye contact, and state their name and the name of their company clearly. Any one of these seemingly innocent actions can end in brutal failure and loss of rapport due to ill timing. What if the client just had hand surgery? What would happen if the client were internally visualizing a previous salesperson that they distrusted while the present salesperson is trying to keep eye contact? What would happen if the salesperson stated his name and the name of his company clearly while the client is experiencing the state of impatience, confusion, or frustration? What would we expect the result to be if the salesperson presented her product, service, or neurocognitive intervention plans while the client was feeling skeptical? These are not uncommon situations. In fact, most clients are lost long before the salesperson or practitioner is ever aware of it. Understanding some of the fundamental principles of information processing makes the job of influencing cognitive neurodynamics and, by implication, the human mind quite a bit easier.

PRINCIPLES OF BRAIN FUNCTION AND INFORMATION PROCESSING

Regardless of the fact that in all its complexity the human brain is a highly complex information processing system, its function for the purposes of persuasion and influence can be broken down into some simple principles.

BRAIN ACTIVATION AND INFORMATION PROCESSING

During the course of an average day the human brain moves through many levels of activation. These levels are subserved by the frequency of electrophysiological activity transmitted or propagated by the brain's neural circuitry at any given moment (Hobson, 1994). These activation levels should be thought of as being more fundamental than emotional states or phases, since they actually determine how these attractors will organize into attractor landscapes.

Landscapes subserved by low activation states/phases may aggregate attractors like depression, sadness, confusion, and laziness, while those subserved by high levels of activation may aggregate attractor states like excitement, enthusiasm, passion, fear, and motivation. In short, brain activation levels are fundamental phase transitions that agglomerate and organize other attractors, and attractor basins, into polytopological landscapes.

The brain's activation level critically determines its information processing speed, accuracy, motivation, and direction of attention (internal/external). If we want to control the brain's information processing, we must first be able to match the information input rate to the present activity level of the brain, and then finally entrain the nervous system to operate at the activity level necessary to optimally package the intended information (i.e., coordinate internal states with the appropriate external objects and events). In neurophysics this process is referred to as *phase locking* and *frequency locking* (Haken, 1988, 1996; Kelso, 1995; Kissin, 1986), which allows the change agent to *select* the appropriate attractor landscape in the subject's neurocognitive system by entrainment.

IDENTIFYING BRAIN RHYTHMS

The rhythms of human behavior reflect the intrinsic rhythms of the brain itself. Communication and change agents, including NLP practitioners, are taught to build rapport by pacing, mirroring, and leading, but are rarely taught which elements of these categories to use and for which purposes. Quite often the most important elements of pacing, phase locking and frequency-locking, are overlooked. This process involves pacing the exact "rhythm" of the neurocognitive system so that the brain may optimally process information at any given moment. That is, the most useful thing to pace is anything that gives you an indication of the frequency (speed or rhythm) at which the neurocognitive system is operating, including eye blink rate, rate of eye ball movement (ocular-motor impulses), breathing rate, speaking rate, sentence length, pause length, rate of limb movement, rate of body shifting, speed, and frequency of gestures. When the communicator matches these rhythms and paces speech rate to the subject's breathing, it is likely that the optimal conditions are engaged for the transmission and propagation (processing) of the communicator's message. When the communicator is not matching these rates, it is likely that information transmission and incorporation rates will be out of synch (out of phase and out of frequency), resulting in confusion, discomfort, and inhibition of comprehension. However, this kind of pacing does not necessarily mean that the communicator is in a state of rapport with the recipient of his/her communication. One can be in perfect rhythm and still violate a client's beliefs or values. Synchronization with regard to phase locking and frequency locking is merely a way of facilitating information transfer between neurocognitive systems.

INFLUENCING BRAIN CHEMISTRY

Transmitting information to the neurocognitive system is of little value without first entraining neural activity, which prepares the information/pattern processing systems of the brain to accept (incorporate) new information. Once neurocognitive activity is entrained by some region of an information field, that field is capable of either increasing pattern (order) or increasing entropy (disorder). If the information source increases entropy, it does so by destabilizing atomic and molecular level energy barriers, resulting in a weakening of neocortical connectivity patterns. This in turn makes the neurocognitive system more fluid, flexible, and ready to change. Entrainment proves to be very useful for selecting a given attractor landscape and making landscape-bound attractors available. These are some of the essential neurophysical mechanics subserving trance induction.

Neuromodulators in the brain stem are in constant dynamic interplay. When the activity level of brain circuitry increases, the brain stem releases a greater amount of *norepinephrine* (NE) into the neocortex (Hobson, 1994). This increase in NE causes excessive strengthening of synapses subserving neocortical connectivity patterns (deepening of involved attractors), which in turn causes interference between previously incorporated patterns and new incoming sensory patterns (Hasselmo and Barkai, 1995).

This increase in order and attractor stability results in a corresponding macroscopic behavioral pattern of inflexibility and resistance to the new information. The effect of excessive synaptic strengthening can be avoided by inducing *activity-dependent depression* of synaptic strength. Reduction in the strength of synaptic connections (shallowing of attractors) and greater neurocognitive fluidity are believed to be facilitated in the presence of acetylcholine (ACh). The deliberate entrainment of intrinsic neural rhythms by effective use of phase locking and frequency locking is a valuable tool for influence and change, which can be used to induce a fundamental phase transition of the neurocognitive system from a more ordered to a less ordered state/phase.

Once phase locking of neurocognitive or biophysical rhythms occurs, an effective communicator can begin to lead the client to a lower activity level by gradually slowing his or her own rate of behavior (breathing, speech rate, gestures, etc.). This is believed to cause the brain stem to produce a greater level of ACh to be released in the neocortex, thus reducing the strength of synaptic connections and allowing for a more fluid acceptance of new sensory patterns (e.g., the communicator's suggestions).

This optimal state for neurocognitive information transfer has also been shown to enhance the induction of *long-term potentiation* (LTP) in the hippocampal region of the brain, resulting in long-term replication of the newly incorporated sensory patterns (e.g., posthypnotic suggestions). Cholinergic enhancement of synaptic modification is believed to play an important role in the permanent storage of information in cortical structures (Hasselmo and Barkai, 1995; Hobson, 1994; Kissin, 1986).

Important behavioral evidence can be found by studying the neocortical dynamics of hypnosis and trance. It is a well-known fact that subjects in trance (phase *trans*ition) are found to have greater recall and susceptibility to suggestion. The neurophysical and neurochemical dynamics discussed above are in large part responsible for this ability. It has also been found that excessive strengthening of intrinsic synapses can lead to simultaneous activation of all stored patterns, resulting in a loss of recall discrimination between patterns (Hasselmo and Barkai, 1995). This neurocognitive state can cause devastating resistance in the face of an otherwise persuasive suggestion. This is similar to trying to tune in a radio or television signal without having an antenna. The state vector is unable to settle into any stable attractor. Excessive strengthening of synaptic connections within the cortex can also result in a loss of specificity of response to newly incorporated patterns. This is commonly seen in the confused, overwhelmed, and vacillating client who needs some time to "think about it" and seems unsure of everything suggested. Extreme strengthening of synaptic connections can sometimes result in cortical activity patterns resembling those characteristic of seizures, giving rise to severe headaches and disorientation.

It is also noteworthy that "accelerated learning" occurs best during this state of *cholinergic neuromodulation* and is associated with Class II theta waves (Kandel et al., 1991, 1995). Entraining the nervous system by phase locking and frequency locking with the intent to lower activity level is essential for reducing resistance (interference between information field and internal patterns) to suggestion, as well as various therapeutic procedures.

INTENSIFYING EMOTION

Another benefit of reducing the activity level of the neurocognitive system prior to suggestion is that the relationship between hemispheric activation begins to change. Neuroanatomical evidence suggests that the right cortical hemisphere has richer connections with subcortical structures including those of the *limbic system (amygdaloid-hippocampal complex)*, which plays a critical role in coordinating emotional states with the appropriate incoming sensory patterns created by external objects and events (Davidson and Hugdahl, 1995). Combined with the association areas, the *amygdala* is also essential for the interpretation and expression of the emotional component of language and, therefore, can be activated by the emotional tone of a therapeutic suggestion.

During activity-dependent depression (decrease in neural activation and loosening of neuronal coupling), there is a shift in dominant activity from the left to the right hemisphere and a corresponding increase in the intensity of emotion experienced by the client, as well as a strong excitation of the limbic reward system of the brain (Kissin, 1986). This shift in hemispheric activity affords the change agent far greater control over the emotional and motivational states necessary for the client to make desired changes. Studies suggest that there is also a corresponding increase of dopamine in the thalamus and visual cortex accompanying this shift in hemispheric dominance, giving rise to more vivid imagery (Davidson and Hugdahl, 1995). Here again we see another fundamental phase transition, this time from left hemisphere dominance to right hemisphere dominance, which gives the state vector access to numerous landscape-bound attractors and their corresponding neurocognitive properties.

The last major benefit of inducing an entropic phase transition by deliberately entraining neurocognitive activity is a decrease in general attention and an increase in sustained focused attention. This type of sustained narrowed focus is typical of trance state.

All of these changes in brain function promote greater receptivity to suggestion and influence. From this point of neuronal entrainment (phase locking), it is also possible to lead a client's state kinesthetically by the communicator subtly changing his/her own state. As the communicator determines the optimal state for receptivity to suggestion and initiation of the desired behavior, the client can be led into that state by the communicator initially entering it. Thus, if motivation is the desired state for the client to be in, previously having calibrated that the desired behaviors will occur during that state, then we must phase lock with the client's present state and then gradually become motivated ourselves. By influencing the nervous system directly at this level, the change agent gains access to one of the core self-organizing principles of the brain and control over several other neurocognitive properties. Entraining intrinsic brain rhythms gives the change agent more direct control over the following:

- Appropriate state-dependent emotions, memories, and behaviors
- Heightened emotional response
- Synchronized information transfer and propagation rate
- Long-term memory-encoding mediums subserved by Class II theta waves, which facilitates effortless incorporation and long-term replication of therapeutic suggestions

In short, spontaneous reorganization of neurocognitive topology can be facilitated by influencing the fundamental organization of a system — its physiological states and by implication its state-bound behaviors. To venture a metaphor, one can greatly reduce the complexity of maze walking by directly influencing the organization of the maze. This is just one of many essential tools that can be utilized for stabilizing or destabilizing neurocognitive patterns in design space. In the next chapter we shall explore the utility of submodal behaviors (submodalities) for influencing cognitive neurodynamics.

REFERENCES

Davidson, R. J. and Hugdahl, K. (1995). *Brain Asymmetry.* Cambridge, MA: MIT Press.

Fuster, J. M. (1995). *Memory in the Cerebral Cortex: An Empirical Approach to Neural Networks in the Human and Nonhuman Primate.* Cambridge, MA: MIT Press.

Furman, M. E. (1996). Neocortical dynamics of persuasion and influence. *Anchor Point,* April, 30–37.

Haken, H. (1988). *Information and Self-Organization: A Macroscopic Approach to Complex Systems.* New York: Springer-Verlag.

Haken, H. (1996). *Principles of Brain Functioning: A Synergetic Approach to Brain Activity Behavior and Cognition.* New York: Springer-Verlag.

Hasselmo, M. E. and Barkai, E. (October 1995). Cholinergic modulation of activity-dependent synaptic plasticity in the piriform cortex and associative memory function in a network biophysical simulation. *Journal of Neuroscience,* 15(10), 6592–6604.

Hobson, J. A. (1994). *The Chemistry of Conscious States: How the Brain Changes its Mind.* New York: Little, Brown & Company.

Houk, J. C., Davis, J. L., and Beiser, D. G. (1995). *Models of Information Processing in the Basal Ganglia.* Cambridge, MA: MIT Press.

Kandel, E. R. and Hawkins, R. D. (September 1992). The biological basis of learning and individuality. *Scientific American.* New York: W. H. Freeman and Company.

Kandel, E. R., Schwartz, J. H., and Jessell, T. M. (Eds.). (1991). *Principles of Neural Science (3rd Edition).* East Norwalk, CT: Appleton & Lange.

Kandel, E. R., Schwartz, J. H., and Jessell, T. M. (Eds.). (1995). *Essentials of Neural Science and Behavior.* Norwalk, CT: Appleton & Lange.

Kelso, J. A. S. (1995). *Dynamic Patterns.* Cambridge, MA: MIT Press.

Kissin, B. (1986). *Psychobiology of Human Behavior (Volume I): Conscious and Unconscious Programs in the Brain.* New York: Plenum Press.

Nunez, P. L. (1995). *Neocortical Dynamics and Human EEG Rhythms.* New York: Oxford University Press.

Roland, P. E. (1993). *Brain Activation.* New York: Wiley-Liss.

Rugg, M. D. and Coles, M. G. (1995). *Electrophysiology of Mind: Event-Related Brain Potentials and Cognition.* New York: Oxford University Press.

14 Submodalities, Cognitive Neurodynamics, and Change

CONTENTS

Submodalities are among the most subtle and elusive neurocognitive properties that the brain uses to construct internal models of the world. Submodalities are influenced by the incorporation of language and they control the transmission of language from individual words to syntax chosen by the communicator. When we communicate, even casually without specific intent, we cannot help but influence submodalities in our listeners and ourselves. Bio-architecturally, they are one of the most effective tools for influencing the structure of thought, emotions, and behavior itself. While submodal behaviors are not new to the neuroscience community, what is new is the deliberate use of these microcognitive behaviors as control parameters that influence neurocognitive dynamics.

HOW SUBMODALITIES INFLUENCE BRAIN FUNCTION

Although scientists have been aware of submodalities for some time, we have only just begun to explore their ubiquitous effects on bio-architecture and resulting cognitive neurodynamics. In order to elucidate their utility as an intervention tool, it is necessary to have a sense of how different submodalities affect brain function.

REENTRANT CIRCUITRY

The first principle of brain function necessary to understand the effects of submodalities is that of *reentrant circuitry* or *reentrant neural cell connectivity* (Edelman, 1987). Reentrant circuitry is a basic bio-architectural principle of nerve cell organization which gives rise to cybernetic-like information loops of which all submodalities make use. Reentrant circuits are circuits in the brain, which are linked cybernetically in elaborate, continuously updating feedback structures. Whatever change that is made in one circuit of nerve cell assemblies is immediately fed back to all participating circuits and vice versa. This principle alone is what gives submodalities their pervasive ability to affect thoughts, emotions, behaviors, and the very maps by which we organize our experience of the world outside.

A simple example of the use of reentrant circuitry can be found in the submodality distinction of *size*. We notice that as we increase the size of an internal image, we simultaneously increase the intensity of the corresponding emotion. Upon deeper examination of this cybernetic link, we find that as the emotional intensity itself decreases, so does the size of the image. This particular submodality depends not only on reentrant circuitry, but also on a property of the nervous system called *population coding*.

POPULATION CODING

Population coding is one way in which the nervous system expresses intensity of a stimulus in the information field (Ferster and Spruston, 1995; Kandel et al., 1991). The larger the population of neurons activated by the stimulus, the more intense the feeling, emotion, or tactile sensation. The principle of population coding is certainly not limited to the visual system. This principle is more simply understood in terms of pain intensity relayed through the somatosensory system. If a person accidentally burns his entire hand, a more intense pain signal would travel through the somatosensory system than if he had burnt only one finger, even if both were considered to be second-degree burns. In this case, the mechanism that encodes the intensity of the pain is the population or number of neurons (transmitting mediums) involved in transmitting the message.

This principle also applies to the visual system or mode. When we perceive (incorporate) an object or an event through our eyes, it is transmitted to the visual cortex to as many as 30 discrete circuits. One such circuit is V1. This area of the visual cortex is spatiotopically mapped (Goldman-Rakic, 1987, 1992; Houk et al., 1995; Kandell et al., 1991). That is, the pattern of electrophysiological activity in this part of the brain spatially matches the pattern of activation that a visual image creates on the retina of the eye itself via photon incorporation. This is one area in the visual cortex where the spatial relationships of a visual image are preserved in much the same way that the pixels of a TV monitor preserve the spatial relationships encoded by the video camera that initially captured the image.

This information is also conveyed to the *prefrontal association cortex*, which is located behind the forehead. It is here that location is encoded by special neurons called pyramidal cells (Damasio, 1994; Goldman-Rakic, 1987, 1992). In this association cortex, information about one's physiological state is linked with information about the visual image (Damasio, 1994). The somatosensory system, which carries physiological information about emotional state and body sensations, actually indexes the visual image being encoded by these pyramidal cells. That is, as a visual image is called up in working memory via the prefrontal cortex, the coordinate spatial location of all active images is marked by a physiological state encoded by the somatosensory system (Damasio, 1994). This function provides one of the means by which the meaning of a previously stored event can be changed simply by changing the location of a visual image (i.e., mapping over). The neural circuitry that subserves this function is so elaborate and far reaching that it makes possible higher-order brain processes such as reasoning, decision making, problem solving, future planning, and the encoding of time. Without this circuitry, these functions would cease to exist.

The principle of reentrant circuitry acts in conjunction with coordinate (spatiotopic) encoding so that as the size of an internal visual image increases, an increase in population of neurons propagating that information occurs in both V1 and the prefrontal cortex via pattern transmission. Both of these circuits convey an intensity shift to all participating reentrant circuits, one region of which is the somatosensory system that engenders feelings and emotions connected with that image. As in the example of the burnt finger or hand, an increased population or number of neurons propagating information will result in an increased intensity signal to the somatosensory system, resulting in an increase in whatever emotions or sensations that were originally encoded or transmitted.

When we influence the size of an image, we transmit desired effects to other target systems via reentrant circuitry and population coding in the brain. The same is true when we influence intensity kinesthetically and perceive the corresponding shift in the size of a linked visual image. This kind of link is commonly referred to as *synesthesia*.

The same principle of reentrant circuitry makes it possible for language to influence a submodality and for a submodality to influence the selection of words that describe the experience. When we ask someone to "take a closer look" at something, if the object is external and they move closer to it, the object will fill up a greater portion of their visual field, retina, and area V1 (Kandel et al., 1991). This is made possible by the mode-neutral behavior of population coding. As we look at something external from a distance, in turn it will require a smaller neuronal population in these same regions. The same is true for an internal image with the addition of another set of reentrant circuits in the prefrontal cortex, which subserves elaborate manipulation of internal imagery (working memory) (Goldman-Rakic, 1987, 1992). Thus, when we ask someone to "take a closer look" at an internal image, the *closer* becomes an instruction to the nervous system to increase population coding and thus intensity. The converse is true when we ask a person to see things from a distance. What makes this possible is the reentrant circuit design between the areas that subserve language (Broca's and Wernicke's area) and those in the visual cortex (occipital cortex), as well as those between the somatosensory and visual cortices. Apparently we cannot help but influence submodal behaviors when we communicate.

FREQUENCY CODING

Another type of neural response to information, which gives rise to manipulatable submodal behavior, is *frequency coding*. Basically, frequency coding refers to the number of times a neuron or neuronal pathway fires (propagates an impulse) in a given period of time. The higher the number of impulses per second, the greater the intensity of the stimulus in the information field. For example, when lightly touching a table top with a finger, the thermal and mechano-receptors of the somatosensory system encode the intensity of the transaction by sending approximately one impulse per second to the brain. However, similarly making contact with a stove burner of 600°F results in a message being transmitted at a frequency in excess of 500 impulses (spikes) per second (Kandel et al., 1991).

This same principle of information intensity transmission operates in the visual system (Kandel et al., 1991). When our eyes perceive the light from a single candle in a dark room, the brain propagates only a few impulses per second via visual pathways, while the retina is capable of transmitting several hundred impulses per second through visual pathways during photon impact from direct observation of the sun. Thus, when we ask someone to brighten an internally visualized image, instructions have been given to the nervous system to increase frequency coding of that image and therefore to increase the intensity of the emotions and sensations encoded with that image. The converse is also true when we ask for an internally visualized image to be made dim or dark. When a person uses these terms while describing an experience, he or she is again making use of the reentrant connections between linguistic and visual circuits in the brain. Frequency coding can also be found transmitting signal intensity in auditory circuits of the temporal lobes. This mechanism subserves the perception of volume. The louder an external stimulus or internal auditory image seems the higher the frequency of propagating impulses through auditory circuits of the brain.

The continued exploration of submodalities has given rise to extensive information that allows us to model human performance and modify cognitive neurodynamics at the neurocognitive level. Such a narrow scope tool as submodalities proves invaluable for the delicate task of modifying the stability distribution of attractors and trajectories in design space. Both stabilization and destabilization are possible by scaling a submodality distinction in one direction or another. For example, one effective NLP technique for the reduction of kinesthetic intensity, characteristic of phobias and traumas, is called visual-kinesthetic dissociation or V/KD. This method requires the subject to recall the experience that was responsible for the development of a phobia or emotional trauma and make an internal visual image of it. Normally, when a subject is asked to do this, the image is reported to be "bigger than life," occupying their entire internal visual field in their working memory. The first part of the technique simply requires the subject to reduce the size (population coded) and brightness (frequency coding) of the image, which is immediately followed by a corresponding involuntary reduction of kinesthetic (somatosensory) intensity of the connected emotion (mostly fear or anxiety). Thus the subject is taught to destabilize the "fear" attractor by altering the population coding (size) and frequency coding (brightness) of the internal stimulus (visual memory of the trauma or phobia) that triggers the "fear" or "anxiety" attractor, thus shallowing it significantly.

A relatively new branch of neuroscience referred to as *functional human brain mapping* has provided extensive validation of much of what was originally only hypothesized 20 years ago. Functional human brain mapping and the various available mediums for brain imaging can now help us to learn how to use these submodal behaviors for therapeutic intervention with greater effectiveness and precision (Kosslyn, 1994; Nunez, 1995; Roland, 1993; Rugg and Coles, 1995; Smith et al., 1995). In our next chapter we discuss how the neurocognitive system organizes cognitive categories from the dynamic interaction between incorporated and replicated information patterns.

REFERENCES

Behrmann, M., Kosslyn, S.M., and Jeannerod, M. (1995). *Neuropsychologia: An International Journal in Behavioural and Cognitive Neuroscience. Special Issue: The Neuropsychology of Mental Imagery*, 33(11).

Damasio, A. R. (1994). *Descartes' Error: Emotion, Reason, and the Human Brain*. New York: G.P. Putnam's Sons.

Davidson, R. J. and Hugdahl, K. (1995). *Brain Asymmetry*. Cambridge, MA: MIT Press.

Dominery, P., Arbib, M., and Joseph, J. P. (1995). A model of corticostriatal plasticity for learning oculomotor associations and sequences. *Journal of Cognitive Neuroscience*, 7(3).

Edelman, G. M. (1987). *Neural Darwinism: The Theory of Neuronal Group Selection*. New York: Basic Books.

Ferster, D. and Spruston, N. (1995). Cracking the neuronal code. *Science*, 270, 756–757.

Furman, M. E. (1996). Submodalities through the eyes of a neuroscientist. *Anchor Point*, May, 19–23.

Fuster, J. M. (1995). *Memory in the Cerebral Cortex: An Empirical Approach to Neural Networks in the Human and Nonhuman Primate*. Cambridge, MA: MIT Press.

Goldman-Rakic, P.S. (1987). Circuitry of primate prefrontal cortex and regulation of behavior by representational memory. In *V.B. Mountcastle Handbook of Physiology*, Plum, F. and Geiger, S. R. (Eds.), New York: Oxford University Press, 5(1), 373–417.

Goldman-Rakic, P.S. (1992). Working memory and the mind. In *Mind and Brain*. New York: W. H. Freeman and Company.

Haken, H. (1988). *Information and Self-Organization: A Macroscopic Approach to Complex Systems*. New York: Springer-Verlag.

Haken, H. (1996). *Principles of Brain Functioning: A Synergetic Approach to Brain Activity Behavior and Cognition*. New York: Springer-Verlag.

Houk, J. C., Davis, J. L., and Beiser, D. G. (1995). *Models of Information Processing in the Basal Ganglia*. Cambridge, MA: MIT Press.

Kandel, E. R., Schwartz, J. H., and Jessell, T. M. (Eds.). (1991). *Principles of Neural Science (3rd Edition)*. East Norwalk, CT: Appleton & Lange.

Kandel, E. R., Schwartz, J. H., and Jessell, T. M. (Eds.). (1995). *Essentials of Neural Science and Behavior*. Norwalk, CT: Appleton & Lange.

Kelso, J. A. S. (1995). *Dynamic Patterns*. Cambridge, MA: MIT Press.

Kosslyn, S. M. (1994). *Image and Brain*. Cambridge, MA: MIT Press.

Nunez, P. L. (1995). *Neocortical Dynamics and Human EEG Rhythms*. New York: Oxford University Press.

Roland, P. E. (1993). *Brain Activation*. New York: Wiley-Liss.

Rugg, M. D. and Coles, M. G. (1995). *Electrophysiology of Mind: Event-Related Brain Potentials and Cognition*. New York: Oxford University Press.

Smith, E. E., Jonides, J., Koeppe, R. A., Awh, E., Schumacher, E. H., and Minoshima, S. (1995). Spatial versus object working memory: PET investigations. *Journal of Cognitive Neuroscience*, 7(3).

15 Neocortical Dynamics of Metaphor

CONTENTS

WHAT IS METAPHOR?

Metaphor is generally thought of as a conceptual tool of linguistics used for story telling, poetry, and the clarification of meaning. Additionally, metaphor is typically viewed as a characteristic of language alone, rather than thought or action (Lakoff and Johnson, 1980). The purpose of this chapter is to elucidate the cognitive neurodynamics of metaphor and discuss its utility as a fundamental tool of change.

A critical function of language itself is to organize incorporated sensory information into functional categories in the human brain. Through metaphor we make use of patterns and relationships that are obtained in our physical experience to order and organize our most abstract understanding. Metaphor is an entropy-decreasing intervention tool with a broader scope than that of submodalities. Since communication (the transference of pattern) is subserved by the same neural substrates as the concept-forming systems of thought, emotions, and behavior, language is an important source of evidence that can be used to elucidate how that system is organized and directed. As we take a deeper look into human communication, linguistic evidence supports the notion that most of our conceptual system is metaphorical in nature. It is hardly possible to utter a single sentence without the use of metaphor (Gernsbacher, 1984; Johnson, 1987; Lakoff, 1987; Lakoff and Johnson, 1980). In its most basic linguistic form, the essence of metaphor is understanding and experiencing one thing in terms of another.

A simple example of this would be the metaphor of the brain as a computer. This metaphorical concept has the ability to structure and organize our sensory perceptions, thoughts, emotions, and behaviors relating to the brain in terms of an object more familiar to our sensory experience, and in terms of a structure more commonly incorporated by a human being's sensory systems in everyday life.

One can obtain linguistic evidence that a basic metaphorical structure is in operation by paying attention to idiomatic expressions. Although a person may not be consciously aware that his/her nervous system is organizing "brain" in terms of "computer," one's language will reflect this organization through utterances such as "He needs some positive programming" or "Can you repeat that, I didn't compute it the first time." Since these linguistic patterns are very common, we tend not to pay attention to the underlying structure responsible for generating the linguistic pattern itself. Consider phrases like the following: "I need some down time," "I need to file that away in my database," "I need time to process that information," "I'll store that in the back of my mind," "I can't take in any more information, I'm on overload," and "I am unable to access that memory." These simple phrases slip through our awareness every day and structure the experience of those hearing it without us ever explicitly laying down the foundation that the brain is a computer. How is this possible?

One way of understanding this phenomenon is to explore its effects at the level of anchoring. Auditory symbolic anchors instantiate tremendous flexibility and discriminatory power for the process of recategorizing replicable sensory information into new structures. An auditory symbolic anchor such as the word *computer* has the ability to release the system from the initial conditions of all related sensory experience, giving the state vector access to a rich array of attractor landscapes through all sensory modalities and across several contexts. Each of these landscapes can be a subset of another set of categories and reference experiences. This is sometimes referred this to as a *4-tuple*.

Few classically conditioned patterns have the precision and discriminatory richness to access as broad a set of conceptual experiences and sensory information as a simple phonetic sequence called a word. To understand the dynamics of this linguistic tool we must descend temporarily to the level of neocortical dynamics and examine more closely the cortical processes that occur in the neurocognitive system during perception (incorporation), replication, and subsequent category formation of incoming sensory information.

HOW THE NERVOUS SYSTEM PROCESSES SENSORY INFORMATION

Perception and Pattern Encoding

Perception is an active, creative process (Kosslyn, 1994) in constant dynamic fluctuation, as compared to *sensing*, which is a more passive process. Internal imagery, via all our representational systems, is active during the process of perception. An elegant example of this is our ability to perceive a partial or degraded image, such as being able to recognize a friend we have not seen for several years. It is astonishing that the brain is able to compare newly incorporated sensory patterns with internally replicated representations, resulting in spontaneous recognition of the friend from nearly any angle, distance, style of walking, voice (even when distorted by telephone transmission), the face that has aged 15 years, or even from the remaining half of a photograph which was torn in two.

This extraordinary neurocognitive capability would not be possible if we stored and retrieved only "exactly" what we experienced.

Compelling evidence suggests that an act of perception is not the copying of an incoming stimulus from the information field. Instead, we "store" relationships between things in a dynamically "soft-coupled" manner through the formation of trajectories between attractors originally established when a stimulus is first sensed (Erickson, 1980, 1989; Kohonen, 1995; Smith and Thelen, 1993, 1994). The recognition of any image sensed by the brain requires simultaneously activated internal imagery (Kosslyn, 1994). One way to think of perception is as that of a matching process between presently incorporated and previously incorporated replicable information/pattern.

As our five sensory systems sample regions of the information field in the external world, replicated patterns of electrochemical activity triggered by those initial conditions attempt to capture the state vector. The resulting process on the neurocellular level is referred to by neuroscientists as *synaptic modification*, a notion first posited in 1945 by Donald Hebb and later validated by the field of molecular biology. The critical notion that was missed for some time by many neuroscientists was the understanding that while the nervous system is sampling the outside world, it is simultaneously sampling the internal world of physiological response via an elaborate network of interoreceptors referred to as the somatosensory system (Damasio, 1994).

The somatosensory system transmits changes in physiological pattern, which occur in response to every event occurring in the outside stimulus field. The pattern of electrochemical activity propagated by the somatosensory system is recombined into a single activation pattern by the formation of trajectories. The trajectories connect, order, and organize widely distributed processes throughout the brain, in turn connecting these patterns by trajectories with the patterns of electrochemical activity resulting from the incorporation of sensory information. This results in the state vector taking a significantly different phase path than it would have otherwise. That is, as we experience our world, our nervous system not only records the events from outside, but also records our bodies' reaction to those events so that at a later date we can formulate an adaptive physiological and biomechanical response to any similarly perceived event in the information field. This is why we are able to evoke a visual or auditory image of an event that happened several years ago and simultaneously experience the feelings connected with that event as if it were occurring in the present. Damasio (1994) discusses this very process as being the neural correlate of human reasoning and decision making.

The incorporation and encoding of activation patterns are essential for the process of perception and memory. As Donald Hebb posited in 1945, it seems that cells that fire together, wire together. Nerve cell assemblies are groups of interconnected neurons whose synapses become mutually and simultaneously strengthened by the propagation of impulses by input neurons during perception and learning. These are referred to as *Hebbian synapses*.

In this regard, experience selects a certain pattern of cell connections (attractor landscape), selectively strengthened for a particular event. But because the connections are widely distributed, when any subset of nerve cell assemblies receives familiar input the entire assembly rapidly responds. The attractor captures the state vector. This is a morphogenetic field phenomenon, namely, *regulation* (Edelman, 1987; Kendrew and Lawrence, 1994). Regulation is a fundamental property of morphogenetic fields, indicating that a portion of the field can regenerate the whole field. Holographic images possess a similar property.

For example, neurons that participate in the recognition of the smell of sawdust are also affected by the history of the neurons encoding the smell of banana (Briggs, 1992). History has preeminence over static representation of a stimulus. This is called *hysteresis* (Haken, 1983, 1988). The act of perception consists of an explosive leap (escape from an attractor) of the neurocognitive system from the basin of one chaotic attractor to another.

The *basin of attraction* is the set of initial conditions from which the state vector is released and predictably captured by the specified attractor (e.g., thought, emotion, or behavior). The basin for each attractor would be defined by the receptor neurons that were activated during perception to form the nerve cell assemblies resulting from impulse propagation. When an external event is first sensed, it forms an attractor. When an external event first becomes meaningful in some way, the attractor formed during sensing becomes connected by phase path or trajectory to another attractor or attractor landscape. On the other hand, if an external event is immediately meaningful, it has caused the state vector to follow a previously formed phase path directly to existent, relevant attractors or attractor landscapes (e.g., thought, emotion, or behavior).

In essence, perception enables the brain to plan and prepare for subsequent action on the basis of past action, sensory incorporation, and perceptual synthesis (modification of neurocognitive topology in design space). An act of perception is not the passive copying of an incoming stimulus.

Patterns of connectivity are spatial and temporal in nature. Both conditions must be met to activate a nerve cell assembly. The time-locked nature of activation patterns simply cannot be violated. This is the essence of what underlies the most effective therapeutic procedures available, namely, pattern interruption.

How Does the Brain Construct Categories Out of Incoming Information?

Categories of perception and action are assembled from multiple brain regions and interconnections primarily on the basis of temporal codes, which give rise to flexibility and uniqueness. Although cell groups coexist and overlap spatially, their activity can be recognized as distinct because of a unique temporal code (Ferster and Spruston, 1995). The same cell or group of cells may participate in different assemblies by changing their temporal relationships (degeneracy), giving rise to extreme context sensitivity of perceptual response (Edelman, 1987). Herein lies the advantage of coupled neuronal oscillations that are not too tightly phase locked.

An example of time coding as a phenomenon is illustrated in a common experiment done by neuroscientists using electrophysiological recordings from the cortex of a cat. In the cat cortex, neuron assemblies of a receptive field oscillate synchronously when a light bar is moving in one direction. The same field fires in an uncorrelated fashion if the light bar moves in the opposite direction, thus violating the time locked nature of the activation pattern of a particular spatial receptor site (Crick, 1994). The notion of *strategies* illustrates this concept on a more macroscopic scale. For example, there are qualitative differences between a visual-to-auditory-to-kinesthetic (V → A → K) neurocognitive strategy and an A → K → V strategy. The different sequences may not be variously employed (significant change in phase path) to create the same state or behavior.

A category is created in context by the formation of trajectories that connect neurocognitive attractor landscapes spatiotemporally (in space and in time). The resulting phase path is always a complex product of the immediate context (internal and external) and the evolved cell assembly's history of reentrant connectivity mapping between heterogeneous, previously incoherent processes. To accomplish metaphor-induced perceptual categorization dynamically, the neurocognitive system must simultaneously incorporate and replicate two or more independent abstracting neural network connectivity patterns in response to a stimulus and then interact reentrantly to modify the state vector's phase path, resulting in "higher order" linkage of their representations to the same attractor landscape. In essence, this is the neocortical dynamic that subserves the process of metaphor-induced category formation.

The function of a metaphor in terms of neocortical dynamics is to simultaneously activate (by incorporation, replication, or transmission) and join two or more previously discrete attractors, which encode two or more different perception–action sequences into a single activation pattern (either spatially or temporally by phase path). For example, when first heard together, the symbolic anchors *computer* and *brain* simultaneously activate nerve cell assemblies, which formerly were discrete attractor landscapes of widely distributed cell connectivity patterns. This leads the state vector to follow a new phase path, giving rise to a new way of thinking about and acting in accordance with both *brain* and *computer.*

An effective metaphor forms a deep and highly stable trajectory or set of trajectories between attractors or attractor landscapes capable of modifying the phase path of the state vector, which previously could only be captured by either one landscape attractor or the other. Subsequently, either symbolic anchor is capable of acting as the initial conditions necessary for specifying the same attractor landscape of thought, emotion, and behavior attractors, hence "unlocking" previously state-bound or landscape-bound resources (e.g., thoughts, emotions, macro-, and microbehaviors). This process allows the neurocognitive system to respond to either brain or computer in terms of the other. Used with precision and creativity, metaphor can be employed to produce significant modifications in the design space phase path through which the state vector travels. There are few tools that are capable of making such profound modifications to the phase path of a neurocognitive system. Effective use of metaphor as a tool for intervention necessarily requires knowledge of existing neurocognitive topology including phase path and attractor-stability distribution, which can be mapped via NeuroPrint.

The success of intervention by metaphor is dependent upon loosely coupled neuronal connectivity. Therefore it follows that the use of metaphor will always be most effective when coupled with hypnotherapeutic patterns that lower brain activation levels and destabilize biophysical energy barriers that were previously established. In this regard, metaphor is able to create a spatiotemporal modification in collective nerve cell assemblies resulting in spontaneous neuronal map reorganization (Kohonen, 1995).

Temporally correlated nerve cell activity among many neurons is the crucial force behind map reorganization. Neurocognitive map modification is believed to involve a spatiotemporal shift in collective, coherent activity. Herein lies the neural correlate of the process of anchoring or classical conditioning itself. This suggests that on the level of neocortical dynamics, linguistic metaphor construction is an elaborate and elegant form of auditory symbolic anchoring, the conversational effects of which are virtually invisible.

A common experience that we all share is the power of metaphor to spontaneously self-organize neuronal mapping so that we may experience that elusive *ah-ha* feeling when a new way of understanding something hits us like a ton of bricks. "Wow, I never thought of it *that way*! This changes everything." A profound shift in the way we perceive or understand something instantiates an equally profound shift in the way we behave with respect to that thing, both microscopically and macroscopically.

This chapter has elaborated the neurophysics of metaphor. With this insight, it is not difficult to understand the far-reaching potential that metaphor possesses for orators, hypnotherapists, educators, trainers, parents, etc. One of metaphor's most elusive effects is that the suggestions that it carries for the unconscious mind can be "time-released." How is this possible? While the brain can process several trillion bits of information per second, it appears that we are perhaps only consciously aware of 50 bits of information per second at any given time. In order to form perceptual filters, the brain must process every bit of that information prior to determining which portions will reach the threshold of consciousness at any given moment (Roland, 1993). Therefore, when a sequence of perceptual events or even a single static event releases the state vector from its initial conditions, reentrant internal processes are maintained for a significant period of time without conscious recognition of the ongoing neocortical dynamics or their neurocognitive result. This is illustrated by the common experience of struggling to think of the name of someone or something, and 3 hours later the information pops into conscious awareness seemingly out of nowhere.

Hypnotic phenomenon and common everyday experiences like this one suggest that several coexisting trajectories may be set in motion simultaneously without conscious awareness (a parallel process). Since conscious awareness itself is a serial process, these trajectories may remain out of awareness for some time, competing for territorial representation in the association cortices. Or they may possibly never enter conscious awareness but instead affect subcortical processes without ever necessitating neocortical representation. The confluence of these previously discrete trajectories is capable of coalescing attractors into a macroscopically detectable phase path (by employing the tool of NeuroPrint) and influencing thought or behavior at some later time. Once the state vector has been captured by a stable phase path resulting from this confluence, the resulting cognition may reach a sufficient threshold of territorial representation in the association cortices, allowing for conscious awareness. Herein lies the neocortical dynamic that underlies the use of posthypnotic suggestion via metaphor.

Metaphors influence, persuade, create understanding, shape perceptions and perceptual filters, and specify appropriate state-bound thoughts, emotions, and behaviors. Linguistic metaphor is a narrow-scope intervention tool, which can be used to either increase order/pattern or increase entropy. Its pattern behavior can be classified as primarily recombination, since its major function is to recombine existing discrete attractors by rearranging phase path. In the next chapter, we discuss the neurophysics of hypnosis and the mechanisms of nonlinear, nonequilibrium phase transition.

REFERENCES

Briggs, J. (1992). *Fractals: The Patterns of Chaos*. New York: Simon and Schuster.

Crick, F. (1994). *The Astonishing Hypothesis: The Scientific Search for the Soul*. New York: Macmillan.

Damasio, A. R. (1994). *Descartes' Error: Emotion, Reason, and the Human Brain*. New York: G.P. Putnam's Sons.

Edelman, G. M. (1987). *Neural Darwinism: The Theory of Neuronal Group Selection*. New York: Basic Books.

Erickson, M. H. (1980). *Hypnotic Alteration of Sensory, Perceptual and Psychophysiological Processes: The Collected Papers of Milton H. Erickson on Hypnosis Volume II*. New York: Irvington.

Erickson, M. H. (1989). *The Nature of Hypnosis and Suggestion: The Collected Papers of Milton H. Erickson on Hypnosis Volume I*. New York: Irvington.

Ferster, D. and Spruston, N. (November 3, 1995). *Science*, 270, 756–757.

Furman, M. E. (1996). Neocortical dynamics of metaphor. *Anchor Point*, February, 44–50.

Gernsbacher, M. A. (1994). *Handbook of Psycholinguistics*. San Diego: Academic Press.

Haken, H. (1983). *Synergetics: An Introduction to Nonequilibrium Phase Transitions and Self-Organization in Physics, Chemistry and Biology (3rd Edition)*. New York: Springer-Verlag.

Haken, H. (1988). *Information and Self-Organization: A Macroscopic Approach to Complex Systems*. New York: Springer-Verlag.

Johnson, M. (1987). *The Body in the Mind: The Bodily Basis of Meaning, Imagination, and Reason*. Chicago, IL: University of Chicago Press.

Kendrew, S. J. and Lawrence, E. (Eds.). (1994). *The Encyclopedia of Molecular Biology*. London: Blackwell Science.

Kohonen, T. (1995). *Self-Organizing Maps*. New York: Springer-Verlag.

Kosslyn, S. M. (1994). *Image and Brain*. Cambridge, MA: MIT Press.

Kosslyn, S. M. and Koenig, O. (1992). *Wet Mind: The New Cognitive Neuroscience*. New York: Free Press.

Lakoff, G. (1987). *Women, Fire, and Dangerous Things: What Categories Reveal about the Mind*. Chicago, IL: University of Chicago Press.

Lakoff, G. and Johnson, M. (1980). *Metaphors We Live By*. Chicago, IL: University of Chicago Press.

Nunez, P. L. (1995). *Neocortical Dynamics and Human EEG Rhythms*. New York: Oxford University Press.

Roland, P. E. (1993). *Brain Activation*. New York: Wiley-Liss.

Smith, L. B. and Thelen, E. (1993). *A Dynamic Systems Approach to Development: Applications*. Cambridge, MA: MIT Press.

Smith, L. B. and Thelen, E. (1994). *A Dynamic Systems Approach to the Development of Cognition and Action*. Cambridge, MA: MIT Press.

16 The Neurophysics of Hypnosis

CONTENTS

Since the beginning of recorded history the human mind has been the source of unending curiosity, drawing the interest of a myriad of interdisciplinary fields. These fields have all attempted to model the enormous complexity of the neurocognitive system in the hopes of resolving some of the mysteries that have riddled human intellect. One of the most enduring of all of these mysteries is hypnosis.

The first serious attempt to investigate hypnotic phenomena in the western world was conducted by Franz Anton Mesmer (1734–1815), who was later dubbed the founding father of modern hypnosis. While his theory of *animal magnetism* was not well received at the time of his writings, his empirical findings led to modern research on hypnosis and its widespread use in medical practice, dentistry, psychiatry, and clinical psychology. In 1779, Mesmer discussed animal magnetism at length in a paper entitled *Dissertation on the Discovery of Animal Magnetism* (Bloch, 1980). He described animal magnetism as the attraction and repulsion of bodies through gravity, mineral magnets, and static electricity. Mesmer's hope was to be able to explain animal magnetism according to the accepted model of physical science of his time, that is, Newtonian physics.

Later, in 1799, he published another paper entitled *Dissertation by F. A. Mesmer, Doctor of Medicine, on His Discoveries*. It seems clear from both of these papers that Mesmer possessed an intuitive sense that something was happening with regard to the pattern and order of the human brain when his described methods of animal magnetism were successful. Mesmer believed that all properties of man and nature were the result of their organization. "... I will prove that 'all properties are the combined result of the organization of bodies and of the motion of the fluid in which they are immersed' " (Bloch, 1980, p. 100). Throughout both papers, Mesmer formulated the belief that physical phenomena and properties of man are interconnected, sharing a common source.

In 1825, Jean Philippe Francois DeLeuze, a French naturalist, published his treatise on animal magnetism entitled *Practical Instruction in Animal Magnetism*, which was later translated by Thomas C. Hartshorn (Deleuze, 1843). The purpose of this text was to formalize many of the principles and methods of animal magnetism.

In November 1841, a Scottish physician by the name of James Braid witnessed a mesmeric demonstration by Lafontaine, the Swiss magnetist. Braid became convinced that the phenomenon was genuine and he immediately began his own experiments. Two years later, he published his first text on the subject entitled *Neurypnology: Or the Rationale of Nervous Sleep, Considered in Relation with Animal Magnetism* (Braid, 1843). Braid's work marked the beginning of a clear division between animal magnetism and what he referred to as hypnosis or *nervous sleep*.

By the 1880s, it became quite clear that the two opposing schools of thought would not be rejoined, with Alfred Binet and Charles Fere carrying on the work of Mesmer and Deleuze in their famous *Animal Magnetism* (1887).

Two years later, Bernheim (1889) published his most famous work entitled *Suggestive Therapeutics: A Treatise on the Nature and Uses of Hypnotism*. Bernheim maintained the belief, originally adopted from Braid, that the state of hypnosis could be attained by suggestion alone.

With the publishing of *Hypnosis and Suggestibility: An Experimental Approach* (1933), Clark Hull, a professor of psychology at Yale University, began to combine techniques of suggestion with the early methodologies of behavioral science. This interdisciplinary integration was continued by Andre Weitzenhoffer, a professor of psychiatry and behavioral sciences at the University of Oklahoma. In *General Techniques of Hypnotism* (1957), he detailed many of the fundamental innovations developed in the field while at the same time warning his readers of some of the possible dangers of hypnotism.

Much of our present understanding of hypnosis, as well as its widespread acceptance by the medical community, is owed to Milton H. Erickson (1901–1980), who was probably the most creative and imaginative contemporary worker in the area of hypnosis and hypnotherapy. Erickson continued to develop therapeutic suggestion as both a science and an art form until his death. During his lifetime, he attracted many ardent students, two of whom were responsible for a majority of the literature that elucidates Erickson's lifetime of creative innovations, namely, Jay Haley (1967) and Ernest Rossi (1980).

By the early 1980s, the field of neurolinguistic programming (NLP) had already adopted many of the unique principles and methods perfected by Erickson. Developers Richard Bandler and John Grinder distilled many of the basic principles and techniques and organized them into verbal and nonverbal models. The verbal induction patterns came to be known as the Milton model.

MODELING NEUROCOGNITIVE DYNAMICS

Recently, impressive developments in the fields of functional brain imaging, dynamical systems science, and biophysics have enabled the discernment of key correlations between microscopic behavior (neurodynamics) and observable, macroscopic behavior. With the development of these tools, we are more firmly positioned than ever before in history to be able to probe the mysteries of hypnotherapy and change.

In order to appreciate the daunting task of modeling neurocognitive behavior, it would be useful to review a few basic facts about brain function. The average human brain has approximately 100 billion brain cells (neurons) which are present from birth, after nearly 100 billion others are pruned away through early developmental processes. Virtually no new brain cells are added throughout one's life. Until recently, these neurons were believed to be the most fundamental unit of information processing and communication in the brain. Each of these neurons will connect with between 1000 and 200,000 other neurons to form a minimum of 100 trillion points of potential connection or synapses.

It is easy to see that through "wiring" potential alone, the task of tracking information patterns transmitted and translated between cells would be difficult, if not impossible. To further complicate the tracking, a message can pass from one neuron to another in just a few thousandths of a second. That message may have arisen out of the confluence of tens of thousands of incoming signals from other neurons widely distributed over the entire nervous system. Here, speed of transmission becomes another challenge. If that's not enough, it's also important to note that the brain maintains its flexibility because its 100 trillion connections are not hard-wired like a computer. Instead, the small synapse allows electromechanical patterns/information to be converted into biochemical patterns/information, which is transmitted from cell to cell and again converted back to electromechanical messages.

Modeling complications increase further when we consider that to date more than 60 different protein-based neuroactive chemical messengers have been identified which are believed to carry qualitatively different instructions depending upon the region of the brain and the target receptor site (Kandel et al., 1991). From this raw material alone, without taking into consideration the numerous subneural, endocrine, and paracrine pattern-encoding mediums (discussed in Section I), the brain is capable of creating an astronomical number of discrete dynamic patterns.

A cutting-edge brain imaging technique, called the SQUID (superconducting quantum interference device), is capable of detecting *intracellular magnetic fields* resulting from brain cell activity that are 1 billion times smaller than the earth's magnetic field (Kelso, 1995; Kruger, 1991). This device has helped neuroscientists in the last 5 years to track correlations between electromagnetic, spatiotemporal patterns, and macroscopic human behaviors such as learning, memory, visualization, auditory imaging, language recognition, and problem solving. Modern brain imaging technology now provides a window for probing many of the mysteries of the neurocognitive system.

We are now far more capable of solving some of the mysteries of the brain than ever before. Many scientists agree that a comprehensive understanding of the brain can only be reached by an interdisciplinary integration of the many disparate branches of research extant today. A major obstacle to such a task is that each sequestered subfield has unwittingly developed its own specialized vocabulary, specifically targeted to its own unique scope of inquiry.

UNDERSTANDING HYPNOSIS

There appears to be no shortage of questions surrounding the field of hypnotherapy. It has long been one of the great enigmas of human behavior, riddled with more than its fair share of myths and mysteries. Some of the most common questions asked by mental health practitioners are "What is hypnosis?" "What is trance?" "What is an altered state?" "How does hypnotic phenomenon work?" "Why is the brain more susceptible to suggestion during hypnosis?" and "Why can we get profound change to occur during trance that seems impossible under other conditions?"

One might ask the question, "Why would it be useful to be able to describe and understand what is happening to the neurocognitive system during trance?" Certainly one reason is that many individuals require a firmly grounded scientific description in order to be convinced of its efficacy as a tool for change. To explore these and other questions, we have to start by considering some new beliefs, presuppositions, and conceptual vocabulary.

Throughout the ages, hypnosis has been explained by invoking terms such as conscious, unconscious, preconscious, subconscious, levels of consciousness, and depth of trance. These metaphors interfere with, rather than enhance, our understanding. To have a clearer understanding of what hypnosis is and what it does, we need to describe it with a new conceptual vocabulary. We must start by denominalizing all of the words we currently use to describe hypnosis, trance, and mind. Linguistic denominalization is a process of turning a noun back into a process (a verb). The following presuppositions are useful to consider:

- Mind and behavior are emergent properties arising out of the incorporation, replication, cleaving, recombination, and transmission of pattern within and between our biophysical systems and the external environment.
- Information/pattern is either accessible to our conscious awareness or not accessible. It cannot exist in any other mode. Presuppositions involving *levels* of consciousness and *depth* of trance must be abandoned in order to develop a deep understanding of the cognitive neurodynamics of hypnosis and hypnotherapy.
- Information/pattern encoded (incorporated) by stable biological substrates that can be reliably replicated is called memory.
- All learning, memory, cognitive faculties, and behavior that are properties of the organization of encoding aggregates (pattern) are state/phase-dependent. The information that is available at any given moment depends on the state/phase of the brain.

The next step is to denominalize the word hypnosis. Hypnosis has long been thought of as a *state* that one *goes into* and *comes out of*. As long as we consider hypnosis and trance as a state, it precludes us from understanding how it works and what is actually taking place in the neurocognitive system.

Consider the fact that hypnosis is a *process* which is constantly evolving — a process of interacting with pattern. To establish a useful model to explain hypnosis, we have to start at the level of spatiotemporal pattern behavior between interacting neuronal assemblies. The process of hypnosis affects these patterns in at least three basic ways:

- Activity level
- Source of information incorporated
- Ambiguity-induced intermittency

ACTIVITY LEVEL

Cell connectivity patterns propagate at certain frequencies in the brain. Each collective frequency gives rise to different emergent properties (Haken, 1988, 1996; Haken and Koepchen, 1991; Hobson, 1994; Kruger, 1991; Nunez, 1995). One of the main goals of any hypnotic induction is to reduce the activity level or frequency by which these patterns are generated and propagated (Bernheim, 1889; Bloch, 1980; Braid, 1843; Deleuze, 1843; Erickson and Rossi, 1981; Hull, 1933; Rossi, 1980; Weitzenhoffer, 1957). Slowing voice tempo and lowering voice volume affect the activity level of the brain. Two principles of neurophysics help to explain this *forced resonance* and *phase locking*.

Forced resonance is a process by which the frequency of a driving force (voice tempo) matches the natural frequency of a structure (the brain) by a process sometimes referred to as pacing. As discussed earlier, the process of pacing in physics is known as phase locking. The driving force (voice tempo) can lead the activity level or frequency of the brain in an inhibitory (into trance) or excitatory (out of trance) direction once phase locked with the brain's current frequency (Haken, 1988; Haken and Koepchen, 1991; Kelso, 1995; Kruger, 1991).

To illustrate this, consider two clocks with pendulums on the same wall. If we start one clock by swinging the pendulum and then do the same to the other, the pendulums will remain out of sync for a short period of time. The vibrations propagated through the wall will provide information/pattern by which the pendulums become phase locked. Later we will find both pendulums swinging in unison.

The mechanism referred to as forced resonance can lead the brain from *beta-wave* activity (approximately 12 to 30 Hz) down through *theta-wave* activity (approximately 4 to 8 Hz). This change in brain activity can be readily seen on an EEG, which measures extracellular magnetic waves (Nunez, 1995). The significance of changing the brain's activity level from beta to theta is that all of the faculties of the brain and mind are dependent upon and subject to these fundamental activity levels. Slowing the brain's activity level gives rise to many trance phenomena such as catalepsy, time distortion, expanding and contracting conscious awareness, perceptual distortion, spontaneous memories, etc.

Thus, faculties such as visualization, memory, attention, and volition will be accessible at one activity level (frequency) and not at another. This can be thought of as a fundamental *phase transition* from one activity level to another where each activity level represents a discrete attractor landscape and its state-bound attractors (Haken and Koepchen, 1991; Haken, 1983, 1988). As the activity level in the brain stem decreases, there is a switch in dominance between norepinephrine (NE) and acetylcholine (ACh). As NE decreases and ACh increases, visual images become more vivid and the ability to voluntarily direct our own attention and exert volition decrease (Hobson, 1989, 1994; Kissin, 1986).

SOURCE OF INFORMATION

As the activity level in the brain stem decreases, the source of pattern propagation (transmission) to the sensory systems of the brain switches in dominance from external to internal (Hobson, 1989, 1994; Kissen, 1986). As we pay less attention to information coming through our senses externally, we are able to pay more attention to internal information resulting from the dynamic interplay between internally replicated patterns in the brain called memory, which are influenced in part by the external environment.

This restriction in external pattern propagation results in an increase in neurocognitive entropy. Hypnotic techniques generally evolve to accomplish both tasks simultaneously. Practitioners of hypnotherapy are instructed to slow voice tempo and breathing (reduce activity level) and direct the subject's attention gradually from a narrowed external focus to an internal one at the same time.

AMBIGUITY-PRODUCED INTERMITTENCY

Hypnotherapists have long known that ambiguity leads people into trance while specificity leads people out of trance (Rossi, 1980). We now have the theoretical foundation that enables us to understand why. Use of ambiguity — syntactic, phonological, and symbolic or otherwise — is the fastest way to move an individual into trance. To understand how this occurs, we need to redefine what trance actually is. Trance is an internal transition from one state/phase or attractor landscape to another — the process of transitioning. During transitioning the state vector is freely moving, unable to be captured by any discrete attractor. When this occurs nerve cell connectivity patterns are destabilized, and once the state vector is captured by an attractor, the patterns are dynamically reorganized (Haken, 1983, 1996; Kruger, 1991). The process of phase transition allows the neurocognitive system to shift smoothly between its collective states.

Hence, trance is a *trans*ition from one neurocognitive state (pattern, phase, or landscape) to another. It is made possible by the temporary destabilization of existing energy barriers (on the atomic, molecular, and cellular levels) that subserve the existing state or phase, allowing spontaneous formation and reorganization of new ones (Haken, 1988; Kruger, 1991). The information available (patterns that are replicable) in each neurocognitive state or phase will be qualitatively different, thus resulting in state-bound information (Kruse and Stadler, 1995; Rossi, 1993). This is why hypnotheraputic processes that induce and maintain transition make state-bound information more accessible. Trance facilitates the brain's ability to shift more quickly and smoothly between collective, coherent, neurocognitive states (patterns, attractors, and attractor landscapes). Consequently, old memories, experiences, and unusual state-bound cognitive and behavioral faculties seemingly emerge from nowhere.

The effect of ambiguity on the brain can be understood by exploring the transition process from one state or phase to another. In the language of physics, this process is referred to as *nonequilibrium phase transition* (Horsthemke and Lefever, 1984). This is the most important concept to grasp in order to understand the process of hypnosis. All nonlinear complex systems in the known universe maintain their flexibility by changing rapidly from state to state. They do this through the process of nonequilibrium phase transition. The operating presupposition is that the state to be exited must be destabilized in order to switch to another state. In short, the main mechanism utilized by any complex system to enter or exit a state, including the brain, is that of a nonequilibrium phase transition.

Any parametric influence, such as hypnotic language patterns or monotonous, rhythmic stimuli, that can keep the state vector near or between attractors rather than in them, results in hypnotic trance. Since the state vector experiences a dramatic increase in degrees of freedom during this time, trance provides the system with greater flexibility and fluidity.

It is easy to see why during trance there is an increased susceptibility to suggestion. When the state vector is in transition, far less perturbation is necessary to specify a desired attractor or attractor landscape in comparison to the much more significant perturbation needed when the state vector must first be released from a deep attractor that has previously captured it. By keeping a complex system operating at points near instability, we are giving the system more rapid, energy-efficient access to its collective states, resources, and information, which are normally trapped by state-dependent encoding. It follows, then, that the more stable a pattern or state is, the less flexibility the system has.

How, then, does the process of hypnosis destabilize existing states/phases in order for the state vector to be released and able to transition into new ones? It does this in two ways. The first is through ambiguity. When a stimulus presented to one of the sensory systems is ambiguous, it causes the neurocognitive system to be released from two or more sets of initial conditions simultaneously. This, in turn, causes the state vector to switch back and forth between two or more possible interpretations — patterns/state or mode attractors (Kruse and Stadler, 1995). The switching is facilitated by noise-induced nonequilibrium phase transition (trance) (Horsthemke and Lefever, 1984).

Rather than the system settling into a recognizable, stable pattern of interpretation, as happens when we employ specificity in our linguistic patterns, the state vector rapidly switches between the possible interpretations — patterns/state or mode attractors. This condition is *intermittency*, a hallmark of all nonlinear complex systems (Kruse and Stadler, 1995). Thus, the more difficult it is to resolve the conflict, as is the case with a phonological, syntactic, or symbolic ambiguity, the longer the system remains in intermittency. In the case of intermittency, the system lives near or between attractors (patterns), rather than in them. In other words, the system lives in transition (Kelso, 1995). A common perceptual experience illustrating intermittency occurs when we focus our attention on a two-dimensional drawing of a three-dimensional Necker cube. The longer we stare at it, the more we realize that it is impossible to resolve whether the opening to the cube is up to the right or down to the left. In this case, the state vector is forced to continually transition between two visual mode attractors, unable to be captured by either of them for any period of time (Kruse and Stadler, 1995).

The second way in which hypnosis destabilizes existing patterns is through forced resonance and *frequency locking*, a type of phase locking. All patterns and resulting information in the brain are frequency dependent. As we gradually slow our voice tempo, breathing, and movements, and focus on monotonous, unchanging stimuli, we lead the brain by forced resonance, through ever-slowing frequencies of activity. At a certain point, the brain is unable to maintain a given pattern for any period of time and begins switching rapidly between patterns — transitioning. This phenomenon can also be seen in REM sleep. During dreaming, the brain switches quite readily through random states, patterns, and contexts virtually uncontrollable by the sleeping subject.

DEPTH OF TRANCE

One of the most common sensations experienced by a subject in trance is the feeling of depth. We can now describe on a neurobiological level that depth of trance is experienced as a result of the change in dynamic balance between NE and ACh in the brain stem. As NE levels drop, a hypnotic subject feels sensations similar to that experienced when moving from waking to sleeping, the two most fundamental brain states. The greater the drop in NE, the deeper the trance feels. Thus, levels of trance experienced subjectively correspond to levels of NE in the brain stem (Hobson, 1989, 1994).

Another common property of the neurocognitive system when in trance is conservation or economy of movement. As the level of ACh in the brain stem increases, the *pons*, a region of the brain stem, sends a command to the skeletal muscles that inhibits movement of limbs. This feature of our neurobiology is particularly useful when we enter REM sleep. It enables us to experience vivid dreams without the corresponding neuromuscular activity, which could endanger us and others nearby. Incidentally, during REM sleep, neuromuscular activity is shut off completely by this mechanism (Hobson, 1989, 1994).

As a hypnotic subject awakens from trance, we observe what we call a reorienting response, which consists of moving from a motionless state to that of normal range and speed of motion. Evidence suggests that this reorienting response means that the balance of neurochemicals in the brain stem has changed in dominance again.

It should be clear by now that the process of change itself is reliant upon the brain's built-in mechanisms for naturally and spontaneously shifting between possible collective modes, phases, or states. Therefore the appearance of persistence of any problem state can now be conceptualized as the deepening or stabilizing of subserving attractors in the neurocognitive attractor landscape. Any change from a present problem state to a desired state necessarily requires the destabilization or weakening of the present attractor or attractor landscape so as to allow the neurocognitive system to explore new collective modes or phases. Trance is the neurocognitive system's built-in process for freeing the state vector and increasing its degrees of freedom.

As an intervention tool, therapeutic trance is broad in scope. It operates normally at the system level, although its scope can be narrowed by implementation of specific techniques such as anchoring and submodalities. Therapeutic trance is entropy increasing and has a destabilizing and disorganizing effect on the neurocognitive system. Its systemic pattern behavior is cleaving in nature. The effectiveness of therapeutic trance as an intervention tool is greatly increased when used in conjunction with linguistic metaphor, anchoring, and submodal behaviors (submodalities). In the remaining chapters, we discuss in great depth the specific use of anchoring as a tool of intervention.

REFERENCES

Bernheim, H. (1889). *Suggestive Therapeutics: A Treatise on the Nature and Uses of Hypnotism.* New York: G. P. Putnam's Sons.

Binet, A. and Fere, C. (1887). *Animal Magnetism.* London: Kegan Paul, Trench & Co.

Bloch, G. (Ed.). (1980). *Mesmerism: A Translation of the Original Scientific and Medical Writings of Franz Anton Mesmer.* Los Altos, CA: William Kaufmann.

Braid, J. (1843). *Neurypnology: Or the Rationale of Nervous Sleep Considered in Relation with Animal Magnetism.* London: Adam & Charles Black.

Brown, P. (1991). *The Hypnotic Brain: Hypnotherapy and Social Communication.* New Haven: Yale University Press.

Deleuze, J. P. F. (1843). *Practical Instruction in Animal Magnetism.* New York: D. Appleton & Co.

Dengrove, E. (Ed.). (1976). *Hypnosis and Behavior Therapy.* Springfield, IL: Charles C Thomas.

Edmonston, W. E. (1986). *The Induction of Hypnosis.* New York: John Wiley & Sons.

Erickson, M. H. and Rossi, E. L. (1979). *Hypnotherapy: An Exploratory Casebook.* New York: Irvington.

Erickson, M. H. and Rossi, E. L. (1981). *Experiencing Hypnosis: Therapeutic Approaches to Altered States.* New York: Irvington.

Erickson, M. H., Rossi, E. L., and Rossi, S. I. (1996). *Hypnotic Realities: The Induction of Clinical Hypnosis and Forms of Indirect Suggestion.* New York: Irvington.

Furman, M. E. (1995). The neurophysics of hypnosis. *Anchor Point,* October, 10–15.

Haken, H. (1981). *The Science of Structure: Synergetics.* New York: Van Nostrand Reinhold Company.

Haken, H. (1983). *Synergetics: An Introduction (3rd Edition).* New York: Springer-Verlag.

Haken, H. (1988). *Information and Self-Organization: A Macroscopic Approach to Complex Systems.* New York: Springer-Verlag.

Haken, H. (1996). *Principles of Brain Functioning: A Synergetic Approach to Brain Activity Behavior and Cognition.* New York: Springer-Verlag.

Haken, H. and Koepchen, H. P. (1991). *Rhythms in Physiological Systems.* New York: Springer-Verlag.

Haley, J. (Ed.). (1967). *Advanced Techniques of Hypnosis and Therapy: Selected Papers of Milton H. Erickson, M.D.* New York: Grune & Stratton.

Hobson, J. A. (1989). *Sleep.* New York: Scientific American Library.

Hobson, J. A. (1994). *The Chemistry of Conscious States: How the Brain Changes its Mind.* New York: Little, Brown & Company.

Horsthemke, W. and Lefever, R. (1984). *Noise-Induced Transitions: Theory and Applications in Physics, Chemistry, and Biology.* New York: Springer-Verlag.

Hull, C. L. (1933). *Hypnosis and Suggestibility: An Experimental Approach.* New York: Appleton-Century-Crofts.

Kandel, E. R., Schwartz, J. H., and Jessell, T. M. (Eds.). (1991). *Principles of Neural Science (3rd Edition).* East Norwalk, CT: Appleton & Lange.

Kelso, J. A. S. (1995). *Dynamic Patterns.* Cambridge, MA: MIT Press.

Kissin, B. (1986). *Psychobiology of Human Behavior (Volume I): Conscious and Unconscious Programs in the Brain.* New York: Plenum Press.

Kroger, W. A. (1963). *Clinical and Experimental Hypnosis in Medicine, Dentistry and Psychology.* Philadelphia: J. B. Lippincott Company.

Kruger, J. (1991). *Neuronal Cooperativity.* New York: Springer-Verlag.

Kruse, P. and Stadler, M. (1995). *Ambiguity in Mind and Nature.* New York: Springer-Verlag.

Nunez, P. L. (1995). *Neocortical Dynamics and Human EEG Rhythms.* New York: Oxford University Press.

Rossi, E. L. (Ed.). (1980). *The Collected Papers of Milton H. Erickson on Hypnosis (Vol. I): The Nature of Hypnosis and Suggestion.* New York: Irvington.

Rossi, E. L. (Ed.). (1980). *The Collected Papers of Milton H. Erickson on Hypnosis (Vol. II): Hypnotic Alteration of Sensory, Perceptual and Psychophysiological Processes.* New York: Irvington.

Rossi, E. L. (Ed.). (1980). *The Collected Papers of Milton H. Erickson on Hypnosis (Vol. III): Hypnotic Investigation of Psychodynamic Processes.* New York: Irvington.

Rossi, E. L. (Ed.). (1980). *The Collected Papers of Milton H. Erickson on Hypnosis (Vol. IV): Innovative Hypnotherapy.* New York: Irvington.

Rossi, E. L. (1993). *The Psychobiology of Mind-Body Healing: New Concepts of Therapeutic Hypnosis (Revised Edition).* New York: W. W. Norton & Company.

Rossi, E. L. and Cheek, D. B. (1988). *Mind-Body Therapy: Methods of Ideodynamic Healing in Hypnosis.* New York: W. W. Norton & Company.

Spiegel, H. and Spiegel, D. (1978). *Trance and Treatment: Clinical Uses of Hypnosis.* Washington, D.C.: American Psychiatric Press.

Watkins, J. G. (1987). *The Practice of Clinical Hypnosis (Vol. I): Hypnotherapeutic Techniques.* New York: Irvington.

Watkins, J. G. (1992). *The Practice of Clinical Hypnosis (Vol. II): Hypnoanalytic Techniques.* New York: Irvington.

Weitzenhoffer, A. M. (1957). *General Techniques of Hypnotism.* New York: Grune & Stratton.

Yapko, M. D. (1990). *Trancework: An Introduction to the Practice of Clinical Hypnosis (2nd Edition).* New York: Brunner/Mazel.

Section III — Part 4

Performance Engineering: Tools and Principles of Neurocognitive Pattern Modification

17 Life, Death, and Anchoring

CONTENTS

THE FAR-REACHING EFFECTS OF ANCHORING

A now-famous series of experiments performed by researchers in the field of psychoneuroimmunology clearly illustrates the long arm of an anchor. The first series of these experiments was performed on rats and mice. The object was to establish the fact that immune system patterns could be trained to respond predictably to a unique stimulus in an information field when presented to the appropriate sensory system.

The rats were first divided into two groups. Group 1 was injected with cyclophosphamide, a known immunodepressant. Group 2 was injected with PolyI:C, an immunostimulant. These injections were used to induce the desired immunophysiological target states (attractor landscaped). When the expected immune system reactions were calibrated, both groups of rats were *selection anchored* to the same stimulus. The researchers used camphor as an olfactory stimulus and saccharine water as a gustatory stimulus, both stimuli presented simultaneously. This uniquely paired anchor insured that the previously calibrated immunophysiological state could not be accidentally triggered by commonly contacted environmental elements. By this point, the researchers had established a reliable selection anchor or conditioned stimulus that, when presented, would activate an immunodepressed state in Group 1 and an immunostimulated state in Group 2. That is, the complex olfactory-gustatory anchor became the initial conditions in both cases, necessary to release the state vector along one phase path in Group 1 and an entirely different phase path in Group 2. In the first case the stimulus specified an immunodepressing attractor landscape and in the second case the same stimulus was used to specify an immunostimulating attractor landscape.

In stage 2 of the experiment, once immunophysiological indicators returned to normal, both groups of rats were injected with a substantial dose of a common pneumonia virus. The olfactory-gustatory selection anchor (a conditioned stimulus in this case) was then presented. Within 48 hours, the immunostimulated rats in Group 2 showed little sign of the virus and within approximately 7 days they showed none. Within the same time frame, all the immunodepressed rats in Group 1 died (Ader et al., 1991). A later research project duplicated these results substituting live cancer cell injections for the pneumonia virus. For these experimental rats, what made the difference between the most fundamental physiological states of life and death was simply a selection anchor that was intentionally conditioned to select different phase paths and their connecting attractor landscapes in design space.

Imagine the biomedical applications of such a tool. In a later research experiment, anchoring was used to help cut the cost of an expensive life-saving cancer therapy. The object of the research was to train the immune system of a young girl to increase production of Interleukin-2 (IL-2) which would be necessary in large quantities to battle her advancing liver cancer. At that time, one course of IL-2 was obtainable at an average cost of approximately $40,000. Upon administration of a single course of IL-2, the same modes were used to establish a complex selection anchor. Researchers were not surprised to find that the process resulted in the presence of higher levels of IL-2 than were normally contained in a single course.

Evidence suggests that the only limitation concerning the use of anchoring to specify a desired attractor or attractor landscape is the accuracy of calibration. The effective use of anchoring to specify a desired phase path and attractor landscape necessarily requires an appropriate method of calibrating when the state vector has been successfully captured by the target landscape. Modern brain imaging technology certainly has no shortage of such methods. These experiments and many others conducted in the field of psychoneuroimmunology strongly support the premise that information can influence the quality of our life and in some cases our very existence (Ader et al., 1991).

DESIGNING NEW METHODS OF CALIBRATION

Mental health practitioners are normally trained to calibrate with mainly their sensory organs. This method severely limits us from calibrating and capturing discrete physiological states (attractors) we may want to specify for intervention. As we seriously explore the use of state-of-the-art calibration tools, such as PET scanning, fMRI, and SQUID, we will significantly extend the range and scope of anchor-based interventions. However, it is not necessary to become this elaborate in order to develop many unique and effective anchor-based interventions. What is necessary is a thorough NeuroPrint and a principle-based understanding of anchoring and its effects on neurocognitive topology and resulting cognitive neurodynamics.

Considering the aforementioned research, the reader may be compelled to search for more commonly available calibration tools a practitioner could use to explore new intervention possibilities. The following example will illustrate how more commonly found calibration tools can be employed for intervention.

Let us consider a client who is suffering with the adverse effects of blood pressure medication and is interested in learning to influence his blood pressure on his own (Chase, 1974). To accomplish this, we needed two tools: an indicator to calibrate ongoing blood pressure and a tool to influence his state (specify an appropriate attractor landscape). Since the brainwave states of Alpha, Theta, and Delta are strongly associated with lower blood pressure, we needed to find a tool that would act as a *control parameter* to lead the neurocognitive system through these collective states. Many companies today produce an auditory-based tool that can be used for this purpose.

For this example we choose a cassette tape that is prerecorded — in the left ear, a pure 100-Hz tone and in the right ear, a 104-Hz tone. As expected through the principle of forced resonance, background brain activity will begin to oscillate at 4 Hz (Delta wave) within a few minutes, which can be verified with an EEG scan. A corresponding decrease in blood pressure will follow in many cases almost immediately, which can be verified via the blood pressure indicator.*

A client who listens to this type of tape will notice a very distinct auditory oscillating at a range of approximately 2 to 4 cycles per second. Since having the cassette tape readily available is restrictive, we can employ the use of a similar bi-modal selection anchor to specify the appropriate attractor landscape in the absence of a cassette tape player. The most important consideration in designing the initial conditions necessary to release the state vector is that the selection anchor be both readily available and unique or discrete (Bechterev, 1928). Those selection anchors that are not unique may specify two or more competing phase paths, causing an increase in entropy and an increased likelihood of the state vector being captured by the wrong attractor.

It should be noted here that background brain activity changes from Beta to Alpha in many cases simply by closing your eyes for a few seconds, which restricts the flow of information the brain must process in any given period. This decrease in information flow through visual pathways causes a temporary decrease in background oscillation frequency (Kandel et al., 1991). Thus, temporary eye closure can be added to the bi-modal selection anchor to form a larger set of initial conditions specifying the appropriate attractor or landscape. This procedure creates a wider basin of attraction and easier access to the desired target attractor state of lower blood pressure. This is only one of many ways in which a creative change agent can begin to explore new intervention possibilities with NeuroPrint. This example shows how an anchor can specify different attractor landscapes of brain activity level. It should be noted here that an added dimension of NeuroPrint-based intervention could be explored through the operant conditioning of brain activity as well (Chase, 1974).

HOW ANCHORS WORK

How are anchors able to influence the human nervous system? The more we understand about how anchors work, the more flexibility we have in designing new and effective uses. Many practitioners in NLP training have heard the phrase, "You cannot, not anchor." This is true. As a human brain samples its external environment via its five senses, a representation of these external events is encoded by *synaptic modification* (Kuno, 1995).

* Casio produces a watch which is very useful for indicating blood pressure.

Synaptic modification is a process by which the nervous system strengthens certain neural pathways and weakens others, resulting in unique electrochemical and electromechanical patterns of activation. That is, a pattern of electrochemical and electromechanical activation encodes the simultaneous (parallel) activity of all five senses as if it were one piece of information (due to the formation of connecting trajectories). At the same time, the brain is propagating information from the internal environment, encoding the body's entire physiological response to the outside event in the same activation pattern. Anything that reactivates this unique pattern of activation (reliably replicates the pattern) also activates the physiological response of the body that was encoded with that pattern (Damasio, 1994). The resulting pattern of activation is an attractor. The more effective the synapses become through modification, the deeper and more stable the attractor. Each sensory system's encoding of this simultaneous event is connected by trajectories to form a basin of attraction. Any of the connected attractors may be capable of causing the entire pattern to be replicated, since each of them can act as the initial conditions necessary to specify the phase path of the state vector. The resulting phenomenon is characteristic of a *morphogenetic field*, capable of spontaneous self-organization. The most important property of this sort of field is that it is capable of regulation. That is, any part of the field can activate the entire field.

What is the relevance of this to the change agent? What is evident from the research of psychoneuroimmunology is that every physiological response, including the expression of DNA (resulting in IL-2 production), is encoded in that activation pattern. The more unique the selection anchor, the more likely the phase path it causes the state vector to travel will lead to the intended activation pattern-attractor state, state-bound behavior, mode-bound behavior, or landscape.

VISUAL MODE ANCHORS AND HUMAN DECISION MAKING

An easy way to tie all of this information together is to understand how anchors are used discretely by the neurocognitive system in the process of decision making. The frontal and prefrontal cortex of the human brain subserves a function referred to as working memory, an internal information field used for thinking, problem solving, decision making, etc. (Goldman-Rakic, 1987, 1992). Patterns encoded by stable substrates throughout the brain are replicated and then transmitted to the prefrontal cortex. These patterns are then indexed by location, as are all morphogenetic fields. This positionally dependent field is subserved by specialized pyramidal cells in the prefrontal cortex (Goldman-Rakic, 1987, 1992).

This bio-architecture helps to explain why submodalities appear to be location-dependent. The location of an image is extremely important in determining its meaning. Recall our discussion of cyto-architectonics and cortical field activation cues in Part II and submodalities in Part III. The reason for this is that the location of the image also appears to be indexed by the somatosensory system (Damasio, 1994). Thus if we access a memory (replicate a pattern) and it is transmitted to the frontal and prefrontal cortex (working memory), the location of that image in the visual information field becomes linked by trajectory (indexed) to the physiological response originally encoded as part of the activation pattern representing that memory (Goldman-Rakic, 1987, 1992).

Let us illustrate this principle of attractor/phase path organization. Imagine something that you are highly motivated to do. As you internally visualize a representation of this, notice the location of the image. If you do not feel that you are proficient at making visual images, then notice where it feels like it is or where it sounds like it is. Now, imagine something that you are not motivated to do, but know you need to do. Notice how working memory organizes these two images (whether visual, auditory, or somatosensory) in different locations. Also notice that as you look back and forth between the two possibilities that there is a kinesthetic difference that becomes easily discernible and lets you know which one you want to do.

This is possible because each location in the visual, as well as auditory or tactile, field now encoded by pyramidal cells in the frontal cortex has become discretely anchored by phase path to your physiological response (discrete attractors or attractor landscapes in the somatosensory system). This is referred to as a *somatosensory* (or somatic) *marker*. The potential intervention applications for this are staggering. Not only is it possible to internally move the unmotivating image to the motivated position in the visual field and feel *differently* about the same image, but one can also discretely influence another person's physiological state from the outside by anchoring the location of each image with a hand gesture (hand position). Richard Bandler first illustrated this. By discretely gesturing to the appropriate location in the visual field where a person has transmitted these images, it is possible to trigger reactivation (replication) of his or her physiological response. The subject's eye fixation while visualizing will indicate the position of the images in the visual field. Rather than establishing artificial anchors, in this case we are utilizing the neurocognitive system's own somatic marking system and precisely pacing a person's internal model of the world. In this case the hand gestures become the initial conditions necessary to release the state vector on the appropriate phase path, thus increasing the basin of attraction of the desired target state.

Anchoring is an extremely versatile intervention tool. Used with creativity it can be designed to increase or decrease neurocognitive entropy. Anchoring can be used for broad-scope interventions, such as specifying a theta wave attractor landscape or causing sustained phase transition (Chase, 1974). It can also be used for narrow-scope applications, such as specifying an attractor, state-bound behavior, mode, or mode-bound behavior (Honig and Fetterman, 1992). For this reason we will devote the remaining chapters to making some highly refined distinctions about various ways in which anchors can be configured for different purposes.

REFERENCES

Ader, R., Felten, D. L., and Cohen, N. (1991). *Psychoneuroimmunology (2nd Edition)*. San Diego: Academic Press.

Bechterev, V. M. (1928). *General Principles of Human Reflexology: An Introduction in the Objective Study of Personality*. New York: International.

Bellack, A. S., Hersen, M., and Kazdin, A. E. (Eds.). (1990). *International Handbook of Behavior Modification and Therapy (2nd Edition).* New York: Plenum Press.

Cautela, J. R. and Kearney, A. J. (1986). *The Covert Conditioning Handbook.* New York: Springer.

Chakotin, S. (1940). *The Rape of the Masses: The Psychology of Totalitarian Political Propaganda.* New York: Fortean Society.

Chase, M. (Ed.). (1974). *Perspectives in the Brain Sciences (Volume 2): Operant Control of Brain Activity.* Los Angeles: University of California, Brain Information Service/Brain Research Institute.

Damasio, A. R. (1994). *Descartes' Error: Emotion, Reason, and the Human Brain.* New York: G.P. Putnam's Sons.

Furman, M. E. (1995). Life, death and anchoring. *Anchor Point*, December, 36–39.

Goldman-Rakic, P.S. (1987). Circuitry of primate prefrontal cortex and regulation of behavior by representational memory. In *Handbook of Physiology*, Mountcastle, V. B., Plum, F., and Geiger, S. R. (Eds.), New York: Oxford University Press, 5(1), 373–417.

Goldman-Rakic, P.S. (1992). Working memory and the mind. In *Scientific American.* New York: W. H. Freeman and Co.

Hayakawa, S. I. and Hayakawa, A. R. (1990). *Language in Thought and Action (5th Edition).* New York: Harcourt Brace & Company.

Honig, W. K. and Fetterman, J. G. (Eds.). (1992). *Cognitive Aspects of Stimulus Control.* Hillsdale, NJ: Lawrence Erlbaum Associates.

Kandel, E. R., Schwartz, J. H., and Jessell, T. M. (Eds.). (1991). *Principles of Neural Science (3rd Edition).* East Norwalk, CT: Appleton & Lange.

Kuno, M. (1995). *The Synapse: Function Plasticity, and Neurotrophism.* New York: Oxford University Press.

Salter, A. (1949). *Conditioned Reflex Therapy: The Direct Approach to the Reconstruction of Personality.* New York: Creative Age Press.

Sechenov, I. M. (1965). *Reflexes of the Brain: An Attempt to Establish the Physiological Basis of Psychological Processes.* Cambridge, MA: MIT Press.

18 Neurophysics and the Principles of Anchoring

CONTENTS

WHAT ARE ANCHORS?

Anchors are information or pattern. In a most fundamental sense anchors are news of difference or change — configurations of matter and energy in an information field that specify a particular target for the state vector of a neurocognitive system. Anchors are *informed* matter and energy that, when incorporated by a neurocognitive system, specify the initial conditions from which that system is released, resulting in qualitative differences in the behavior produced by the system as well as the information available to it (Haken, 1988). Anchors can significantly alter cognitive neurodynamics by modifying the stability distribution of attractors and trajectories, and by rearranging the phase path of the state vector through design space. Anchors can be configured to facilitate incorporation of other patterns. They can be used to replicate, cleave, recombine, and transmit pattern throughout the neurocognitive system. Anchors can be used as sensory markers — *neurocognitive bookmarks* for experience, memories, desired states, state-bound and mode-bound behaviors, and thoughts.

CLASSES OF ANCHORS

In order to develop a deeper understanding of how anchors can be configured for purposes of intervention, it is useful to make even finer distinction than we have thus far. In this chapter we divide anchors into several distinct classes.

SIGNAL/SELECTION ANCHORING

A *signal anchor* acts as the initial conditions necessary to release the state vector of the neurocognitive system on a trajectory to a desired target. In this regard, a signal anchor can select a particular neurocognitive target for its action (attractor, attractor landscape, state/phase, state-bound, or mode-bound behavior thought, memory, etc.) (Haken, 1988). For this reason it is sometimes referred to as a *selection anchor*. But thinking of anchors only in this way imposes severe limitations on the flexibility that anchoring could otherwise provide.

Another common limitation is that anchoring is normally taught and utilized explicitly through the kinesthetic/tactile system, employing only one of several possible sensory systems that exist both externally and internally in the human nervous system. Even though visual and auditory anchors are used implicitly, the principles and conditions for their accurate development have not as yet been formally arranged in a teachable and reproducible form. Signal or selection anchors are one of two major classes of anchors (Bechterev, 1928; Pavlov, 1927). The other is referred to as *confirmation* or *reinforcement anchors* (Hull, 1943, 1952; Skinner, 1938). We will discuss the various classes of signal/selection anchors first.

APPEARANCE ANCHORS

Appearance anchors can be considered as any set of initial conditions that can be perceived by any of the internal or external sensory systems of the brain. In short, any agent that can be cleanly differentiated by a sensory receptor can act as an appearance anchor. In NLP training, appearance anchors are the only major class of anchors that are dealt with explicitly. In all cases, it is defined as the appearance of some object, sound, touch, or chemical to the sensory receptors that clearly specify a particular target attractor, landscape, or phase path.

It is important to note that there are more than five sensory systems that are capable of responding to appearance anchors (Haken, 1988). One of them is the proprioceptive sense, which keeps track of the position and velocity of limbs in space. Another is the vestibular system, which keeps track of balance and position of the body with respect to gravity (Kandel et al., 1991).

DISAPPEARANCE ANCHORS

An entire class of anchors, almost as vast and versatile as appearance anchors, is *disappearance anchors*. Disappearance anchors occur everywhere in life but they are rarely seen as anchors (Pavlov, 1927). As such, they become an illusive and covert method of anchoring. One example of a disappearance anchor involves the feeling of homesickness that a person might experience while away from home. In this case the initial condition specifying the homesickness attractor is not an object that can be seen, a sound, or a touch, but instead the disappearance of possibly any or all of those things.

Another example of this is the state that occurs when you hear your car door slam firmly behind you, you look in your hand, and you notice your keys are not there. Here, again, it is not the sound of the door alone that is the initial condition necessary to lead to that physiological state attractor of panic. The initial conditions necessary to enter that state are totally dependent upon the sound plus the disappearance of the keys from their expected location — your hand. This particular type of anchor is also referred to as a *complex* or *compound bi-modal* anchor (i.e., affecting two sensory modes). This compound anchor requires both conditions to be met in order to trigger the state of panic. Condition 1 is the *auditory appearance* of the sound of the door locking behind you. Condition 2 is the *visual disappearance* of the keys from their expected place — your hand.

Trace Anchors

A *trace anchor* is possible when a stimulus is very strong. The strength of the stimulus creates an ongoing oscillation inside the nervous system. This sustained trajectory can itself become the initial condition necessary to trigger a specific physiological state or behavior (Hull, 1943, 1952; Pavlov, 1927, 1928, 1941). One way to notice whether the stimulus or the trace of the stimulus is the anchor is to track the period of delay between the strong stimulus and the response. If the delay is long, then the trace is the necessary initial condition. If the delay is short, then the strong stimulus itself is the initial condition. Trace anchors are sometimes referred to as *paused* or *delayed* anchors.

Time Interval Anchors

The basal ganglia of the human brain coordinate many functions by sequencing sensory motor activity. They also have a function by which the brain can keep track of time intervals. Dopamine is utilized by this complex circuit in much the same way that sand is used in an hourglass (Houk et al., 1995). As a result of this precise function in the brain, time itself can be used as an anchor for any physiological state or behavior desired (Pavlov, 1927).

One subtle example of this in daily life is the experience of setting an alarm clock for 7:00 a.m. and finding yourself wide-awake at 6:59 a.m. In this case, a time interval itself was responsible for shifting the neurobiological dominance of the brain stem from cholinergic to aminergic activity (Hobson, 1994; Pavlov, 1927), therefore transitioning the neurocognitive system from sleeping to waking state (the most fundamental of all physiological brain-state transitions). Another example of this in everyday life is when you find your stomach grumbling for food every day at a specific time even without looking at a clock. In this case, the time interval triggers digestive system messenger molecules that result in feeling hungry, rather than an external stimulus such as the smell of food, the sound of a Burger King jingle, or the site of a hamburger.

Time interval anchors provide the practitioner with a distinct intervention advantage over the other types of anchors previously mentioned. The modifications to the neurocognitive system they engender are not dependent upon an external information field to maintain the changes made. Select regions of design space can be continually stabilized or destabilized by a precise time interval anchor that is carried around within the subject's neurocognitive system, independent of a specially arranged external stimulus field.

Complex/Compound Anchors

Compound or *complex anchors* utilize either more than one sensory system's receptor circuits and/or more than one class of anchors (Hull, 1943, 1952). One of the most important reasons to utilize compound anchors is to maintain uniqueness of the initial conditions so that the nervous system can easily differentiate between signals that select or specify particular attractors or attractor landscapes. The greater the differentiation, the greater the number of possible attractors that can be cleanly specified without interference. Utilization of compound anchors allows us to have far greater flexibility.

Another advantage of utilizing compound anchors is that when an anchor is designed to utilize a sensory system specifically, one can eliminate uncontrollable interference that might have occurred in that sensory system, which would normally occlude the accuracy and weaken the effect of the anchor (Haken, 1988; Bechterev, 1928). One example of a *compound signal anchor* naturally occurring every day would be the sound of someone's voice combined with the facial expression accompanying his or her words. There are many physiological states that can only be triggered by the combination of a facial expression and a certain voice tone. We come to expect these two things to occur together. As a result, comedians are able to generate uproarious laughter by making use of this phenomenon by combining a facial expression with an unexpected voice tone or vice versa.

Compound or complex anchors can be designed for two or more sensory system encoding circuits (bi-modal, tri-modal), as well as for two or more different classes of anchors. Any combination of appearance, disappearance, trace, and time interval anchors can be considered a complex or compound anchor giving rise to great versatility and specificity.

Analogical Variation and Threshold Anchors

Variation anchors depend upon submodal behaviors (submodalities). This is another very rich source of anchors. A variation anchor is any external or internal control parameter that can be scaled analogically for submodalities such as brightness, speed, size, volume, temperature, etc. Within the design of a variation anchor, one of two things can be used to specify a target attractor — either the variation of the stimulus itself or the reaching of a specific threshold for that stimulus (Pavlov, 1927, 1928, 1941).

Threshold anchors are a subclass of variation anchors (Haken, 1988). We have all experienced physiological states and behaviors that are dependent upon threshold.

A common example of a variation anchor is the luminescence of a room. As we scale the light of a room from very bright to nearly that of candlelight, one may become romantic. Thus, either the variation itself or a certain threshold is recognized by the nervous system as the initial condition necessary to specify the state/phase attractor of "romantic." Variation and threshold anchors are also highly prevalent in the parts of the nervous system that transmit pain messages.

Rate of Change/Acceleration Anchors

An *acceleration anchor* is another subclass of variation anchors. In this case the initial condition that specifies a particular attractor is the rate of change of a pattern, not its velocity alone (Haken, 1988). For example, a person walking alone on a quiet, dark street might be in a calm emotional state when the sound of distant footsteps behind him match that of his own, and a state of fear if the footsteps rapidly accelerate in relationship to his own footsteps. In this case, the accelerating footsteps may in turn also specify the state-dependent behavior of running for safety.

This same person may experience no major change in neurocognitive dynamics in the case where footsteps are heard at a constant velocity, even if that velocity is significantly greater than his own footsteps. A third possibility is that the same neurocognitive system's state vector may only be captured by the fear attractor if a certain variation in the pattern exists, such as the sound of the footsteps moving toward or away from him, or if a particular threshold is met, such as the proximity of the sound of the footsteps in relationship to his own.

Still other possibilities exist in this same scenario. The state vector may only be captured by the fear attractor in the event that the footsteps are heard for a certain time interval, matching that of a stored time interval (previously encoded during a traumatic experience) that reliably acts as the initial conditions to the memory (attractor) of a previous crime-related, traumatic experience.

In order to understand the structure of a problem or to craft an effective intervention, it is critical that the practitioner make a clear distinction as to the exact initial conditions (type of pattern) necessary to specify a particular attractor affecting the quality of functioning of the neurocognitive system. These initial conditions may include the appearance, disappearance, time interval, variation, velocity, or the acceleration of a pattern in the information field that directs the state vector.

It is certainly possible for the initial conditions to be a compound of two or more types of these anchors. In such a case the state vector may move from the calm attractor to the caution attractor when the person hears the footsteps moving toward him (variation) for a particular period of time (time interval). From the state of caution, the state vector may trigger the state-dependent behavior of checking for car keys in the attempt to find a safe place quickly. At this point, the absence of the car keys from their familiar place (disappearance anchor) may force the state vector to be captured by the fear attractor. This in turn could cause the state vector to initiate the state-dependent behavior of walking fast. If the person then perceives the footsteps to accelerate (acceleration anchor) in relationship to his own, the state vector may then be captured by the terror attractor, perhaps initiating the state-dependent behaviors of running, crying, and screaming. In this example, note that all of the anchors that specify phase path are submodal.

For an intervention in this case to be successful, it requires that the practitioner cleanly differentiate the anchors. To know that a particular subject relives a traumatic experience intermittently when he hears footsteps is worthless. We must be able to distinguish the precise initial conditions that cause the neurocognitive system to transition from one state to another or from one state-bound behavior to another. This is important because regions of phase path making up the strategy must remain in place or be designed to be more adaptive, while other regions of phase path that are unnecessary for the adaptive response to a real problem should be destabilized.

Inhibitory Anchors

Inhibitory anchors are a class of anchors (patterns) whose neural activity interferes with that of another signal/selection anchor (Bechterev, 1928; Dagenbach and Carr, 1994; Dempster and Brainerd, 1995; Hull, 1943, 1952; Luria, 1932; Pavlov, 1927, 1928, 1941; Salter, 1949). Inhibition anchors can be external or internal. To create an external inhibition anchor, we can anchor an incompatible state or utilize a particularly strong stimulus. In this way it is possible to inhibit or collapse the attractor prescribed by the signal anchor. An internal inhibition anchor is effective when a signal anchor no longer leads the state vector to its previously linked response (attractor). Inhibition anchors are used to cleave and rearrange specified regions of phase path. In this case, the original signal anchor itself becomes the inhibition anchor, permanently specifying a change in phase path (Salter, 1949). All other anchors discussed prior to this point are considered excitatory (Pavlov, 1927).

Embedded Anchors

An *embedded anchor* is a signal anchor that triggers a state or behavior specified by a *linguistic embedded command.* This type of anchoring structure is very useful for situations that call for covert anchors and undetectable influence. An embedded command (not consciously recognized) can be attached to a signal anchor, which consistently triggers its performance (Benson, 1994; Cautela and Kearney, 1986; Kazdin, 1994; Kissin, 1986). Embedded anchors work particularly well with metaphor and therapeutic trance.

Confirmation/Reinforcement Anchors

Confirmation or *reinforcement anchors* differ from signal anchors in that they do not set the state vector on a trajectory to a specific state or behavior, or other attractor. Instead, they confirm or reinforce a specific state, behavior, or other attractor that the state vector has been captured by, thereby increasing the possibility of reoccurrence by reinforcing it (Hull, 1943; Skinner, 1938). The confirmation/reinforcement anchor specifies a particular attractor or attractor landscape that the subject finds pleasurable. By continually directing the state vector to this pleasurable attractor or landscape directly after being captured by the target attractor (in most cases a behavior or thought), the frequency of recurrence for the target attractor is increased. This is known as instrumental or operant conditioning when macroscopic behaviors are being reinforced/stabilized (Hull, 1943; Kazdin, 1994; Pryor, 1995; Skinner, 1953).

On a molecular level, many confirmation anchors utilize the dopamine reward system of the brain to increase the likelihood that a particular nerve cell assembly (attractor) will become strengthened and reoccur (Haken, 1988, 1996; Kandel and Hawkins, 1992; Kazdin, 1994; Kosslyn, 1994; Partridge and Partridge, 1993; Pryor, 1995).

Timing is the critical factor that differentiates the function of a specific stimulus in the nervous system. If the stimulus starts before and overlaps with a given state or behavior, it is a signal/selection anchor. If the stimulus occurs only during the state or behavior (specifying a pleasure attractor or landscape), it is a confirmation/reinforcement anchor. The stimulus must begin and end very rapidly and leave little or no physiological trace in order to effectively mark a state, behavior, or other attractor for reoccurrence; otherwise, it may be inhibitory — perturbing rather than stabilizing the target attractor. While the main function of an inhibitory anchor is to destabilize and cleave, the main function of a signal/selection anchor or conformation anchor is to stabilize. The three types can be effectively used together to cleave and recombine specified regions of phase path in design space.

PRINCIPLES AND CONDITIONS

There are five basic principles and conditions for the effective establishment of an anchor. They are as follows:

1. *Stimuli Indifference*. In order to be effective, the information pattern we wish to transform into an anchor must not already prescribe a particular state, behavior, or other attractor. If it does, it will become an inhibition anchor rather than a signal/selection anchor (Bechterev, 1928). Effective anchors are those that are novel to the nervous system when first specifying an attractor. The exception to this is when a strong stimulus or response is involved, which may simultaneously countercondition another relationship between attractors resulting in a destabilization of the previous phase path. Strong stimuli and strong responses can sometimes have a cleaving effect (Dagenbach and Carr, 1994; Dempster and Brainerd, 1995).
2. *Health*. The health of the brain and connecting biophysical system is critical in the formation of activity-dependent synapses underlying stable changes in the neurocognitive system. Activity-dependent patterns of neuronal connections underlie the formation of all physiological states, behaviors, and thoughts. If the state of the brain or body is not one of good health, it will take many more repetitions of stimulus presentation to create an effective and lasting link (Pavlov, 1927).
3. *Interference Free*. Foreign stimuli (i.e., uncontrolled stimuli in the information field) can and will interfere with the accurate and consistent accessing of the desired target state, behavior, or other specified attractor.

Complex or compound anchors that utilize some or all of our sensory system's encoding circuits (modes) will substantially eliminate this interference (Bechterev, 1928; Dagenbach and Carr, 1994; Dempster and Brainerd, 1995; Pavlov, 1927).

4. *Brain State*. Less repetitions will be required to couple stimuli with target attractors when the brain state is alert. Use of confirmation/reinforcement anchors increases alertness by creating a state of reafference — an increased blood flow to the early sensory cortices of the brain resulting in faster and more accurate coupling (Fuster, 1995; Hobson, 1994; Kandel et al., 1991; Nunez, 1995; Roland, 1993; Rugg and Coles, 1995).
5. *Timing*. A stimulus that begins before and overlaps with a particular state or behavior is transformed into a signal/selection anchor. A stimulus that occurs only during a specified state, behavior, or thought is transformed into a confirmation/reinforcement anchor. The selected attractor can be increased (stabilized) or decreased (destabilized) in frequency by pairing the selection anchor with either reinforcing or aversive stimuli, respectively (Pavlov, 1927).

The scope of this book can only allow for a preliminary discussion of the vast science and art form of anchoring. Thus far we have only laid out some major classes of anchors as well as some basic principles and conditions of the process of anchoring. Mastery of this skill must also take into account procedures such as *transferring, fading, collapsing,* and *chaining* (states, behaviors, thoughts, modes, etc.), to name just a few. Each one of these procedures has a unique set of principles and conditions which affect its neurophysiological outcome. Both neurophysics and neurophysiology have shed light on thousands of new ways that these classes of anchors can be used to engineer change.

Anchors can be used to eliminate trauma, improve memory and learning, maintain health, facilitate change, influence and persuade, transfer models of human excellence, and more. In Chapter 19, we shall discuss some anchoring traps that can be avoided with some basic knowledge of nonlinear dynamical systems and neurophysics.

REFERENCES

Baddeley, A. (1986). *Working Memory: Oxford Psychology Series — II*. New York: Oxford University Press.

Bechterev, V. M. (1928). *General Principles of Human Reflexology: An Introduction in the Objective Study of Personality*. New York: International.

Bellack, A. S., Hersen, M., and Kazdin, A. E. (Eds.). (1990). *International Handbook of Behavior Modification and Therapy (2nd Edition)*. New York: Plenum Press.

Benson, D. F. (1994). *The Neurology of Thinking*. New York: Oxford University Press.

Cautela, J. R. and Kearney, A. J. (1986). *The Covert Conditioning Handbook*. New York: Springer.

Chakotin, S. (1940). *The Rape of the Masses: The Psychology of Totalitarian Political Propaganda*. New York: Fortean Society.

Dagenbach, D. and Carr, T. H. (Eds.). (1994). *Inhibitory Processes in Attention, Memory, and Language*. San Diego: Academic Press.
Dempster, F. N. and Brainerd, C. J. (Eds.). (1995). *Interference and Inhibition in Cognition*. San Diego: Academic Press.
Furman, M. E. (1996). Neurophysics and the principle of anchoring. *Anchor Point*, June, 30–40.
Fuster, J. M. (1995). *Memory in the Cerebral Cortex: An Empirical Approach to Neural Networks in the Human and Nonhuman Primate*. Cambridge, MA: MIT Press.
Haken, H. (1988). *Information and Self-Organization: A Macroscopic Approach to Complex Systems*. New York: Springer-Verlag.
Haken, H. (1996). *Principles of Brain Functioning: A Synergetic Approach to Brain Activity, Behavior and Cognition*. New York: Springer-Verlag.
Hayakawa, S. I. and Hayakawa, A. R. (1990). *Language in Thought and Action (5th Edition)*. New York: Harcourt Brace & Company.
Hobson, J. A. (1994). *The Chemistry of Conscious States: How the Brain Changes its Mind*. New York: Little, Brown & Company.
Houk, J. C., Davis, J. L., and Beiser, D. G. (1995). *Models of Information Processing in the Basal Ganglia*. Cambridge, MA: MIT Press.
Hull, C. L. (1943). *Principles of Behavior: An Introduction to Behavior Theory*. New York: Appleton-Century-Crofts.
Hull, C. L. (1952). *A Behavior System: An Introduction to Behavior Theory Concerning the Individual Organism*. New Haven: Yale University Press.
Kandel, E. R. and Hawkins, R. D. (September 1992). The biological basis of learning and individuality. In *Scientific American*. New York: W. H. Freeman and Company.
Kandel, E. R., Schwartz, J. H., and Jessell, T. M. (Eds.). (1991). *Principles of Neural Science*, 3rd ed. Norwalk, CT: Appleton & Lange.
Kandel, E. R., Schwartz, J. H., and Jessell, T. M. (Eds.). (1995). *Essentials of Neural Science and Behavior*. Norwalk, CT: Appleton & Lange.
Kazdin, A. E. (1994). *Behavior Modification in Applied Settings*. Pacific Grove, CA: Brooks/Cole.
Kelso, J. A. S. (1995). *Dynamic Patterns*. Cambridge, MA: MIT Press.
Kissin, B. (1986). *Psychobiology of Human Behavior (Volume I): Conscious and Unconscious Programs in the Brain*. New York: Plenum Press.
Kosslyn, S. M. (1994). *Image and Brain*. Cambridge, MA: MIT Press.
Kruse, P. and Stadler, M. (1995). *Ambiguity in Mind and Nature*. New York: Springer-Verlag.
Luria, A. R. (1932). *The Nature of Human Conflicts: An Objective Study of Disorganization and Control of Human Behavior*. New York: Grove Press.
Nunez, P. L. (1995). *Neocortical Dynamics and Human EEG Rhythms*. New York: Oxford University Press.
Paillard, J. (1991). *Brain and Space*. New York: Oxford University Press.
Partridge, L. D. and Partridge, L. D. (1993). *The Nervous System: Its Function and Its Interaction with the World*. Cambridge, MA: MIT Press.
Pavlov, I. P. (1927). *Conditioned Reflexes*. New York: Oxford University Press.
Pavlov, I. P. (1928). *Lectures on Conditioned Reflexes (Vol. I): Twenty-five years of Objective Study of the Higher Nervous Activity (Behavior) of Animals*. New York: International.
Pavlov, I. P. (1941). *Lectures on Conditioned Reflexes (Vol. II): Conditioned Reflexes in Psychiatry*. New York: International.
Pribram, K. H. (1991). *Brain and Perception: Holonomy and Structure in Figural Processing*. Hillsdale, NJ: Lawrence Erlbaum Associates.
Pryor, K. (1995). *On Behavior: Essays & Research*. North Bend, WA: Sunshine Books.

Roland, P. E. (1993). *Brain Activation*. New York: Wiley-Liss.

Rugg, M. D. and Coles, M. G. (1995). *Electrophysiology of Mind: Event-Related Brain Potentials and Cognition*. New York: Oxford University Press.

Salter, A. (1949). *Conditioned Reflex Therapy: The Direct Approach to the Reconstruction of Personality.* New York: Creative Age Press.

Sechenov, I. M. (1965). *Reflexes of the Brain: An Attempt to Establish the Physiological Basis of Psychological Processes*. Cambridge, MA: MIT Press.

Skinner, B. F. (1938). *The Behavior of Organisms: An Experimental Analysis*. New York: Appleton-Century-Crofts.

Skinner, B. F. (1953). *Science and Human Behavior*. New York: Free Press.

19 Neurophysics, Self-Organization, and Anchoring

CONTENTS

A BRIEF HISTORY OF ANCHORING

The first known paper giving a more or less systematic description of "natural" conditioned reflexes, referred to as *psychic reflexes* at the time, was published by Wolfson in the form of a thesis entitled "Observations upon Salivary Secretion" (Petrograd, 1899). Tolochinov, who completed his experiments in 1901 and communicated the results at the Congress of Natural Sciences in Helsingfors in 1903, first used the term "conditioned reflex" in print. Of course the story goes back long before the turn of the 20th century; however, Wolfson, Tolochinov, and their huge team of research scientists were the first to formalize this discovery.

At this time, Ivan Pavlov was a physiologist interested in the physiology of digestion. He found the work done by Wolfson and Tolochinov so fascinating that he devoted the rest of his life to the research of conditioned reflexes. Pavlov won the Nobel Prize in 1904, and finally in the spring of 1924 he delivered a series of lectures at the Military Medical Academy in Petrograd to an audience of medical practitioners and biologists. These lectures were taken down by a stenographer and subsequently published. Pavlov admitted at that time that the original idea of the reflex, as he knew it, had evolved 300 years before him by the work of Rene Descartes.

Descartes' conception of the reflex stopped short at the cerebral cortex where Pavlov's began. Pavlov's belief was that psychological states such as playfulness, fear, and anger could be demonstrated as reflex activities of the subcortical parts of the brain. A bold attempt to apply the idea of the reflex to the activities of the hemispheres was made by Russian physiologist Ivan Mikhailovich Sechenov in his pamphlet entitled *Reflexes of the Brain*, published in Russian in 1863. Sechenov, considered the founder of Russian physiology, strongly influenced the life's work of his devoted pupil Pavlov. All major landmarks in the history of world and Russian physiology are closely linked with Sechenov's historic research and theories. He was the first to advance the idea that the initial cause of any action always lies in external sensory stimulation, and that without such stimulation, thought and psychological life as we know it would be inconceivable. He further posited that the initial cause of any human action actually lies outside the person, that all acts of conscious and unconscious life are reflexes.

It is from this notion investigated by Sechenov and Pavlov that the NLP concept of anchoring was born. While the intention of Pavlov's lectures was to enlighten the scientific medical community as to the extraordinary positive value of his findings to humanity, this valuable information inadvertently slipped into the hands of a few whose intentions were ill fated.

Behavioral science, NLP, and psychoneuroimmunology have been the first to implement this technology for the intention of healing since 1924. Each of these fields has applied the technology to a limited extent. Behavioral science applies it predominantly to the sensory motor system. NLP applies these principles predominantly to the autonomic nervous system and neuroendocrine system. And psychoneuroimmunology targets the molecular behavior of the neuroimmune system. All of these fields make use of the unspoken rules of neurocybernetics when their applications are successful.

ASSEMBLING ANCHORS FOR SPECIFIC APPLICATION

In the previous chapter, we discussed various types of anchors such as signal/selection anchors, confirmation/reinforcement anchors, inhibition anchors, etc. In order to effect lasting change, all these types of anchors must be combined and sequenced with respect to the principles and conditions that govern their establishment. Let us explicate some correct and incorrect uses of these intervention tools.

CHAINING ANCHORS

The phenomenon of chaining was first discovered by Pavlov (1927), and then extensively used by Bechterev (1928), Salter (1949), Hull (1943, 1952), and Skinner (1938, 1953), in order to establish various types of response chains. Chaining was later adopted by the field of NLP for use in chaining (ordering) cognitive behaviors sequentially at the mode level (strategies) as well as for physiological states.

Chaining has been found to be an excellent way for creating a phase path between two radically different physiological states. This procedure facilitates smoother transition between present and desired target attractors or landscapes. In addition to being one of the most commonly documented applications available in NLP literature, with the aid of a NeuroPrint this chaining procedure provides the practitioner with an effective way of rearranging the phase path of the neurocognitive system.

One area of applied value is that of the installation of *behavioral chains* or *strategies*. For this type of installation to be successful, a minimum of three conditions must be present:

- A signal/selection anchor must be established, specifying each behavior required in the chain.
- A confirmation anchor, which could be the same for each behavior in the chain, must be established to increase response potential (frequency of recurrence).
- The behavioral chain must be built backward or it will fall apart. The chain or strategy must be designed such that the highly established or stable behavior always follows a less well-established or unstable behavior, thus acting as a confirmation/reinforcement anchor for the weaker behavior. This condition creates a self-reinforcing behavioral chain or strategy, such that removal of the first confirmation/reinforcement anchor leaves the strategy (behavior chain) intact. Installation of a strategy that falls apart suggests that it probably did not meet one or more of these conditions.

Chaining is a very effective way to stabilize a series of states/phases, modes, state-bound or mode-bound behaviors, and thoughts along a phase path. This is one of the mechanisms responsible for the formation of Markov chains.

Generalization vs. Discrimination

Much of the behavior modification literature discusses the distinction between generalization and discrimination. What does getting something to generalize actually mean? And how is generalization accomplished using anchors? This is an important distinction to make when designing a neurocognitive intervention.

Generalization is the tendency for a learned response to occur in the presence of stimuli that were not present during the incorporation of the new pattern. Thus more than one stimulus in the information field can act as the initial conditions necessary to specify that attractor or landscape. This type of cognitive design can stabilize a new attractor quickly (Hull, 1943; Skinner, 1938).

Conversely, *discrimination* is the tendency for a learned response to occur only in the presence of stimuli presented during the initial incorporation of the pattern and not in the presence of stimuli absent during the incorporation. This means that the state vector will only be captured by the target attractor when a precise set of initial conditions is met (Hull, 1943; Skinner, 1938).

If we intend for a new behavior or state to generalize, we must use a confirmation/reinforcement anchor when the desired response occurs in the absence of the original signal/selection anchor that established the response. If we want the new behavior or state to occur only at specific times or in specific situations, we must utilize a confirmation/reinforcement anchor only when the desired response occurs in the presence of the signal anchor. This procedure guarantees that the new behavior or state remains under stimulus control from the information field (Honig and Fetterman, 1992).

From the point of view of neurophysics, when a behavior or state generalizes, *the basin of attraction is widened*. Conversely, when we *discrimination-train* a behavior or state, we are *narrowing the basin of attraction*. Posthypnotic suggestion, which is a form of future pacing in NLP, narrows the basin of attraction and brings the new behavior or state under stimulus control (discrimination).

In posthypnotic suggestion, a signal/selection anchor is proposed verbally. This is a higher-order form of anchoring in that both the stimulus itself (first order) and the verbal instruction (second order) are now capable of eliciting the response (Pavlov, 1927; Skinner, 1957). For example, if you were told (convinced) that reading this chapter will make you curious about the potential advantages of anchoring for your next client's intervention plan, both the visual stimulus of reading this chapter (first order) and the verbal instruction (second order) have the capability of eliciting the state of curiosity (Platonov, 1959).

Stacking Anchors

Stacking is normally described as a procedure by which a signal anchor is created, which triggers several different emotional states to be used later as a powerful resource. The instructions involve anchoring many resource states with the same signal anchor, assuming that the signal anchor will become increasingly more powerful with each added resource. In fact, this procedure creates just the opposite. Neurophysiologically, a signal anchor that specifies more than one physiological state will cause *internal inhibition, response suppression*, and finally *extinction* of the states we are attempting to stack (Bechterev, 1928; Dagenbach and Carr, 1994; Dempster and Brainerd, 1995; Pavlov, 1927, 1928, 1941; Salter, 1949). This accounts for why a stacked anchor weakens rather than strengthens over time. Each physiological state, which is triggered by the identical signal/selection anchor, competes for territorial representation in the association cortices, to become the *order parameter* of the neurocybernetic system. The state that wins control is always the state with the greatest rate of increase. Eventually even that state becomes inhibited by the competition of the others.

Stacking is a procedure that is most well suited to cause a learned behavior or state to generalize. However, the procedure must be modified in the following way in order to be effective. To get something to generalize, we must first of all increase the number of signal anchors capable of eliciting the response, not the number of responses capable of being elicited by the signal anchor. Second, we must utilize confirmation/selection anchoring when the state or behavior is occurring in the absence of the original signal anchor. In this way we are able to widen the basin of attraction for a new state or behavior, thus causing it to generalize.

Collapsing Anchors

The last procedure we will discuss is the proper use of *collapsing anchors*. This notion was the largest and most detailed topic discussed by Pavlov in his lectures to the medical and scientific community on *conditioned inhibition* (Dagenbach and Carr, 1994; Dempster and Brainerd, 1995; Luria, 1932; Pavlov, 1927, 1928, 1941).

The topic became extensive due to the fact that there were more conditions necessary to inhibit a previously conditioned reflex than for the formation of any other procedure. Much of his efforts were trial and error because he did not have the advantage of imaging neurons at work at the molecular level, as we do today.

The first type of conditioned inhibition is similar to what we refer to today as *pattern interruption*. The method consists of utilizing a signal anchor to trigger a state or behavior and interrupting its activation pattern (replication) with a strong sensory stimulus. Those familiar with EMDR may notice striking similarities. This type of anchor is used for destabilizing an attractor.

Another type of conditioned inhibition requires a strong stimulus to be presented first, and the anchored state or behavior to be inhibited must be triggered second. In this case, the trace activity of the first sensory stimulus interrupts spontaneous formation of the previously conditioned state, behavior, or thought attractor. This type of anchor is used when cleaving a trajectory between a stimulus and the attractor it specifies.

A third major type of conditioned inhibition was referred to by Pavlov as *internal inhibition*. The conditions for this are simple. If a signal anchor that normally triggers a state or behavior is prevented from triggering its preconditioned response, each time the anchor is fired the activation pattern undergoes further *conditioned suppression,* until finally *extinction* occurs. This case is particularly interesting in that the anchor, which originally elicited and strengthened a given response, now is responsible for suppressing the response. This type of anchoring is used to cleave a trajectory between a stimulus and an attractor by simply presenting the stimulus several times, each time preventing the state vector from being captured by the attractor it normally specifies.

Once established, an anchor must reliably produce the intended response every time or suppression begins to occur. By acting well within the guidelines of the explicit principles and conditions necessary to form various types of anchors, it is possible to control exactly how long an anchor will last and whether or not it has the capability of influencing the neurocognitive system without the practitioner being present.

The most important notion that needs to be dispelled is that by collapsing anchors the two anchored states are caused to integrate. This is not true. Try this experiment for verification: Anchor one visual perception of a *Necker cube* tactilely on your right shoulder and the opposite perception of the Necker cube on your left shoulder. Once each anchor produces a reliable shift in visual perception, use both anchors simultaneously. If integration were possible in the nervous system, you would look at a Necker cube from now on and see both perceptual positions simultaneously or an integration forever. However, this is not the case. Instead what you will find is that each visual perceptual state will compete for dominance and will oscillate back and forth until the perceptual state with the greatest rate of increase (greatest phase velocity) finally dominates.

Collapsing anchors results in either temporary or permanent inhibition of one of the two states, this being dependent upon conditions such as timing, pairing order, number of presentations, intertrial interval, and rate of increase (phase velocity) of each anchored state, and is subject to spontaneous recovery (regulation of the morphogenetic field) if done incorrectly. This type of anchor is used to create a broad scope increase in entropy or when destabilizing both of the attractors specified.

Collapsing anchors and conditioned inhibition can also be accomplished in other creative ways. Pavlov experimented for many years with the idea of establishing a *compound anchor* and then producing conditioned inhibition by extracting one sensory component (mode) of the original compound anchor to produce inhibition.

Sensitization, Pseudoconditioning, and Backward Conditioning

Many practitioners who are new to the notion of anchoring find that they have difficulty in establishing effective anchors that last. This is most commonly due to errors in timing and pairing. It is possible for a stimulus to temporarily elicit a desired response (specify a target attractor) even though it has not become an effective anchor. This is referred to as *pseudoconditioning*, made possible by the nervous system phenomenon of *sensitization* (Hull, 1943; Pavlov, 1927; Skinner, 1938). Sensitization is the tendency of a stimulus to elicit a reflex response following the elicitation of that response by a different stimulus. Thus the linguistic elicitation (auditory symbolic selection anchor) of a physiological state can be incorrectly paired with a touch (tactile selection anchor) on the shoulder and still temporarily produce the kinesthetic state even though it is incorrectly timed.

It is also possible for an anchor to temporarily trigger a desired state via *backward conditioning* (Hull, 1943). This occurs when an already occurring state is anchored, thus perturbing the state vector instead of specifying it. In this case, the state precedes the anchor by just fractions of a second. Under laboratory conditions, Pavlov attempted to backward condition salivation using 427 trials before concluding that the resulting conditioned reflex could only be a temporary one, due to the fact that the stimulus actually interrupted and inhibited the very state it was supposed to anchor (Pavlov, 1927).

The importance of understanding anchors extends beyond intervention flexibility. Anchors hold regions of design space together. Understanding how they work, how they stabilize particular neurocognitive patterns, as well as being able to identify their effect on cognitive neurodynamics, is essential for successful cognitive intervention planning. It will be efficacious for the practitioner to utilize the information discussed in the last three chapters during the final steps of NeuroPrint elicitation and diagramming when identifying control parameters in the information field that affect cognitive neurodynamics.

REFERENCES

Baddeley, A. (1986). *Working Memory: Oxford Psychology Series — II.* New York: Oxford University Press.

Bechterev, M. (1928). *General Principles of Human Reflexology: An Introduction to the Objective Study of Personality.* New York: International.

Bellack, A. S., Hersen, M., and Kazdin, A. E. (Eds.). (1990). *International Handbook of Behavior Modification and Therapy (2nd Edition).* New York: Plenum Press.

Benson, D. F. (1994). *The Neurology of Thinking*. New York: Oxford University Press.

Cautela, J. R. and Kearney, A. J. (1986). *The Covert Conditioning Handbook*. New York: Springer.

Chakotin, S. (1940). *The Rape of the Masses: The Psychology of Totalitarian Political Propaganda*. New York: Fortean Society.

Chance, P. (1994). *Learning and Behavior (3rd Edition)*. Pacific Grove, CA: Brooks/Cole.

Dagenbach, D. and Carr, T. H. (1994). *Inhibitory Processes in Attention, Memory, and Language*. San Diego: Academic Press.

Dempster, F. N. and Brainerd, C. J. (1995). *Interference and Inhibition in Cognition*. San Diego: Academic Press.

Furman, M. E. (1996). Neurocybernetics, self-organization and anchoring. *Anchor Point*, July, 31–37.

Fuster, J. M. (1995). *Memory in the Cerebral Cortex: An Empirical Approach to Neural Networks in the Human and Nonhuman Primate*. Cambridge, MA: MIT Press.

Haken, H. (1988). *Information and Self-Organization: A Macroscopic Approach to Complex Systems*. New York: Springer-Verlag.

Haken, H. (1996). *Principles of Brain Functioning: A Synergetic Approach to Brain Activity, Behavior and Cognition*. New York: Springer-Verlag.

Hayakawa, S. I. and Hayakawa, A. R. (1990). *Language in Thought and Action (5th Edition)*. New York: Harcourt Brace & Company.

Hobson, J. A. (1994). *The Chemistry of Conscious States: How the Brain Changes its Mind*. New York: Little, Brown & Company.

Honig, W. K. and Fetterman, J. G. (Eds.). (1992). *Cognitive Aspects of Stimulus Control*. Hillsdale, NJ: Lawrence Erlbaum Associates.

Houk, J. C., Davis, J. L., and Beiser, D. G. (1995). *Models of Information Processing in the Basal Ganglia*. Cambridge, MA: MIT Press.

Hull, C. L. (1943). *Principles of Behavior: An Introduction to Behavior Theory*. New York: Appleton-Century-Crofts.

Hull, C. L. (1952). *A Behavior System: An Introduction to Behavior Theory Concerning the Individual Organism*. New Haven: Yale University Press.

Kandel, E. R. and Hawkins, R. D. (1992). The biological basis of learning and individuality. *Scientific American*, September.

Kandel, E. R., Schwartz, J. H., and Jessell, T. M. (Eds.). (1991). *Principles of Neural Science*, 3rd ed. Norwalk, CT: Appleton & Lange.

Kandel, E. R., Schwartz, J. H., and Jessell, T. M. (Eds.). (1995). *Essentials of Neural Science and Behavior*. Norwalk, CT: Appleton & Lange.

Kazdin, A. E. (1994). *Behavior Modification in Applied Settings*. Pacific Grove, CA: Brooks/Cole.

Kelso, J. A. S. (1995). *Dynamic Patterns*. Cambridge, MA: MIT Press.

Kissin, B. (1986). *Psychobiology of Human Behavior (Volume I): Conscious and Unconscious Programs in the Brain*. New York: Plenum Press.

Kosslyn, S. M. (1994). *Image and Brain*. Cambridge, MA: MIT Press.

Kruse, P. and Stadler, M. (1995). *Ambiguity in Mind and Nature*. New York: Springer-Verlag.

Luria, A. R. (1932). *The Nature of Human Conflicts: An Objective Study of Disorganization and Control of Human Behavior*. New York: Grove Press.

Nunez, P. L. (1995). *Neocortical Dynamics and Human EEG Rhythms*. New York: Oxford University Press.

Paillard, J. (1991). *Brain and Space*. New York: Oxford University Press.

Partridge, L. D. and Partridge, L. D. (1993). *The Nervous System: Its Function and Its Interaction with the World*. Cambridge, MA: MIT Press.

Pavlov, I. P. (1927). *Conditioned Reflexes*. New York: Oxford University Press.
Pavlov, I. P. (1928). *Lectures on Conditioned Reflexes (Vol. I): Twenty-Five Years of Objective Study of the Higher Nervous Activity (Behavior) of Animals*. New York: International.
Pavlov, I. P. (1941). *Lectures on Conditioned Reflexes (Vol. II): Conditioned Reflexes in Psychiatry*. New York: International.
Platonov, K. (1959). *The Word as a Physiological and Therapeutic Factor*. Moscow: Foreign Languages Publishing House.
Pribram, K. H. (1991). *Brain and Perception: Holonomy and Structure in Figural Processing*. Hillsdale, NJ: Lawrence Erlbaum Associates.
Pryor, K. (1995). *On Behavior: Essays & Research*. North Bend, WA: Sunshine Books.
Roland, P. E. (1993). *Brain Activation*. New York: Wiley-Liss.
Rugg, M. D. and Coles, M. G. (1995). *Electrophysiology of Mind: Event-Related Brain Potentials and Cognition*. New York: Oxford University Press.
Salter, A. (1949). *Conditioned Reflex Therapy: The Direct Approach to the Reconstruction of Personality*. New York: Creative Age Press.
Sechenov, I. M. (1965). *Reflexes of the Brain: An Attempt to Establish the Physiological Basis of Psychological Processes*. Cambridge, MA: MIT Press.
Skinner, B. F. (1938). *The Behavior of Organisms: An Experimental Analysis*. New York: Appleton-Century-Crofts.
Skinner, B. F. (1953). *Science and Human Behavior*. New York: Free Press.
Skinner, B. F. (1957). *Verbal Behavior*. New York: Appleton-Century-Crofts.

Epilogue

In the ancient Greek dialog known as *Theaetetus*, Plato writes about the nature of knowledge and wisdom. By the end of the dialog, however, no definition of knowledge is reached, or so they think. We are only told what it is not. Albeit the fact that no formal conclusion is reached, the dialog reveals deep insight into the relationship between brain, mind, behavior, and information that would take nearly 2300 years of neuroscientific probing to validate. It appears that even then man's relationship to nature, and the properties of their interaction, were strongly suspected to have emerged out of conversations with her in the universal language of pattern.

> **Socrates:** Imagine, then, for the sake of argument, that our minds contain a block of wax, which in this or that individual may be larger or smaller, and composed of wax that is comparatively pure or muddy, and harder in some, softer in others, and sometimes of just the right consistency.
>
> **Theaetetus:** Very well.
>
> **Socrates:** Let us call it the gift of Muses' mother, Memory, and say that whenever we wish to remember something we see or hear or conceive in our own minds, we hold this wax under the perceptions or ideas and imprint them on it as we might stamp the impression of a seal ring. Whatever is so imprinted we remember and know so long as the image remains; whatever is rubbed out or has not succeeded in leaving an impression we have forgotten and do not know.
>
> **Theaetetus:** So be it.

Later on in the dialog, Socrates and Theaetetus discuss how false judgment and its related disorders of mind and behavior are born out of the interaction between sense and impression. Socrates first makes reference to the importance of the impression in the waxen block coinciding with sense. Here he illustrates a deep insight for the implicit necessity that an organism and its environment correspond. He then likens the act of perception to the proper fitting of a new sensory pattern or form into one of a myriad of patterns previously incorporated by the encoding substrate of mind.

> **Socrates:** It remains, then, that false judgment should occur in a case like this — when I, who know you and Theodorus and possess imprints of you both like seal impressions in the waxen block, see you both at a distance indistinctly and am in a hurry to assign the proper imprint of each to the proper visual perception, like fitting a foot into its own footmark to effect a recognition, and then make the mistake of interchanging them, like a man who thrusts his feet into the wrong shoes, and apply the perception of each to the imprint of the other. Or my mistake might be illustrated by the sort of thing that happens in a mirror when the visual current transposes right to left. In that case mistaking or false judgment does result.

> **Theaetetus:** I think it does, Socrates. That is an admirable description of what happens to judgment.

As soon as Theaetetus is seen to have apprehended the inseparable relationship between sensing, perceiving, and impressions in mind, the dialog moves to a telling description of how false judgments can be born of the variation in the stability, of the encoding (incorporating) substrate of mind — the waxen block. Socrates reveals to Theaetetus his suspicion that learning, memory, and forgetting are inextricably unified properties born of the dynamically shifting balance of pattern and entropy in the incorporating substrate, and that in some instances this will give rise to intrinsic interference with perception, and the failure to produce properly corresponding behavior.

> **Socrates:** When a person has what the poet's wisdom commends as a 'shaggy heart,' or when the block is muddy or made of impure wax, or oversoft or hard, the people with soft wax are quick to learn, but forgetful, those with hard wax the reverse. Where it is shaggy or rough, a gritty kind of stuff containing a lot of earth or dirt, the impressions obtained are indistinct; so are they too when the stuff is hard, for they have no depth. Impressions in soft wax also are indistinct, because they melt together and soon become blurred. And if, besides this, they overlap through being crowded together into some wretched little narrow mind, they are still more indistinct. All these types, then, are likely to judge falsely. When they see or hear or think of something, they cannot quickly assign things to their several imprints. Because they are so slow and sort things into the wrong places, they constantly see and hear and think amiss, and we say they are mistaken about things and stupid.

The history of life is the history of pattern and entropy — order and disorder. We as human beings mark the genesis of a new epoch in life's history, with the unprecedented ability to perceive our own pattern and influence our own design. It seems ironic that western thought, as far back as 2300 years ago, was riddled with many of the same questions extant today. Yet, even then the seed of the answer was planted firmly on the tip of humanity's tongue. For it was as far back as the philosophical musings found in the collected dialogs of Plato, that written record revealed the name that the scholars of the day had unwittingly assigned, to describe all manner of human maladies of mind by the term "dis-order."

> **Socrates:** Then let us not leave it incomplete. There remains the question of dreams and disorders, especially madness and all the mistakes madness is said to make in seeing or hearing or otherwise misperceiving.

While the informational bio-architecture revealed by neuro-imaging bears little physical resemblance to Plato's cleverly conceived, metaphorical waxen block, the fundamental principles of nature's interaction endure. As it is recounted by the dialogs, the motion of nature penetrates all things. To say that a thing exists is to say that it is constantly changing, for that which does not change will cease to exist.

It is our sincere hope that in the generations to come, NeuroPrint and the Standard Theory of Pattern-Entropy Dynamics find their way into the able hands of those agents of change who possess the depth of mind to apply the principles contained therein for the good of humanity.

REFERENCE

Hamilton, E. and Cairns, H. (Eds.). (1994). *The Collected Dialogues of Plato including the Letters*. Princeton, NJ: Princeton University Press.

Section III — Part 5

Glossary of Neurophysics, Nonlinear Science, and Dynamical Systems Terminology

Glossary

Adaptation: A naturally occurring process of interaction between brain and environment that acts on the dynamics of the brain in the form of changing the relative strength of different synapses to form dynamically evolving spatiotemporal thought/action patterns in the brain.

Anticipation: A system anticipates upcoming stable solutions. The system spends more time near a particular phase as it approaches a critical point in its present state or phase, giving rise to an enhanced phase density that specifies the locus of the upcoming state (stabilization of desired state must occur before fluctuating/destabilizing existing state. This enhances phase density and prevents random spontaneous switching). Critical slowing, the increased recovery time of a state or phase, is a predictor of upcoming phase transitions. This aspect of coordination dynamics is an anticipatory dynamical system (ADS).

Attractor: The region of the state space of a dynamical system toward which trajectories travel as time passes. As long as the parameters are unchanged, if the system passes close enough to the attractor, it will never leave that region without significant perturbation. Attractors are ordered states of high stability surrounded by instability (apparently random activity). These attractors are surrounded both temporally and spatially by chaotic activity in the brain. Brain cell assemblies are more tightly functionally coupled in attractor regions.

> *Note:* **The Birth of an Attractor:** An order parameter, once established, has a backward effect on the activity of all the elements from which it has emerged. This is the so-called slaving process. Once the collective behavior is in a state of high stability it demonstrates hysteresis prior to change.
>
> *Note:* Meanings are generated through the emergent relations between attractor neural networks. Meanings are a function of an attractor network. (Melody does not consist of tones, which are the sensory data for musical experience, but of the intervals between the tones and the relations between the intervals. So it is possible for melody to be transposed in such a way that it seems to remain the same even though every individual tone has been changed.) We classify stimuli entering the attractor neural network as meaningful if they lead the network quickly to an attractor. Otherwise, sensory input is classified as meaningless and ignored. This is the foundation of perception. Through associative learning, meanings can be attached or detached from attractors.
>
> *Note:* Sensitization will correspond to an opening, habituation to a closing, of attractors. Such an opening or closing of perceptual attractors can be achieved by changing attention parameters.

Related Terms:

Attractor Depth: Relative depth indicates the stability of one attractor over another. Relative depth is measured by switching behavior. A system switches more quickly from a shallow attractor to a deep one. Depth can also be measured by dwelling time and recovery speed after perturbation.

Attractor Width: A relative marker. Refers to the width of valleys in a potential well. Attractor width indicates variability inherent in the attractor space. This differs from the width of a basin of an attractor, which is defined by the set(s) of initial conditions from which the system goes into a particular behavior. The basin for each attractor would be defined by the receptor neurons that were activated during training or perception to form the nerve cell assembly.

Barrier Height: A relative marker of stability indicates amount of push or perturbation the system needs to escape the attractor.

Basin of Attraction: Collection of all points of the state space that tend to the attractor in forward time. The basin of an attractor can also be thought of as the set of initial conditions (anchors, strategies, etc.) from which the system goes to a specific behavior. The basin for each attractor in the brain is defined by the receptor neurons activated during training to form a nerve cell assembly, the whole range of sensory inputs that separately evoke a particular perception or behavior. A basin of attraction can also be thought of as the region within which all trajectories converge on a particular attractor.

Behavioral Attractor: A behavior that is stable within an individual and to which the system returns, when perturbed, acts like a behavioral attractor. Behavioral attractors are always softly assembled (functionally coupled) from interactions between their component elements (activity-dependent synapses) and are always in open energy exchange with their surroundings (other neurons and glial cells). Changes in either components (nerve cells) or in the context (sensory environment) may influence the patterns that emerge and their stability.

Critical Fluctuation: Large fluctuations as instability is approached — fluctuation enhancement (may be said to anticipate an ongoing pattern change).

Destabilized: An attractor is said to be destabilized when the time to recover from perturbation increases.

Development: The evolving and dissolving of sequences of attractors and the relationships between them.

Fixed-Point Attractor: Trajectory evolves toward a fixed stable point in phase space. The activity of a dampened pendulum would trace the trajectory of a fixed-point attractor. Fixed-point attractors are common for dissipative systems.

Flattening of Attractor Well: Indicates attractor instability and variability of behavior.

Instabilities: Provide a special entry point to a system because they allow a clear distinction between one pattern of behavior and another. They demarcate/separate behavioral patterns, enabling us to identify when pattern change occurs.

Landscape: A series of potential attractor wells that evolve and dissolve spatiotemporally over time. If a potential well is steep and narrow, it indicates that the system has few and highly stable behavioral choices. If the potential well is steep with a flat floor, it indicates that the system has several highly stable choices, none of which are preferred. A deep attractor literally "sucks in" other competing organizations of a system — the deeper the more preferred.

Limit Cycle Attractor: Patterns that repeat in time. Collective oscillations (phase locked) of neurons form stable attractors resistant to small perturbations. Habitual patterns of thought or behavior can be thought of as limit cycle attractors. Pattern interruption breaks limit cycles.

Open/Closed Attractors: When a network of cell assemblies in the brain becomes activated, the attractor is said to be open. Open indicates an active attractor basin.

Repeller: A region of state space where the state vector is repelled. Stable states are "attracting." Unstable states are repelling.

Searching Instability: While a system is searching, it is unstable. Instabilities offer a way to find control parameters (you know when you have a control parameter when its variation/scaling causes a qualitative change). All submodalities can act as control parameters.

Stability: A small change in initial conditions leads to only a small change in the trajectory. A system must lose stability during phase shifts. As you weaken functionally coupled bonds, there will be greater behavioral "variability" of the order parameter. Variability is an indicator of strength of a behavioral or perceptual attractor. A second index of strength/stability is resistance to perturbation (see "critical slowing").

State Vector: A point in phase space on the trajectory indicating the current state of the system. State vector describes a single point in state space.

Strange/Chaotic Attractor: Similar to limit cycle attractor. Patterns almost but never exactly repeat. Trajectories nearly but never cross. Most brain activity derived from phase portraits of EEGs depict strange attractors in the brain. A strange attractor's trajectories never exactly repeat twice. A strange attractor is geometrically a fractal.

Trajectory: The history of a system or a state vector. Can indicate strengthening and weakening of an attractor. A trajectory is a solution curve of a vector field. It is the path taken by the state of a system through the state's space as time progresses.

Asymptotic Trajectories: Represent the stable behavior that is seen once initial transient effects of perturbation have died away.

Measurements of Stability:

Barrier Height: The amount of "push" or perturbation the system needs to escape the attractor (i.e., intensity of stimulus).

Critical Slowing: Refers to the ability of the system to recover from perturbation as it nears critical point. Critical slowing is longer the closer the system is to instability. Can also be measured as the time needed for an unstable system (*attractor*) to find a stable state. The time increases the closer it approaches bifurcation. The measurement of time it takes to return to *some* previously observed state (local relaxation time) is an important index of stability and its loss when patterns spontaneously form (can be used to test stability of newly evolved attractors).

Dwelling Time: How long a system spends in the narrow channel of an attractor well before exiting; persistence of a given state before switching occurs (increases near fixed points). The *switching rate* out of or back into a perceptual or behavioral attractor is a measure of relative stability of different attractors. Residence or dwelling time is also a measure of stability. (The "swish pattern" and the "physiological snap" decrease switching time out of and into precepts and behaviors. Anchors can be used to increase residence/dwelling time.)

Phase Velocity: The speed at which a state vector is captured by a particular attractor. The time it takes for a particular state/phase to enslave the collective dynamics of a system once the state vector is released from initial conditions.

Variability: Growing instability of an attractor is detectable by increased measures of variability and behavior; variability indicates the approach of a phase transition. When a system approaches transition, the range of variability around a stable mode is greatly expanded. Variability reflects different developmental trajectories leading to the stabilization or destabilization of a pattern. If you track a course of a behavior over an extended time scale, it allows you to identify places in a trajectory where new forms appear.

Readiness to Change = High Degree of Dimensional Complexity (Variability/Many Degrees of Freedom)

Bifurcation: A sudden qualitative and discontinuous change (transition) in the dynamics of a system when a certain value of a parameter is reached. A point where one of the concurring modes predominates the others by slaving the elementary components of the process. The predominating mode is called the order parameter (i.e., a tornado is an example of a predominating mode of air flow which has a backward effect on the action of the air molecules that make it up. Once the tornado forms its action governs the movement of molecules originally involved in creating it). When bifurcation or phase transition occurs, it causes a qualitative change in action mode. The time needed for a system in a stable state to find a new stable state increases the closer the bifurcation is approached (see "critical slowing"). Normally, bifurcations provide a mechanism that converts one functional state to another.

Hopf Bifurcation: Occurs when a steady state changes to a periodic (oscillating) state. A very common way for periodic oscillation to "switch on." Hopf bifurcation is one in which one stable pattern switches to another stable pattern at a critical value of spatial orientation. Two stable solutions coexist; an exchange of stability occurs at the bifurcation point as in the example of a person who begins walking from a standing-still position after being pushed from the back.

Note: A bifurcation can also manifest as a sudden transition from limit cycle activity to a chaotic (apparently random) activity, as the value of a control parameter is slightly altered.

Period-Doubling Bifurcation: A period is the amount of time it takes for a system to return to its original state. The time it takes for a system to oscillate back to its starting point doubles at certain critical values. After several period doubling cycles, the system will show no predictable period for return to its original state. Its activity will become "apparently random." Continuous period doubling will eventually result in chaotic activity.

Saddle-Node Bifurcation: Occurs when there are attracting (stable) and repelling (unstable) directions in the neural coordination dynamics.

Bistability: The coexistence or simultaneous availability of two behavioral/perceptual attractor states, tendencies, or patterns.

Catastrophe: Sudden change in the state of a continuous system when the dynamics of the system undergo a bifurcation. Change may occur suddenly and discontinuously even though there has only been a small change in one of the system parameters. The magnitude of the change in the system is out of proportion to the change in the control parameter. This type of change is evident in human behavior when in one case information (a control parameter) presented to a person makes no apparent qualitative change in his/her emotions or behaviors, yet at another time the same information creates a sudden discontinuous response whose magnitude is drastically out of proportion to the previously witnessed response.

Chaos: Happens when the future of a trajectory is computationally unpredictable because small errors in the location of the trajectory lead to exponentially larger errors at later times. A chaotic dynamical system generally must meet three conditions: (1) it must be sensitive to initial conditions, (2) have dense periodic points, and (3) a dense orbit. Even simple systems can be chaotic having unpredictable trajectories. At the same time it can be considered "regular" in the sense that it can be completely analyzed and understood.

Chaotic Itinerancy: In the brain migration through state space along a trajectory that is, in part, determined by successive input and, in part, by input by other parts of the brain (see "reentrant").

Circular Causality: (Between microscopic and macroscopic processes.) Macroscopic structure and function emerge from elementary components and in turn organize/govern the microscopic elements of the system they arose from (i.e., a tornado).

Control Parameter: A parameter that when scaled leads a system to explore its collective states. Control parameters in self-organized dynamical systems are nonspecific, moving the system through its collective states, but not prescribing them. Control parameters break symmetry by concentrating energy in a system and inducing phase transition which enables the system to explore its collective variables/states. Some examples of control parameters affecting the human brain are regional blood flow (head/eye movement compresses degrees of freedom). This control parameter causes phase shifts between sensory systems and subsecondary representational systems. Other examples of control parameters are gravity, motion, breathing, diet, attention, intention, training, anchors, practice, biochemistry (hormones, peptides) instantiated by the elicitation of reference experiences, etc.

Coupling: Elements of a system are coupled if they influence one another.

Degeneracy: Brain wiring is so overlapping that any single function can be carried out by more than one pattern of neuronal connections and a single group of neurons can participate in more than one function.

Degrees of Freedom: When degrees of freedom are exposed by dissolution of an old pattern or attractor by perturbation, the system is allowed to explore new, more functional behaviors (degrees of opportunity; see "variability"). The brain exhibits a lower functional dimension when it is in a stable, recognizable state and greater degrees of freedom during phase transition.

Dissipative System: Refers to a system that loses or dissipates energy as a function of time (i.e., a dampened pendulum that loses energy to friction).

Dynamical System: Can be thought of as a set of possible states (its phase space or state space) plus evolution rules which determine sequences of points in that space (trajectories).

Feedback: Positive feedback amplifies while negative feedback regulates (i.e., a TV camera pointed at its own monitor is an example of visual iterative positive feedback).

Fixed Point: A resting point or equilibrium of the system. For example, the pendulum of a clock always eventually stops moving, so hanging vertically downward is a fixed point of this system.

Forced Resonance: When a system is acted upon by an external, periodic driving force, its oscillations become phase locked by the oscillations of the driving force. In forced resonance, the response is greatest when the frequency of the periodic driving force matches the natural frequency of the structure. The resulting oscillations are phase locked (i.e., pacing somebody's speech rate or movement patterns to gain rapport, two clocks with oscillating pendulums will oscillate at the same frequency via vibrations carried by the wall they hang from).

Fractal: An object or process in which patterns occurring on a small spatial or temporal scale are repeated at ever larger/smaller scales.

Fractal Dimension: The self-similarity of a fractal implies that it possesses some fundamental aspect that does not vary as a function of scale. Regardless of how much a fractal is magnified, it continues to look similar in appearance. Fractal structures found in nature (trees, mountains, coastlines, clouds, the structure of the lung, etc.) have what is called statistical self-similarity. In this case, smaller crinkles are not necessarily exact copies of larger crinkles, but have the same qualitative appearance and are the same on "average." Strange attractors also fall into this category.

Functional Synergies: Collective functional organizations that are neuronally based are subserved by soft coupling of nerve cell assemblies that render control of complex multivariable systems. Principles of self-organizing pattern formation govern their assembly.

Hysteresis: The tendency of a system to favor its history. The temporary resistance of one stable state against the dynamics of formation of another state. Hysteresis involves the persistence of a perception, state, or pattern despite settings of the control parameter that would favor the alternative. The presence of competing patterns under gradual parametric change will favor the initially established pattern.

Intermittency: Periods of stability and predictability in the midst of random fluctuation. Intermittency can manifest as periods of order inside randomness or periods of randomness interrupting order.

Intermittent Dynamics: The brain's state vector, rather than residing in attractors of a neural network, dwells for varying times near attractive states where it can switch flexibly and quickly. The probability of switching will always increase as the state vector nears category boundaries. Categories are determined by the stability of attractive states. The brain lives at the edge of instability where it can switch spontaneously among collective states. Rather than requiring an active process to destabilize and switch from one stable state to another, intermittency seems to be an inherent built-in feature of neural machinery that supports perception/behavior and the brain itself. The main mechanism of intermittency is believed to be the coalescence of stable (attracting) and unstable (repelling) directions in neural coordination dynamics.

Iteration: The process of feeding the solution/result of an equation/process back in as "input" into that equation or process. Fractals such as that found in a shoreline are created by the process of repeated iteration. The chaotic activity of the ocean continually subtracts elements in an iterative recursive process. Submodalities can be iterated to affect mental processes in the brain (i.e., compulsion blowout).

Monostability: The existence of a single behavioral/perceptual state.

Morphogen: Any substance thought to impair or alter positional information in a developmental morphogenetic gradient.

Morphogenesis: Spontaneous self-organizing pattern formation.

Morphogenetic Field: A position/location-dependent self-organizing field. Can develop independently without instructive influences. An important property of morphogenetic fields is that it is capable of regulation, which means that any portion of the field is capable of regenerating the whole field.

Multistability: Parallel reentrant processing circuits in the brain give rise to differential perception of the same physical stimulus configuration (ambiguities set up a multistable/intermittent attractor layout).

Nonlinear: Refers to a system governed by nonlinear differential or difference equations.

Nonlinearity: The emergent properties of a system are more than the sum of its parts.

Open System: A system that is free to exchange energy with the surrounding universe.

Order Parameter: A sudden, spontaneous, macroscopic reorganization resulting from the nonlinear behavior of a system where concurring modes reach a bifurcation point and one of the modes predominates the others by slaving the elementary components of the system. An order parameter acts to compress the degrees of freedom available to the elemental components of the system. This results in spontaneous reorganization of connectivity and pattern formation.

Period: The time it takes for a trajectory on a periodic cycle to return to its starting point.

Periodic Point: Point that lies on a periodic cycle, i.e., an orbit that returns to previous values.

Perturbation: Something that perturbs or disturbs a system.

Phase Locking: Collective oscillations of neurons form limit cycles far more stable and resistant to small perturbations than a collection of individual oscillations. Habitual patterns of thought form limit cycles.

Phase Shifts: When systems are undergoing phase shifts, their components are more loosely coupled. While systems are fluid, they are freer to seek new places in their phase space and they do so when any control parameter is changed. Larger perturbations are required when components are tightly coupled.

Phase Space: A term describing the state space of a system that usually implies that at least one axis is a time derivative, like velocity.

Phase Transition: An autonomous reorganization of macroscopic order emerging spontaneously from elementary interactions. Phase transition occurs when a system reaches critical fluctuation. During phase transition the system lies between or near attractors, not in them. The system becomes more fluid and flexible resulting in bifurcation. The spontaneous appearance and disappearance of attractors and the restructuring of the dynamics of a system indicate phase transition.

Reafference: A command issued by the limbic system altering all the sensory systems to prepare to respond to new information.

Reentrant: Two or more abstracting networks are working disjunctively to process the same stimuli. Reentrant circuits communicate with each other in a simultaneous/parallel fashion and continually update representations of the same sampled stimulus.

Self-Organization: New and different forms emerge spontaneously due to instabilities (i.e., pattern interruption). Self-organization theory demonstrates pattern formation and change under nonspecific parametric influences. Self-organizing systems spontaneously form and change patterns due to nonlinear interactions among the components of a system.

Sensitivity to Initial Conditions: This is a fundamental to the unpredictability found in chaotic systems (see "chaos"). It means that a small change in initial condition leads to a large change in the trajectory of a state vector as time progresses. Tiny differences become drastically magnified.

Slaving Principle: In the neighborhood of critical points the behavior of a complex system is completely governed by few collective modes, the order parameters that slave all the other modes and elementary components. (Words emerge from elementary processes and, in turn, govern self-organization of those processes, i.e., pain/thirst perception.)

> *Note:* Within a complex system, long-lasting or slowly changing events govern short-lasting or swiftly changing events.

Spontaneous Pattern Formation: Is caused by symmetry breaking. Symmetry breaking occurs when there are changes in a system's concentration of energy. When local energy levels (concentrations) change, new forms spontaneously self-organize. (Temperature, speed, ion concentration, regional blood flow, blood volume, oxygen, etc. are all components of shifting energy concentrations in the brain. All are capable of breaking symmetry.) Intention and training are specific control parameters also capable of concentrating energy and symmetry breaking.

Stability: Stability means that a small change in initial condition leads to only a small change in the trajectory.

State Space: Also referred to as Phase Space. This is the space of points whose coordinates completely specify the model system. A way of visually representing the dynamics of a system over time. The mathematical description of a dynamical system consists of two parts, the state and the dynamics. The state is a snapshot of the system at a given instant in time, while the dynamics are the set of rules by which the state trajectory evolves over time. The state of a system is represented by the state vector. The state vector represents a point in state space.

Stochastic Resonance: Optimum noise intensity in a system, which maximizes coherent switching. Applying an optimal level of external noise can enhance weak signals. Stochastic resonance can facilitate information transmission by amplifying weak signals as they flow through the nervous system. It is thought to be a self-generated optimization process. Noise is beneficial in probing stability of coordinated states and discovering new ones. Noise provides fluctuation and enhancement, which is necessary for nonequilibrium phase transitions in sensory motor behavior and learning.

Switching Time: A measure of attractor stability. The amount of time it takes to switch from one attractor/pattern to another. Switching time is always faster from a less stable to a more stable attractor.

Symmetry Breaking: Results from shifting energy concentration in a system. Gives rise to spontaneous pattern formation. Symmetry breaking is a basic requisite process for obtaining information from the external environment via the human sensory systems. Symmetry removal and symmetry breaking are both basic features of the process of perception. If two regions of the visual field are related by exchange symmetry, at first they will be relegated automatically to background. The perceptual system undergoes continuous removal of exchange symmetry in order to discern pattern.

The second step of perception is accomplished suddenly, as soon as a critical state is reached where the interiorized figure falls into coincidence with, and is captured by, one of the previously established attractors in the brain system. At this point visual thinking is formed, and the order parameter develops embracing the interiorized figure as a whole unit.

Synergetics: A branch of theoretical physics founded by Hermann Haken. Synergetics is the study of self-organizing systems; the cooperation of individual parts of a system resulting in the spontaneous production of macroscopic, spatial, temporal, or functional structures/patterns.

REFERENCES

Kelso, J. A. S. (1995). *Dynamic Patterns*. Cambridge, MA: MIT Press.
Haken, H. (1983). *Synergetics: An Introduction (3rd Edition)*. New York: Springer-Verlag.

Suggested Readings and Resource Guide

Adams, R. D. and Victor, M. (1994). *Principles of Neurology: Companion Handbook (5th Edition)*. New York: McGraw-Hill.

Ader, R., Felten, D. L., and Cohen, N. (1991). *Psychoneuroimmunology (2nd Edition)*. San Diego: Academic Press.

Adler, A. (1927). *Understanding Human Nature*. New York: Greenberg Publishers.

Aggleton, J. P. (1992). *The Amygdala: Neurobiological Aspects of Emotion, Memory, and Mental Dysfunction*. New York: John Wiley & Sons.

Andreas, S. and Andreas, C. (1987). *Change Your Mind and Keep the Change*. Moab, UT: Real People Press.

Andreas, S. and Andreas, C. (1989). *Heart of the Mind: Engaging Your Inner Power to Change with Neuro-Linguistic Programming*. Moab, UT: Real People Press.

Andreas, S. and Faulkner, C. (1994). *NLP: The New Technology of Achievement*. New York: William Morrow and Company.

Arkes, H. R. and Hammond, K. R. (Eds.). (1986). *Judgment and Decision Making: An Interdisciplinary Reader*. New York: Cambridge University Press.

Aronson, E. (1972). *The Social Animal*. New York: Viking Press.

Ashby, W. R. (1956). *An Introduction to Cybernetics*. London: Methuen & Co.

Ashby, W. R. (1960). *Design for a Brain (2nd Edition)*. New York: John Wiley & Sons.

Atkins, P. W. (1994). *The 2nd Law: Energy Chaos, and Form*. New York: Scientific American Books.

Atmanspacher, H. and Dalenoort, G. J. (1994). *Inside Versus Outside*. New York: Springer-Verlag.

Baddeley, A. (1986). *Working Memory: Oxford Psychology Series — 11*. New York: Oxford University Press.

Baker, G. L. and Gollub, J. P. (1990). *Chaotic Dynamics: An Introduction*. Cambridge, MA: Cambridge University Press.

Bandler, R. (1984). *Magic in Action*. Cupertino, CA: Meta Publications.

Bandler, R. (1985). *Using Your Brain for a Change*. Moab, UT: Real People Press.

Bandler, R. (1993). *Time for a Change*. Cupertino, CA: Meta Publications.

Bandler, R. and Grinder, J. (1975). *Patterns of the Hypnotic Techniques of Milton H. Erickson, M.D. Volume I*. Cupertino, CA: Meta Publications.

Bandler, R. and Grinder, J. (1975). *The Structure of Magic*. Palo Alto, CA: Science and Behavior Books.

Bandler, R. and Grinder, J. (1976). *The Structure of Magic II*. Palo Alto, CA: Science and Behavior Books.

Bandler, R. and Grinder, J. (1979). *Frogs Into Princes: Neuro-Linguistic Programming*. Moab, UT: Real People Press.

Bandler, R. and Grinder, J. (1982). *Reframing: Neuro-Linguistic Programming and the Transformation of Meaning*. Moab, UT: Real People Press.

Bandler, R. and MacDonald, W. (1988). *An Insider's Guide to Sub-Modalities*. Cupertino, CA: Meta Publications.

Bandura, A. (1969). *Principles of Behavior Modification*. New York: Holt, Rinehart and Winston.
Barkow, J. H., Cosmides, L., and Tooby, J. (1992). The *Adapted Mind: Evolutionary Psychology and the Generation of Culture*. New York: Oxford University Press.
Baron-Cohen, S. (1995). *Mindblindness: An Essay on Autism and Theory of Mind*. Cambridge, MA: MIT Press.
Barondes, S. H. (1993). *Molecules and Mental Illness*. New York: Scientific American Library.
Basar, E., Flohr, H., Haken, H., and Mandell, A. J. (1983). *Synergetics of the Brain*. New York: Springer-Verlag.
Bateson, G. (1972). *Steps to an Ecology of Mind*. New York: Ballantine Books.
Bateson, G. (1979). *Mind and Nature: A Necessary Unity*. New York: Bantam Books.
Beaulieu, J. (1987). *Music and Sound in the Healing Arts*. Barrytown, NY: Station Hill Press.
Bechterev, V. M. (1928). *General Principles of Human Reflexology*. New York: International.
Beck, D. E. and Cowan, C. C. (1996). *Spiral Dynamics: Mastering Values, Leadership and Change*. Cambridge, MA: Blackwell.
Becker, J. B., Breedlove, S. M., and Crews, D. (1993). *Behavioral Endocrinology*. Cambridge, MA: MIT Press.
Bellack, A. S., Hersen, M., and Kazkin, A. E. (1990). *International Handbook of Behavior Modification and Therapy (2nd Edition)*. New York: Plenum Press.
Benson, D. F. (1994). *The Neurology of Thinking*. New York: Oxford University Press.
Bernays, E. L. (Ed.). (1955). *The Engineering of Consent*. Norman: University of Oklahoma.
Bernheim, H. (1889). *Suggestive Therapeutics*. New York: Knickerbocker Press.
Bickerton, D. (1995). *Language and Human Behavior*. Seattle, WA: University of Washington Press.
Binet, A. and Fere, C. (1887). *Animal Magnetism*. Patternoster Square, London: Kegan Paul, Trench & Co.
Birbaumer, N. and Ohman, A. (Eds.). (1993). *The Structure of Emotion*. Seattle, WA: Hogrefe & Huber.
Bittle, C. N. (1935). *Logic: The Science of Correct Thinking*. New York: Bruce Publishing Company.
Black, I. B. (1991). *Information in the Brain: A Molecular Perspective*. Cambridge, MA: MIT Press.
Blackmore, S. (1999). *The Meme Machine*. New York: Oxford University Press.
Bloch, G. (Ed.). (1980). *Mesmerism: A Translation of the Original Scientific and Medical Writings of Franz Anton Mesmer*. Los Altos, CA: William Kaufmann.
Bloedel, J. R., Ebner, T. J., and Wise, S. P. (1996). *The Acquisition of Motor Behavior in Vertebrates*. Cambridge, MA: MIT Press.
Bloom, F. and Lazerson, A. (1988). *Brain, Mind, and Behavior, Second Edition*. New York: W. H. Freeman and Company.
Bloom, P., Peterson, M. A., Nadel, L., and Garrett, M. F. (1996). *Language and Space*. Cambridge, MA: MIT Press.
Bohm, D. (1994). *Thought as a System*. New York: Routledge.
Bohm, D. (1995). *Wholeness and the Implicate Order*. New York: Routledge.
Boole, G. (1854). *The Laws of Thought: An Investigation of the Laws of Thought on Which Are Founded the Mathematical Theories of Logic and Probabilities*. New York: Dover.
Braid, J. (1843). *Neurypnology*. Edinburgh: Adam & Charles Black.
Brandt, S. and Dahmen, H. D. (1995). *The Picture Book of Quantum Mechanics (Second Edition)*. New York: Springer-Verlag.
Bregman, A. S. (1990). *Auditory Scene Analysis: The Perceptual Organization of Sound*. Cambridge, MA: MIT Press.

Briggs, J. (1992). *Fractals: The Patterns of Chaos*. New York: Touchstone.
Briggs, J. and Peat, F. D. (1971). *Turbulent Mirror: An Illustrated Guide to Chaos Theory and the Science of Wholeness*. New York: Harper & Row.
Brodie, R. (1996). *Virus of the Mind: The New Science of the Meme*. Seattle: Integral Press.
Buckminster Fuller, R. (1975). *Synergetics: Explorations in the Geometry of Thinking*. New York: Macmillian.
Burke, K. (1989). *On Symbols and Society*. Chicago, IL: University of Chicago Press.
Burt, A. M. (1993). *Textbook of Neuroanatomy*. Philadelphia, PA: W.B. Saunders Company.
Buzan, T. (1989). *Use Both Sides of Your Brain (3rd Edition)*. New York: Penguin Books.
Buzan, T. (1989). *Use Your Perfect Memory (3rd Edition)*. New York: Penguin Books.
Buzan, T. and Buzan, B. (1993). *The Mind Map Book: How to Use Radiant Thinking to Maximize Your Brain's Untapped Potential*. New York: Penguin Books.
Cacioppo, J. T. and Petty, R. E. (1983). *Social Psychophysiology: A Source Book*. New York: Guilford Press.
Cailliet, R. (1993). *Pain: Mechanisms and Management*. Philadelphia: F.A. Davis Company.
Cairns-Smith, A. G. (1985). *Seven Clues to the Origin of Life*. New York: Cambridge University Press.
Calvin, W. H. (1996). *The Cerebral Code*. Cambridge, MA: MIT Press.
Calvin, W. H. and Ojemann, G. A. (1994). *Conversations with Neil's Brain: The Neural Nature of Thought and Language*. Reading, MA: Addison-Wesley Publishing.
Cambel, A. B. (1993). *Applied Chaos Theory: A Paradigm for Complexity*. San Diego: Academic Press.
Cameron-Bandler, L., Gordon, D., and Lebeau, M. (1985). *The Emprint Method: A Guide to Reproducing Competence*. Moab, UT: Real People Press.
Campbell, D. G. (1992). *Introduction to the Musical Brain (2nd Edition)*. St. Louis, MO: MMB Music.
Casti, J. L. (1994). *Complexification: Explaining a Paradoxical World Through the Science of Surprise*. New York: HarperCollins.
Casti, J. L. (1996). *Five Golden Rules: Great Theories of 20th Century Mathematics — and Why They Matter*. New York: John Wiley & Sons.
Cautela, J. R. and Kearney, A. J. (1986). *The Covert Conditioning Handbook*. New York: Springer.
Chakotin, S. (1940). *The Rape of the Masses: The Psychology of Totalitarian Political Propaganda*. New York: Fortean Society.
Chalmers, D. J. (1996). *The Conscious Mind: In Search of a Fundamental Theory*. New York: Oxford University Press.
Chance, P. (1994). *Learning and Behavior (3rd Edition)*. Belmont, CA: Wadsworth.
Charvet, S. R. (1995). *Words that Change Minds: Mastering the Language of Influence*. Dubuque, IA: Kendall/Hunt.
Checkland, P. (1993). *Systems Thinking, Systems Practice*. New York: John Wiley & Sons.
Chierchia, G. (1995). *Dynamics of Meaning: Anaphora, Presupposition, and the Theory of Grammar*. Chicago, IL: University of Chicago Press.
Chopra, D. (1989). *Quantum Healing: Exploring the Frontiers of Mind/Body Medicine*. New York: Bantam Books.
Churchland, P. M. (1995). *The Engine of Reason, the Seat of the Soul: A Philosophical Journey into the Brain*. Cambridge, MA: MIT Press.
Cialdini, R. B. (1993). *Influence: Science and Practice (3rd Edition)*. New York: HarperCollins College Publishers.
Citrenbaum, C. M., King, M. E., and Cohen, W. I. (1985). *Modern Clinical Hypnosis for Habit Control*. New York: W. W. Norton & Company.

Clayman, C. (1995). *The Human Body: An Illustrated Guide to Its Structure, Function and Disorders*. New York: Dorling Kindersley Publishing.

Cohen, J. and Steward, I. (1994). *The Collapse of Chaos: Discovering Simplicity in a Complex World*. New York: Penguin Books.

Conway, F. and Siegelman, J. (1978). *Snapping (2nd Edition)*. New York: Stillpoint Press.

Cooper, J. R., Bloom, F. E., and Roth, R. H. (1991). *The Biochemical Basis of Neuropharmacology (6th Edition)*. New York: Oxford University Press.

Corbett, E. P. J. (1990). *Classical Rhetoric for the Modern Student (3rd Edition)*. New York: Oxford University Press.

Cornoldi, C., Logie, R. H., Brandimonte, M. A., Kaufmann, G., and Reisberg, D. (1996). *Stretching the Imagination: Representation and Transformation in Mental Imagery*. New York: Oxford University Press.

Cornwell, J. (1995). *Nature's Imagination: The Frontiers of Scientific Vision*. New York: Oxford University Press.

Coveney, P. and Highfield, R. (1995). *Frontiers of Complexity: The Search for Order in a Chaotic World*. New York: Fawcett Columbine.

Crapo, L. (1985). *Hormones: The Messengers of Life*. New York: W. H. Freeman and Company.

Crick, F. (1994). *The Astonishing Hypothesis: The Scientific Search for the Soul*. New York: Charles Scribner's Sons.

Crick, F. and Koch, C. (1992). The problem of consciousness. *Scientific American*, September.

Dagenbach, D. and Carr, T. H. (1994). *Inhibitory Processes in Attention, Memory, and Language*. San Diego: Academic Press.

Damasio, A. R. (1994). *Descartes' Error: Emotion, Reason, and the Human Brain*. New York: G.P. Putnam's Sons.

Damasio, A. R. and Damasio, H. (1992). Brain and language. *Scientific American*, September.

Damasio, H. (1995). *Human Brain Anatomy in Computerized Images*. New York: Oxford University Press.

Darwin, C. (1859). *The Origin of Species by Means of Natural Selection of the Preservation of Favoured Races in the Struggle for Life*. London: John Murray.

Das, J. P. (1969). *Verbal Conditioning and Behaviour*. New York: Pergamon Press.

Daudel, R. (1993). *The Realm of Molecules*. New York: McGraw-Hill.

Davidson, R. J. and Hugdahl, K. (1995). *Brain Asymmetry*. Cambridge, MA: MIT Press.

Davis, P. J., Hersh, R., and Marchisotto, E. A. (1981). *The Mathematical Experience: Study Edition*. Cambridge, MA: Birkhauser Boston.

Dawkins, R. (1976). *The Selfish Gene (New Edition)*. New York: Oxford University Press.

Dawkins, R. (1982). *The Extended Phenotype*. New York: Oxford University Press.

Dawkins, R. (1986). *The Blind Watchmaker: Why the Evidence of Evolution Reveals a Universe without Design*. New York: W. W. Norton & Company.

Deacon, T. W. (1997). *The Symbolic Species: The Co-Evolution of Language and the Brain*. New York: W. W. Norton & Company.

DeArmond, S. J., Fusco, M., and Dewey, M. M. (1989). *Structure of the Human Brain: A Photographic Atlas (3rd Edition)*. New York: Oxford University Press.

Deleuze, J. P. F. (1982). *Practical Instruction in Animal Magnetism*. New York: Da Capo Press.

Dempster, F. N. and Brainerd, C. J. (1995). *Interference and Inhibition in Cognition*. San Diego: Academic Press.

Denes, P. B. and Pinson, E. N. (1993). *The Speech Chain: The Physics and Biology of Spoken Language (2nd Edition)*. New York: W. H. Freeman and Company.

Dengrove, E. (Ed.). (1976). *Hypnosis and Behavior Therapy*. Springfield, IL: Charles C Thomas.

Dennett, D. C. (1991). *Consciousness Explained.* New York: Little, Brown & Company.
Despopoulos, A. and Silbernagl, S. (1991). *Color Atlas of Physiology.* New York: Georg Thieme Verlag.
Devaney, R. L. (1992). *A First Course In Chaotic Dynamical Systems: Theory and Experiment.* Reading, MA: Addison-Wesley.
Devlin, K. (1991). *Logic and Information.* Cambridge, MA: Cambridge University Press.
Dichter, E. (1960). *The Strategy of Desire.* New York: Doubleday & Company.
Dichter, E. (1964). *Handbook of Consumer Motivations: The Psychology of the World of Objects.* New York: McGraw-Hill.
Dichter, E. (1971). *Motivating Human Behavior.* New York: McGraw-Hill.
Dilts, R. (1983). *Applications of Neuro-Linguistic Programming.* Cupertino, CA: Meta Publications.
Dilts, R. (1990). *Changing Belief Systems with NLP.* Capitola, CA: Meta Publications.
Dilts, R. (1998). *Modeling with NLP.* Capitola, CA: Meta Publications.
Dilts, R., Hallbom, T., and Smith, S. (1990). *Beliefs: Pathways to Health & Well-Being.* Portland, OR: Metamorphous Press.
Dilts, R. B., Epstein, T., and Dilts, R. W. (1991). *Tools for Dreamers: Strategies for Creativity and the Structure of Innovation.* Cupertino, CA: Meta Publications.
Dilts, R. B. and Epstein, T. A. (1995). *Dynamic Learning.* Capitola, CA: Meta Publications.
Doob, L. W. (1935). *Propaganda: Its Psychology and Technique.* New York: Henry Holt and Company.
Dyson, F. (1985). *Origins of Life.* New York: Cambridge University Press.
Eccles, J. C. (1994). *How the Self Controls Its Brain.* New York: Springer-Verlag.
Eco, U. (1976). *A Theory of Semiotics.* Bloomington: Indiana University Press.
Edelman, G. M. (1987). *Neural Darwinism: The Theory of Neuronal Group Selection.* New York: BasicBooks.
Edelman, G. M. (1992). *Bright Air, Brilliant Fire: On the Matter of the Mind.* New York: BasicBooks.
Edmonston, W. E. (1986). *The Induction of Hypnosis.* New York: John Wiley & Sons.
Einstein, A. (1961). *Relativity: The Special and the General Theory.* New York: Three Rivers Press.
Einstein, A. (1997). *The World As I See It.* Secaucus, NJ: Carol Publishing Group.
Ellul, J. (1965). *Propaganda: The Formation of Men's Attitudes.* New York: Vintage Books.
Engelkamp, J. and Zimmer, H. D. (1994). *The Human Memory: A Multi-Modal Approach.* Kirdland, WA: Hogrefe & Huber.
Epstein, R. (Ed.). (1980). *Notebooks B. F. Skinner.* Englewood Cliffs, NJ: Prentice-Hall.
Erickson, M. H. (1980). *Hypnotic Alteration of Sensory, Perceptual and Psychophysiological Processes: The Collected Papers of Milton H. Erickson on Hypnosis Volume II.* New York: Irvington.
Erickson, M. H. (1989). *The Nature of Hypnosis and Suggestion: The Collected Papers of Milton H. Erickson on Hypnosis Volume I.* New York: Irvington.
Erickson, M. H., Rossi, E. L., and Rossi, S. I. (1976). *Hypnotic Realities: The Induction of Clinical Hypnosis and Forms of Indirect Suggestion.* New York: Irvington.
Erickson, M. H. and Rossi, E. L. (1979). *Hypnotherapy: An Exploratory Casebook.* New York: Irvington.
Erickson, M. H. and Rossi, E. L. (1981). *Experiencing Hypnosis: Therapeutic Approaches to Altered States.* New York: Irvington.
Everaert, M., Van Der Linden, E. J., Schenk, A., and Schreuder, R. (1995). *Idioms: Structural and Psychological Perspectives.* Hillsdale, NJ: Lawrence Erlbaum Associates.
Feldenkrais, M. (1977). *Awareness through Movement.* New York: HarperCollins.

Feldenkrais, M. (1981). *The Elusive Obvious*. Capitola, CA: Meta Publications.
Feldendrais, M. (1984). *The Master Moves*. Capitola, CA: Meta Publications.
Feldenkrais, M. (1985). *The Potent Self: The Dynamics of the Body and the Mind*. New York: HarperCollins.
Feldenkrais, M. (1992). *Body and Mature Behaviour: A Study of Anxiety, Sex, Gravitation & Learning*. Madison, CT: International Universities Press.
Feldenkrais, M. (1993). *Body Awareness as Healing Therapy: The Case of Nora*. Berkeley, CA: Somatic Resources.
Festinger, L. (1957). *A Theory of Cognitive Dissonance*. Stanford, CA: Stanford University Press.
Festinger, L. (1964). *Conflict, Decision, and Dissonance*. Stanford, CA: Stanford University Press.
Feynman, R. P. (1985). *QED: The Strange Theory of Light and Matter*. Princeton, NJ: Princeton University Press.
Finke, R. A. (1989). *Principles of Mental Imagery*. Cambridge, MA: MIT Press.
Fischbach, G. D. (1992). Mind and brain. *Scientific American*, September.
Flood, R. L. and Jackson, M. C. (1991). *Creative Problem Solving: Total Systems Intervention*. New York: John Wiley & Sons.
Fortey, R. (1997). *Life: A Natural History of the First Four Billion Years of Life on Earth*. New York: Alfred A. Knopf.
Franklin, S. (1995). *Artificial Minds*. Cambridge, MA: MIT Press.
Freese, J. H. (1926). *Aristotle XXII "Art" of Rhetoric*. Cambridge, MA: Harvard University Press.
Freud, S. (1913). *The Interpretation of Dreams*. New York: Macmillan.
Friedman, S. (1993). *The New Language of Change: Constructive Collaboration in Psychotherapy*. New York: Guilford Press.
Friedman, W. J. (1990). *About Time: Inventing the Fourth Dimension*. Cambridge, MA: MIT Press.
Furman, M. E. (1995). The neurophysics of hypnosis. *Anchor Point*, October, 10–15.
Furman, M. E. (1995). Life, death and anchoring. *Anchor Point*, December, 36–39.
Furman, M. E. (1996). Neocortical dynamics of metaphor. *Anchor Point*, February, 44–50.
Furman, M. E. (1996). Neocortical dynamics of persuasion and influence. *Anchor Point*, April, 30–37.
Furman, M. E. (1996). Submodalities through the eyes of a neuroscientist. *Anchor Point*, May, 19–23.
Furman, M. E. (1996). Neurophysics and the principle of anchoring. *Anchor Point*, June, 30–40.
Furman, M. E. (1996). Neurocybernetics, self-organization and anchoring. *Anchor Point*, July, 31–37.
Furman, M. E. (1996). The science and practice of human performance modeling and engineering. *Anchor Point*, September, 31–43.
Furman, M. E. (1996). Neuro-Cognitive modeling: the art and science of capturing the invisible. *Anchor Point*, November, 40–42.
Furman, M. E. (1996). Foundations of neuro-cognitive modeling: eye movement – a window to the brain. *Anchor Point*, December, 14–22.
Furman, M. E. (1997). Modeling at the speed of sight. *Anchor Point*, January, 29–33.
Furman, M. E. (1997). Mind, music and miracles. *Anchor Point*, September, 11–18.
Furman, M. E. (1998). NLP and the behavior of information. *Anchor Point*, July, 27–32.
Furman, M. E. (1998). Manufacturing reality: uncovering the hidden forces that control the behavior of information patterns. *Anchor Point*, August, 34–40.

Furman, M. E. (1998). The art and science of designing an information virus. *Anchor Point*, October, 19–25.

Furman, M. E. (1998). Designer information viruses: the art and science of memetic engineering. *Anchor Point*, November, 31–36.

Furman, M. E. (1998). Intelligent learning systems: a student-centered model for reconstructing the human education system. *Anchor Point*, December, 21–29.

Furman, M. E. (1999). Simon says trauma gone. *Anchor Point*, November, 37–44.

Fuster, J. M. (1995). *Memory in the Cerebral Cortex: An Empirical Approach to Neural Networks in the Human and Nonhuman Primate*. Cambridge, MA: MIT Press.

Gallagher, W. (1993). *The Power of Place: How Our Surroundings Shape Our Thoughts, Emotions, and Actions*. New York: HarperCollins.

Gallo, F. (1996). Therapy by energy. *Anchor Point*, June, 46–51.

Gallo, F. (1996). Reflections on active ingredients in efficient treatments of PTSD, Part 1. *Electronic Journal of Traumatology*, 2(1). Available at <http://www.fsu.edu/~trauma/>.

Gallo, F. (1996). Reflections on active ingredients in efficient treatments of PTSD, Part 2. *Electronic Journal of Traumatology*, 2(2). Available at <http://www.fsu.edu/~trauma/>.

Gallo, F. P. (1998). *Energy Psychology: Explorations at the Interface of Energy, Cognition, Behavior, and Health*. Boca Raton, FL: CRC Press.

Gallo, F. (1999). A no talk cure for trauma. In *The Art of Psychotherapy: Case Studies from the Family Therapy Networker*, Simon, R., Markowitz, L., Barrilleaux, C., and Topping, B. (Eds.), New York: John Wiley & Sons, 244–255.

Gallo, F. P. (2000). *Energy Diagnostic and Treatment Methods*. New York: W. W. Norton & Company.

Gallo, F. P. and Vincenzi, H. (2000). *Energy Tapping: How to Rapidly Eliminate Anxiety, Depression, Cravings, and More Using Energy Psychology*. Berkley, CA: New Harbinger.

Gazzaniga, M. S. (1992). *Nature's Mind: The Biological Roots of Thinking, Emotions, Sexuality, Language, and Intelligence*. New York: BasicBooks.

Gazzaniga, M. S. (1995). *The Cognitive Neurosciences*. Cambridge, MA: MIT Press.

Geissler, H. G., Miller, M. H., and Prinz, W. (1990). *Psychophysical Explorations of Mental Structures*. Lewiston, NY: Hogrefe & Huber.

George, F. H. (1965). *Cybernetics and Biology*. San Francisco: W. H. Freeman and Company.

Gernsbacher, M. A. (1994). *Handbook of Psycholinguistics*. San Diego: Academic Press.

Gershon, E. S. and Rieder, R. O. (1992). Major disorders of mind and brain. *Scientific American*, September.

Gilligan, S. and Price, R. (1993). *Therapeutic Conversations*. New York: W. W. Norton & Company.

Glaser, R., Kennedy, S., Lafuse, W., Bonneau, R., Speicher, C., Hillhouse, J., and Kiecolt-Glaser, J. (1990). Psychological stress-induced modulation interleukin-2 receptor gene expression and interleukin-2 production in peripheral blood leukocytes. *Archives of General Psychiatry*, 47, 707–712.

Glasser, W. (1984). *Control Theory: A New Explanation of How We Control our Lives*. New York: Harper & Row.

Gleick, J. (1987). *Chaos: Making a New Science*. New York: Penguin Books.

Gold, P. (1984). *Memory Modulation: Neurobiology of Memory and Learning*. New York: Guilford.

Goldman-Rakic, P. S. (1987). Circuitry of primate prefrontal cortex and regulation of behavior by representational memory. In *Handbook of Physiology*, Vol. 1. Mountcastle, V. B., Plum, F., and Geiger, S. R. (Eds.), New York: Oxford University Press, 373–417.

Goldman-Rakic, P. S. (1992). Working memory and the mind. *Scientific American*, September.

Goldratt, E. M. (1994). *It's Not Luck*. Great Barrington, MA: North River Press.

Goldratt, E. M. and Cox, J. (1992). *The Goal: A Process of Ongoing Improvement (2nd Edition)*. Great Barrington, MA: North River Press.

Goldstein, M. and Goldstein, I. F. (1993). *The Refrigerator and the Universe: Understanding the Laws of Energy*. Cambridge, MA: Harvard University Press.

Goodwin, P. (1988). *Foundation Theory: Report on the Efficacy of the Formal Education Process in Rural Alaska (Vol. I)*. Honolulu, HI: Advanced Neuro Dynamics, Inc.

Goodwin, P. and James, T. (1993). *NLP & Neurophysics*. Honolulu, HI: Advanced Neuro Dynamics, Inc.

Gordon, D. (1978). *Therapeutic Metaphors*. Cupertino, CA: Meta Publications.

Greenfield, S. A. (1995). *Journey to the Centers of the Mind*. New York: W. H. Freeman & Company.

Greenspan, F. S. and Baxter, J. D. (1994). *Basic & Clinical Endocrinology (4th Edition)*. Norwalk, CT: Appleton & Lange.

Gribbin, J. (1995). *Schrodinger's Kittens and the Search for Reality: Solving the Quantum Mysteries*. New York: Little, Brown and Company.

Griffiths, A. J. F., Miller, J. H., Suzuki, D. T., Lewontin, R. C., and Gelbart, W. M. (1993). *An Introduction to Genetic Analysis (5th Edition)*. New York: W. H. Freeman and Company.

Grinder, J., DeLozier, J., and Bandler, R. (1977). *Patterns of the Hypnotic Techniques of Milton H. Erickson, M.D. (Volume 2)*. Cupertino, CA: Meta Publications.

Grinder, J. and Bandler, R. (1981). *Trance-Formations: Neuro-Linguistic Programming and the Structure of Hypnosis*. Moab, UT: Real People Press.

Gutnick, M. J. and Mody, I. (1995). *The Cortical Neuron*. New York: Oxford University Press.

Hall, E. T. (1959). *The Silent Language*. New York: Doubleday.

Hall, E. T. (1976). *Beyond Culture*. New York: Doubleday.

Haken, H. (1981). *The Science of Structure: Synergetics*. New York: Van Nostrand Reinhold Company.

Haken, H. (1983). *Synergetics: An Introduction (3rd Edition)*. New York: Springer-Verlag.

Haken, H. (1988). *Information and Self-Organization: A Macroscopic Approach to Complex Systems*. New York: Springer-Verlag.

Haken, H. (1996). *Principles of Brain Functioning: A Synergetic Approach to Brain Activity Behavior and Cognition*. New York: Springer-Verlag.

Haken, H. and Koepchen, H. P. (1991). *Rhythms in Physiological Systems*. New York: Springer-Verlag.

Hale, J. D. and Kocak, H. (1991). *Dynamics and Bifurcations*. New York: Springer-Verlag.

Haley, J. (1986). *Uncommon Therapy: The Psychiatric Techniques of Milton H. Erickson, M.D.* New York: W. W. Norton & Company.

Haley, J. (1967). *Advanced Techniques of Hypnosis and Therapy: Selected Papers of Milton H. Erickson, M.D.* New York: Grune & Stratton.

Hall, M. (1995). *Meta-States: A Domain of Logical Levels*. Grand Junction, CO: Empowerment Technologies.

Hameroff, S. R. (1987). *Ultimate Computing: Biomolecular Consciousness and NanoTechnology*. New York: North-Holland.

Hammond, D. C. (1990). *Handbook of Hypnotic Suggestions and Metaphors*. New York: W. W. Norton & Company.

Hannon, B. and Ruth, M. (1994). *Dynamic Modeling*. New York: Springer-Verlag.

Harth, E. (1993). *The Creative Loop: How the Brain Makes a Mind*. New York: Addison-Wesley.

Hassan, S. (1988). *Combating Cult Mind Control*. Rochester, VT: Park Street Press.

Havens, R. A. (1989). *The Wisdom of Milton H. Erickson: Hypnosis & Hypnotherapy (Volume I)*. New York: Irvington.
Havens, R. A. (1989). *The Wisdom of Milton H. Erickson: Human Behavior & Psychotherapy (Volume III)*. New York: Irvington.
Hawkins, J. A. and Gell-Mann, M. (1992). *The Evolution of Human Languages*. Reading, MA: Addison-Wesley.
Hayakawa, S. I. and Hayakawa, A. R. (1990). *Language in Thought and Action (5th Edition)*. New York: Harcourt Brace & Company.
Hebb, D. O. (1949). *The Organization of Behavior: A Neuropsychological Theory*. New York: John Wiley & Sons.
Hebb, D. O. and Donderi, D. C. (1987). *Textbook of Psychology (4th Edition)*. Hillsdale, NJ: Lawrence Erlbaum Associates.
Heller, S. and Steele, T. (1991). *Monsters and Magical Sticks*. Los Angeles: Falcon Press.
Hellhammer, D., Florin, I., and Weiner, H. (1988). *Neuronal Control of Bodily Function, Basic and Clinical Aspects (Vol. 2): Neurobiological Approaches to Human Disease*. Lewiston, NY: Hans Huber.
Hellige, J. B. (1993). *Hemispheric Asymmetry: What's Right and What's Left*. Cambridge, MA: Harvard University Press.
Hendriks-Jansen, H. (1996). *Catching Ourselves in the Act: Situated Activity, Interactive Emergence, Evolution and Human Thought*. Cambridge, MA: MIT Press.
Herbert, N. (1993). *Elemental Mind: Human Consciousness and the New Physics*. New York: Penguin Group.
Hilts, P. F. (1995). *Memory's Ghost: The Strange Tale of Mr. M and the Nature of Memory*. New York: Simon & Schuster.
Hinton, G. E. (1992). How neural networks learn from experience. *Scientific American*, September.
Hobson, J. A. (1989). *Sleep*. New York: Scientific American Library.
Hobson, J. A. (1994). *The Chemistry of Conscious States: How the Brain Changes its Mind*. New York: Little, Brown & Company.
Hofstadter, D. R. (1980). *Gödel, Escher, Bach: An Eternal Golden Braid*. New York: Vintage Books.
Hofstadter, D. R. (1985). *Metamagical Themas: Questing for the Essence of Mind and Pattern*. New York: Basic Books.
Holland, J. H. (1995). *Hidden Order: How Adaptation Builds Complexity*. Reading, MA: Addison-Wesley.
Holyoak, K. J. and Thagard, P. (1995). *Mental Leaps: Analogy in Creative Thought*. Cambridge, MA: MIT Press.
Honig, W. K. and Fetterman, J. G. (1992). *Cognitive Aspects of Stimulus Control*. Hillsdale, NJ: Lawrence Erlbaum Associates.
Hoopes, J. (1991). *Peirce on Signs: Writings on Semiotic by Charles Sanders Peirce*. Chapel Hill: University of North Carolina Press.
Horsthemke, W. and Lefever, R. (1984). *Noise-Induced Transitions: Theory and Applications in Physics, Chemistry, and Biology*. New York: Springer-Verlag.
Houk, J. C., Davis, J. L., and Beiser, D. G. (1995). *Models of Information Processing in the Basal Ganglia*. Cambridge, MA: MIT Press.
Hull, C. L. (1933). *Hypnosis and Suggestibility: An Experimental Approach*. New York: Appleton-Century-Crofts.
Hull, C. L. (1943). *Principles of Behavior: An Introduction to Behavior Theory*. New York: Appleton-Century-Crofts.
Hull, C. L. (1951). *Essentials of Behavior*. New Haven: Yale University Press.

Hull, C. L. (1952). *A Behavior System: An Introduction to Behavior Theory Concerning the Individual Organism*. New Haven: Yale University Press.
Hume, D. (1888). *A Treatise of Human Nature*. New York: Oxford University Press.
Hunter, W. S. (1919). *Human Behavior*. Chicago, IL: University of Chicago Press.
Iyengar, S. and McGuire, W. J. (1993). *Explorations in Political Psychology*. London: Duke University Press.
Jackendoff, R. (1994). *Patterns in the Mind*. New York: HarperCollins.
James, T. and Woodsmall, W. (1988). *Time Line Therapy and the Basis of Personality*. Cupertino, CA: Meta Publications.
James, W. (1890). *The Principles of Psychology (Volume I)*. New York: Henry Holt and Company.
James, W. (1890). *The Principles of Psychology (Volume II)*. New York: Henry Holt and Company.
James, W. (1902). *The Varieties of Religious Experience: A Study in Human Nature*. New York: Longmans, Green and Co.
Janis, I. L. and Mann, L. (1977). *Decision Making: A Psychological Analysis of Conflict, Choice, and Commitment*. New York: Free Press.
Johnson, M. (1987). *The Body in the Mind: The Bodily Basis of Meaning, Imagination, and Reason*. Chicago, IL: University of Chicago Press.
Jones, S. (1993). *The Language of Genes: Solving the Mysteries of Our Genetic Past, Present and Future*. New York: Bantam Doubleday Dell Publishing Group.
Kaandorp, J. A. (1994). *Fractal Modelling Growth and Form in Biology*. New York: Springer-Verlag.
Kandel, E. R. and Hawkins, R. D. (1992). *The Biological Basis of Learning and Individuality*. New York: W. H. Freeman and Company.
Kandel, E. R., Schwartz, J. H., and Jessell, T. M. (Eds.). (1991). *Principles of Neural Science (3rd Edition)*. East Norwalk, CT: Appleton & Lange.
Kandel, E. R., Schwartz, J. H., and Jessell, T. M. (Eds.). (1995). *Essentials of Neural Science and Behavior*. Norwalk, CT: Appleton & Lange.
Kane, G. (1995). *The Particle Garden: Our Universe as Understood by Particle Physicists*. Reading, MA: Addison-Wesley.
Katz, M. A. (1976). *Calculus for the Life Sciences*. New York: Marcel Dekker.
Kauffman, S. (1995). *At Home in the Universe: The Search for the Laws of Self-Organization and Complexity*. New York: Oxford University Press.
Kauffman, S. A. (1993). *The Origins of Order: Self-Organization and Selection in Evolution*. New York: Oxford University Press.
Kazdin, A. E. (1994). *Behavior Modification in Applied Settings*. Pacific Grove, CA: Brooks/Cole.
Kelly, G. A. (1963). *A Theory of Personality: The Psychology of Personal Constructs*. New York: W. W. Norton & Company.
Kelso, J. A. S. (1995). *Dynamic Patterns*. Cambridge, MA: MIT Press.
Kendrew, S. J. (1994). *The Encyclopedia of Molecular Biology*. Cambridge, MA: Blackwell Science.
Kenyon, T. (1994). *Brain States*. Naples, FL: United States Publishing.
Key, W. B. (1993). *The Age of Manipulation: The Con in Confidence — the Sin in Sincere*. Lanham, MD: Madison Books.
Kiesler, C. A. (1971). *The Psychology of Commitment: Experiments Linking Behavior to Belief*. New York: Academic Press.
Kissin, B. (1986). *Psychobiology of Human Behavior (Volume I): Conscious and Unconscious Programs in the Brain*. New York: Plenum Press.

Kleinsmith, L. J. and Kish, V. M. (1995). *Principles of Cell and Molecular Biology, Second Edition*. New York: HarperCollins College Publishers.
Klippstein, H. (1991). *Ericksonian Hypnotherapeutic Group Inductions*. New York: Brunner/Mazel.
Klir, G. J. (1985). *Architecture of Systems Problem Solving*. New York: Plenum Press.
Kohonen, T. (1995). *Self-Organizing Maps*. New York: Springer-Verlag.
Kolb, B. and Whishaw, I. (1990). *Fundamentals of Human Neuropsychology, Third Edition*. New York: W. H. Freeman and Company.
Kordon, C. (1993). *The Language of the Cell*. New York: McGraw-Hill.
Korzybski, A. (1933). *Science and Sanity*. Englewood, NJ: Institute of General Semantics.
Korzybski, A. (1950). *Manhood of Humanity (2nd Edition)*. Englewood, NJ: Institute of General Semantics.
Kosslyn, S. M. (1994). *Image and Brain*. Cambridge, MA: MIT Press.
Kosslyn, S. M. and Koenig, O. (1992). *Wet Mind: The New Cognitive Neuroscience*. New York: Free Press.
Kotre, J. (1995). *White Gloves: How We Create Ourselves Through Memory*. New York: Free Press.
Kroger, W. A. (1963). *Clinical and Experimental Hypnosis in Medicine, Dentistry and Psychology*. Philadelphia, PA: J. B. Lippincott Company.
Kruger, J. (1991). *Neuronal Cooperativity*. New York: Springer-Verlag.
Kruse, P. and Stadler, M. (1995). *Ambiguity in Mind and Nature*. New York: Springer-Verlag.
Kuhn, T. S. (1962). *The Structure of Scientific Revolutions (3rd Edition)*. Chicago, IL: University of Chicago Press.
Kuno, M. (1995). *The Synapse: Function Plasticity, and Neurotrophism*. New York: Oxford University Press.
Lakoff, G. (1987). *Women, Fire, and Dangerous Things: What Categories Reveal about the Mind*. Chicago, IL: University of Chicago Press.
Lakoff, G. and Johnson, M. (1980). *Metaphors We Live By*. Chicago, IL: University of Chicago Press.
Lakoff, G. and Turner, M. (1989). *More than Cool Reason: A Field Guide to Poetic Metaphor*. Chicago, IL: University of Chicago Press.
Lappe, M. (1994). *Evolutionary Medicine: Rethinking the Origins of Disease: Strategies to Stop the Advance of Epidemic Drug Resistant Bacteria and Viruses, Tuberculosis, Malaria, Cancer, and AIDS*. San Francisco: Sierra Club Books.
Lewis, B. and Pucelik, F. (1990). *Magic of NLP Demystified: A Pragmatic Guide to Communication & Change*. Portland, OR: Metamorphous Press.
Levine, B. (1991). *Your Body Believes Every Word You Say: The Language of the Body/Mind Connection*. Lower Lake, CA: Aslan Publishing.
Levine, D. S. (1991). *Introduction to Neural & Cognitive Modeling*. Hillsdale, NJ: Lawrence Erlbaum Associates.
Levine, D. S. and Leven, S. J. (1992). *Motivation, Emotion, and Goal Direction in Neural Networks*. Hillsdale, NJ: Lawrence Erlbaum Associates.
Levine, J. and Suzuki, D. (1993). *The Secret of Life: Redesigning the Living World*. Boston, MA: WGBH Educational Foundation.
Lifton, J. (1961). *Thought Reform and the Psychology of Totalism: A Study of "Brainwashing" in China*. New York: W. W. Norton & Co.
Lindzey, G. and Aronson, E. (1985). *The Handbook of Social Psychology Volumes I and II (3rd Edition)*. Hillsdale, NJ: Lawrence Erlbaum Associates.
Llinas, R. (Ed). (1988). *The Biology of the Brain from Neurons to Networks*. New York: W. H. Freeman and Company.

Llinas, R. (Ed.). (1990). *The Workings of the Brain Development, Memory, and Perception.* New York: W. H. Freeman and Company.
Loewenstein, W. R. (1999). *The Touchstone of Life: Molecular Information, Cell Communication, and the Foundations of Life*. New York: Oxford University Press.
Lorenz, K. (1992). *The Natural Science of the Human Species: An Introduction to Comparative Behavioral Research in The "Russian Manuscript" (1944–1948).* Cambridge, MA: MIT Press.
Luce, R. D. and Raiffa, H. (1985). *Games and Decisions: Introduction and Critical Survey.* New York: Dover.
Luer, G., Lass, U., and Shallo-Hoffmann, J. (1988). *Eye Movement Research: Physiological and Psychological Aspects*. Lewiston, NY: C.J. Hogrefe.
Luria, A. R. (1932). *The Nature of Human Conflicts or Emotion, Conflict and Will: An Objective Study of Disorganisation and Control of Human Behaviour.* New York: Grove Press.
Lynch, A. (1996). *Thought Contagion: How Belief Spreads Through Society: The New Science of Memes*. New York: Basic Books.
Mandelbrot, B. B. (1983). *The Fractal Geometry of Nature*. New York: W. H. Freeman and Company.
Margulis, L. and Sagan, D. (1986). *Microcosmos: Four Billion Years of Microbial Evolution.* Berkeley: University of California Press.
Martin, G. and Pear, J. (1996). *Behavior Modification: What It Is and How To Do It (5th Edition).* Upper Saddle River: NJ: Prentice-Hall.
Maslow, A. H. (1987). *Motivation and Personality (3rd Edition).* New York: Harper & Row Publishers.
Masters, R. (1994). *Neurospeak: Transforms Your Body While You Read*! Wheaton, IL: Quest Books.
McClelland, D. C. (1987). *Human Motivation*. Cambridge, MA: Press Syndicate of the University of Cambridge.
McGaugh, J. L., Weinberger, N. M., and Lynch, G. (1990). *Brain Organization and Memory: Cells, Systems, and Circuits*. New York: Oxford University Press.
McGaugh, J. L., Weinberger, N. M., and Lynch, G. (1995). *Brain and Memory: Modulation and Mediation of Neuroplasticity.* New York: Oxford University Press.
McMullin, R. E. (1986). *Handbook of Cognitive Therapy Techniques*. New York: W. W. Norton & Company.
McNeill, D. (1992). *Hand and Mind: What Gestures Reveal about Thought*. Chicago, IL: University of Chicago Press.
Meichenbaum, D. (1977). *Cognitive-Behavior Modification: An Integrative Approach*. New York: Plenum Press.
Meyer, M. and Booker, J. (1991). *Eliciting and Analyzing Expert Judgement: A Practical Guide (Vol. 5)*. New York: Academic Press.
Miller, G. A. (1991). *The Science of Words*. New York: W. H. Freeman and Company.
Miller, G. A., Galanter, E., and Pribram, K. H. (1979). *Plans and the Structure of Behavior.* New York: Adams-Bannister-Cox.
Minsky, M. (1986). *The Society of Mind*. New York: Touchstone.
Mishra, R. K., Maab, D., and Zwierlein, E. (1994). *On Self-Organization: An Interdisciplinary Search for a Unifying Principle*. New York: Springer-Verlag.
Mittenthal, J. E. and Baskin, A. B. (1992). *The Principles of Organization in Organisms*. Reading, MA: Addison-Wesley.
Montanaro, T. (1995). *Mime Spoken Here: The Performer's Portable Workshop*. Gardiner, ME: Tilbury House.

Morowitz, H. J. and Singer, J. L. (1995). *The Mind, The Brain, and Complex Adaptive Systems.* Reading, MA: Addison-Wesley.

Morrison, N. (1994). *Introduction to Fourier Analysis.* New York: John Wiley & Sons.

Mosekilde, E. and Mouritsen, O. G. (1995). *Modelling the Dynamics of Biological Systems: Nonlinear Phenomena and Pattern Formation.* New York: Springer-Verlag.

Moyers, B. (1993). *Healing and The Mind.* New York: Doubleday.

National Research Council. (1991). *In the Mind's Eye: Enhancing Human Performance.* Washington, D.C.: National Academy Press.

National Research Council. (1994). *Learning, Remembering, Believing: Enhancing Human Performance.* Washington, D.C.: National Academy Press.

National Research Council, Baron, S., Kruser, D. S., and Huey, B. M. (Eds.). Panel on Human Performance Modeling, Committee on Human Factors, Commission on Behavioral and Social Sciences and Education. (1990). *Quantitative Modeling of Human Performance in Complex, Dynamic Systems.* Washington, D.C.: National Academy Press.

National Research Council, Druckman, D. and Swets, J. A. (Eds.). Committee on Techniques for the Enhancement of Human Performance, Committee on Behavioral and Social Sciences and Education. (1988). *Enhancing Human Performance Issues, Theories, and Techniques.* Washington, D.C.: National Academy Press.

Nayfeh, A. H. and Balachandran, B. (1995). *Applied Nonlinear Dynamics: Analytical, Computational, and Experimental Methods.* New York: John Wiley & Sons.

Neelakanta, P. S. and DeGroff, D. F. (1994). *Neural Network Modeling: Statistical Mechanics and Cybernetic Perspectives.* Boca Raton, FL: CRC Press.

Nesse, R. M. and Williams, G. C. (1994). *Why We Get Sick: The New Science of Darwinian Medicine.* New York: Random House.

Newell, A. (1990). *Unified Theories of Cognition.* Cambridge, MA: Harvard University Press.

Nolte, J. (1993). *The Human Brain: An Introduction to Its Functional Anatomy (3rd Edition).* St. Louis, MO: Mosby-Year Book.

Nolte, J. and Angevine, J. B. (1995). *The Human Brain: In Photographs and Diagrams.* St. Louis, MO: Mosby-Year Book.

Nonnenmacher, T. F., Losa, G. A., and Weibel, E. R. (1994). *Fractals in Biology and Medicine.* Basel, Switzerland: Birhauser Verlag.

Nunez, P. L. (1995). *Neocortical Dynamics and Human EEG Rhythms.* New York: Oxford University Press.

O'Connor, J. and Seymour, J. (1990). *Introducing Neuro-Linguistic Programming: The New Psychology of Personal Excellence.* Hammersmith, London: HarperCollins.

Ofshe, R. and Watters, E. (1994). *Making Monsters: False Memories, Psychotherapy, and Sexual Hysteria.* New York: Charles Scribner's Sons.

O'Hanlon, W. H. (1992). *Solution-Oriented Hypnosis: An Erickson Approach.* New York: W. W. Norton & Company.

O'Hanlon, W. H. and Hexum, A. L. (1990). *An Uncommon Casebook: The Complete Clinical Work of Milton H. Erickson, M.D.* New York: W. W. Norton & Company.

Olomucki, M. (1993). *The Chemistry of Life.* New York: McGraw-Hill.

Ono, T., Squire, L. R., Raichle, M. E., Perrett, D. I., and Fukuda, M. (1993). *Brain Mechanisms of Perception and Memory from Neuron to Behavior.* New York: Oxford University Press.

Ornstein, R. and Sobel, D. (1987). *The Healing Brain.* New York: Simon & Schuster.

Osherson, D. (1990). *An Invitation to Cognitive Science Vol. I: Language.* Cambridge, MA: MIT Press.

Overdurf, J. and Silverthorn, J. (1995). *Training Trances: Multi-Level Communication in Therapy and Training.* Portland, OR: Metamorphous Press.

Paillard, J. (1991). *Brain and Space.* New York: Oxford University Press.

Parker, S. P. (1993). *McGraw-Hill Encyclopedia of Physics (2nd Edition).* New York: McGraw-Hill.

Parmelee, M. (1913). *The Science of Human Behavior: Biological and Psychological Foundations.* New York: Macmillan Company.

Partridge, L. D. and Partridge, L. D. (1993). *The Nervous System: Its Function and Its Interaction with the World.* Cambridge, MA: MIT Press.

Passingham, R. (1993). *The Frontal Lobes and Voluntary Action: Oxford Psychology Series — 21.* New York: Oxford University Press.

Pavlov, I. P. (1927). *Conditioned Reflexes.* New York: Oxford University Press.

Pavlov, I. P. (1928). *Lectures on Conditioned Reflexes (Volume I).* New York: International.

Pavlov, I. P. (1941). *Lectures on Conditioned Reflexes (Volume II).* New York: International.

Penrose, R. (1991). *The Emperor's New Mind: Concerning Computers, Minds, and the Laws of Physics.* New York: Penguin Group.

Penrose, R. (1994). *Shadows of the Mind: A Search for the Missing Science of Consciousness.* New York: Oxford University Press.

Piattelli-Palmarini, M. (1994). *Inevitable Illusions: How Mistakes of Reason Rule Our Mind.* New York: John Wiley & Sons.

Pinker, S. (1994). *The Language Instinct: How the Mind Creates Language.* New York: HarperCollins.

Pinker, S. (1997). *How the Mind Works.* New York: W. W. Norton & Company.

Platonov, K. I. (1959). *The Word as a Physiological and Therapeutic Factor: The Theory and Practice of Psychotherapy according to I.P. Pavlov.* Moscow: Foreign Languages Publishing House.

Plotkin, H. (1993). *Darwin Machines and the Nature of Knowledge.* Cambridge, MA: Harvard University Press.

Popper, K. R. (1959). *The Logic of Scientific Discovery.* New York: Routledge.

Port, R. F. and Van Gelder, T. (1995). *Mind as Motion: Explorations in the Dynamics of Cognition.* Cambridge, MA: MIT Press.

Porter, P. K. (1995). *Psycho-Linguistics: The Language of the Mind.* Phoenix, AZ: PureLight Publishing.

Posner, M. and Raichle, M. (1994). *Images of Mind.* New York: W. H. Freeman and Company.

Powers, W. T. (1973). *Behavior: The Control of Perception.* New York: Aldine De Gruyter.

Pribram, K. H. (1971). *Languages of the Brain: Experimental Paradoxes and Principles in Neuropsychology.* New York: Brandon House.

Pribram, K. H. (1991). *Brain and Perception: Holonomy and Structure in Figural Processing.* Hillsdale, NJ: Lawrence Erlbaum Associates.

Pribram, K. H. (1993). *Rethinking Neural Networks: Quantum Fields and Biological Data, Proceedings of the First Appalachian Conference on Behavioral Neurodynamics.* Hillsdale, NJ: Lawrence Erlbaum Associates.

Pribram, K. H. (1994). *Origins: Brain & Self Organization.* Hillsdale, NJ: Lawrence Erlbaum Associates.

Prigogine, I. (1996). *The End of Certainty: Time, Chaos, and the New Laws of Nature.* New York: Free Press.

Prochiantz, A. (1989). *How the Brain Evolved.* New York: McGraw-Hill.

Pryor, K. (1975). *Lads Before the Wind: Diary of a Dolphin Trainer.* North Bend, WA: Sunshine Books.

Pryor, K. (1995). *On Behavior: Essays & Research.* North Bend, WA: Sunshine Books.

Ram, A. and Eiselt, K. (1994). *Proceedings of the Sixteenth Annual Conference of the Cognitive Science Society, August 13 to 16, 1994, Georgia Institute of Technology, Atlanta, Georgia.* Hillsdale, NJ: Lawrence Erlbaum Associates.

Rasmussen, H. (Ed.). (1991). *Cell Communication in Health and Disease*. New York: W. H. Freeman and Company.
Reese, M., Reese, E. J., Nagel, C. V., and Siudzinski, R. (1985). *Mega Teaching and Learning*. Portland, OR: Metamorphous Press.
Reese, M. and Yancar, C. L. (1986). *Practitioner Manual for Introductory Patterns of Neuro Linguistic Programming*. Indian Rocks Beach, FL: Southern Institute Press.
Restak, R. M. (1994). *Receptors*. New York: Bantam.
Restak, R. M. (1994). *The Modular Brain*. New York: Charles Scribner's Sons.
Richardson, J. T. E., Engle, R. W., Hasher, L., Logie, R. H., Stoltzfus, E. R., and Zacks, R. T. (1996). *Working Memory and Human Cognition*. New York: Oxford University Press.
Rieke, F., Warland, D., de Ruyter van Steveninck, R., and Bialek, W. (1997). *Spikes: Exploring the Neural Code*. Cambridge, MA: MIT Press.
Rifkin, J. (1980). *Entropy*. New York: Viking Press.
Roederer, J. G. (1995). *The Physics and Psychophysics of Music: An Introduction (3rd Edition)*. New York: Springer-Verlag.
Roland, P. E. (1993). *Brain Activation*. New York: Wiley-Liss.
Rose, S. (1992). *The Making of Memory from Molecules to Mind*. New York: Anchor Books, Doubleday.
Rosen, S. (1982). *My Voice Will Go with You: The Teaching Tales of Milton H. Erickson*. New York: W. W. Norton & Company.
Rossi, E. L. (1989). *Hypnotic Investigation of Psychodynamic Processes: The Collected Papers of Milton H. Erickson on Hypnosis (Volume III)*. New York: Irvington.
Rossi, E. L. (1989). *Innovative Hypnotherapy: The Collected Papers of Milton H. Erickson on Hypnosis (Volume IV)*. New York: Irvington.
Rossi, E. L. (1993). *The Psychobiology of Mind-Body Healing; New Concepts of Therapeutic Hypnosis*. New York: W. W. Norton & Company.
Rossi, E. L. and Cheek, D. B. (1988). *Mind-Body Therapy: Methods of Ideodynamic Healing in Hypnosis*. New York: W. W. Norton & Company.
Rossi, E. L. and Ryan, M. O. (1986). *Mind-Body Communication in Hypnosis: Milton H. Erickson (Volume III): The Seminars, Workshops, and Lectures of Milton H. Erickson*. New York: Irvington.
Rossi, E. L., Ryan, M. O., and Sharp, F. A. (1983). *Healing in Hypnosis: Milton H. Erickson: Volume I — The Seminars, Workshops, and Lectures of Milton H. Erickson*. New York: Irvington.
Rugg, M. D. and Coles, M. G. (1995). *Electrophysiology of Mind: Event-Related Brain Potentials and Cognition*. New York: Oxford University Press.
Rushkoff, D. (1994). *Media Virus: Hidden Agendas in Popular Culture*. New York: Ballantine.
Rywerant, Y. (1983). *The Feldenkrais Method: Teaching by Handling*. New Canaan, CT: Keats Publishing.
Salter, A. (1949). *Conditioned Reflex Therapy: The Direct Approach to the Reconstruction of Personality*. New York: Creative Age Press.
Salter, F. K. (1995). *Emotions in Command: A Naturalistic Study of Institutional Dominance*. New York: Oxford University Press.
Schilpp, P. A. (Ed.). (1949). *Albert Einstein: Philosopher-Scientist — The Library of Living Philosophers Volume VII*. La Salle, IL: Open Court.
Schmoll, H. J., Tewes, U., and Plotnikoff, N. P. (1989). *Psychoneuroimmunology; Interactions between Brain, Nervous System, Behavior, Endocrine and Immune System*. Lewiston, NY: Hogrefe & Huber.
Schrodinger, E. (1944). *What is Life? With Mind and Matter and Autobiographical Sketches*. New York: Cambridge University Press.

Schroeder, M. (1991). *Fractals, Chaos, Power Laws: Minutes from an Infinite Paradise*. New York: W. H. Freeman and Company.

Scott, A. (1995). *Stairway to the Mind: The Controversial New Science of Consciousness*. New York: Springer-Verlag.

Sechenov, I. M. (1965). *Reflexes of the Brain: An Attempt to Establish the Physiological Basis of Psychological Processes*. Cambridge, MA: MIT Press.

Shepherd, G. M. (1994). *Neurobiology (Third Edition)*. New York: Oxford University Press.

Skinner, B. F. (1938). *The Behavior of Organisms: An Experimental Analysis*. New York: D. Appleton-Century Company.

Skinner, B. F. (1953). *Science and Human Behavior*. New York: Free Press.

Skinner, B. F. (1957). *Verbal Behavior*. New York: Appleton-Century-Crofts.

Skinner, B. F. (1968). *The Technology of Teaching*. New York: Appleton-Century-Crofts.

Skinner, B. F. (1972). *Cumulative Record: A Selection of Papers (3rd Edition)*. New York: Appleton-Century-Crofts.

Skinner, B. F. (1974). *About Behaviorism*. New York: Vintage.

Skinner, B. F. (1980). *Notebooks B. F. Skinner*. Englewood Cliffs, NJ: Prentice-Hall.

Skinner, B. F. (1987). *Upon Further Reflection*. Englewood Cliffs, NJ: Prentice-Hall.

Smith, L. B. and Thelen, E. (1993). *A Dynamic Systems Approach to Development: Applications*. Cambridge, MA: MIT Press.

Smith, L. B. and Thelen, E. (1994). *A Dynamic Systems Approach to the Development of Cognition and Action*. Cambridge, MA: MIT Press.

Smolin, L. (1997). *The Life of the Cosmos*. New York: Oxford University Press.

Someren, M. W. V., Barnard, Y. F., and Sandberg, J. A. C. (1994). *The Think Aloud Method: A Practical Guide to Modelling Cognitive Processes*. San Diego: Academic Press.

Spencer-Brown, G. (1994). *Laws of Form*. Portland, OR: Cognizer Co.

Spiegel, H. and Spiegel, D. (1978). *Trance and Treatment: Clinical Uses of Hypnosis*. Washington, D.C.: American Psychiatric Press.

Springer, S. and Deutsch, G. (1993). *Left Brain: Right Brain, Fourth Edition*. New York: W. H. Freeman and Company.

Squire, L. R., Weinberger, N. M., Lynch, G., and McGaugh, J. L. (1991). *Memory: Organization and Locus of Change*. New York: Oxford University Press.

Stapp, H. P. (1993). *Mind, Matter, and Quantum Mechanics*. New York: Springer-Verlag.

Steen, G. (1994). *Understanding Metaphor in Literature*. New York: Longman Group Limited.

Stein, B. E. and Meredith, M. A. (1993). *The Merging of the Senses*. Cambridge, MA: MIT Press.

Steinberg, D. D. (1982). *Psycholinguistics: Language, Mind and World*. New York: Longman Group Limited.

Stemmer, B. and Whitaker, H. (Eds.). (1998). *Handbook of Neuro-linguistics*. New York: Academic Press.

Stewart, I. and Golubitsky, M. (1992). *Fearful Symmetry: Is God a Geometer?* New York: Penguin Books.

Stich, S. P. (1996). *Deconstructing the Mind*. New York: Oxford University Press.

Stites, D. P., Terr, A. I., and Parslow, T. G. (1994). *Basic & Clinical Immunology (8th Edition)*. Norwalk, CT: Appleton & Lange.

Strogatz, S. H. (1994). *Nonlinear Dynamics and Chaos: With Applications to Physics, Biology, Chemistry, and Engineering*. Reading, MA: Addison-Wesley.

Sutton, E. W. and Rackham, H. (1942). *Cicero III De Oratore Books I-II*. Cambridge, MA: Harvard University Press.

Tarski, A. (1941). *Introduction to Logic and to the Methodology of Deductive Sciences*. New York: Dover Publications.

Taylor, G. (1996). *Cultural Selection: Why Some Achievements Survive the Test of Time — and Others Don't.* New York: Basic Books.
Ter Meulen, A. G. B. (1995). *Representing Time in Natural Language: The Dynamic Interpretation of Tense and Aspect.* Cambridge, MA: MIT Press.
Thompson, R. F. (1993). *The Brain: A Neuroscience Primer (2nd Edition).* New York: W. H. Freeman and Company.
Thorndike, E. L. (1931). *Human Learning.* New York: Century Co.
Tilney, F. (1930). *The Structural Basis of Behaviorism.* Philadelphia, PA: J. B. Lippincott Company.
Touyz, S., Byrne, D., and Gilandas, A. (1994). *Neuropsychology in Clinical Practice.* San Diego: Academic Press.
Tschacher, W., Schiepek, G., and Brunner, E. J. (1992). *Self-Organization and Clinical Psychology: Empirical Approaches to Synergetics in Psychology.* New York: Springer-Verlag.
Turner, M. (1991). *Reading Minds: The Study of English in the Age of Cognitive Science.* Princeton, NJ: Princeton University Press.
Valiant, L. G. (1994). *Circuits of the Mind.* New York: Oxford University Press.
Vallacher, R. R. and Nowak, A. (1994). *Dynamical Systems in Social Psychology.* San Diego: Academic Press.
Van Gigch, J. P. (1991). *System Design Modeling and Metamodeling.* New York: Plenum Press.
Volk, T. (1995). *Metapatters: Across Space, Time and Mind.* New York: Columbia University Press.
Von Neumann, J. (1958). *The Computer and the Brain.* New Haven, CT: Yale University Press.
Waldrop, M. M. (1992). *Complexity: The Emerging Science at the Edge of Order and Chaos.* New York: Simon & Schuster.
Wandell, B. A. (1995). *Foundations of Vision.* Sunderland, MA: Sinauer Associates.
Watkins, J. G. (1987). *Hypnotherapeutic Techniques: The Practice of Clinical Hypnosis (Volume I).* New York: Irvington.
Watkins, J. G. (1992). *Hypnoanalytic Techniques: The Practice of Clinical Hypnosis (Volume II).* New York: Irvington.
Watson, J. B. (1919). *Psychology from the Standpoint of a Behaviorist.* Philadelphia, PA: Washington Square Press.
Watson, J. B. (1930). *Behaviorism.* New York: W. W. Norton & Company.
Watzlawick, P. (1978). *The Language of Change: Elements of Therapeutic Communication.* New York: W. W. Norton & Company.
Watzlawick, P., Bavelas, J. B., and Jackson, D. D. (1967). *Pragmatics of Human Communication: A Study of Interactional Patterns, Pathologies, and Paradoxes.* New York: W. W. Norton & Company.
Watzlawick, P., Weakland, J., and Fisch, R. (1974). *Change: Principles of Problem Formation and Problem Resolution.* New York: W. W. Norton & Company.
Webb, T. W. and Webb, D. (1990). *Accelerated Learning with Music: A Trainer's Manual.* Norcross, GA: Accelerated Learning Systems.
Weinberg, G. M. (1975). *An Introduction to General Systems Thinking.* New York: John Wiley & Sons.
Weitzenhoffer, A. M. (1957). *General Techniques of Hypnotism.* New York: Grune & Stratton.
Werner, H. and Kaplan, B. (1963). *Symbol Formation: An Organismic-Developmental Approach to Language and the Expression of Thought.* New York: John Wiley & Sons.
Wiener, N. (1950). *The Human Use of Human Beings: Cybernetics and Society.* Boston: Houghton Mifflin.

Wiener, N. (1962). *Cybernetics: Or Control and Communication in the Animal and the Machine (2nd Edition).* Cambridge, MA: MIT Press.
Wilkie, W. L. (1994). *Consumer Behavior (3rd Edition).* New York: John Wiley & Sons.
Woodsmall, W. (1988). *Strategies.* Arlington, VA: Advanced Behavioral Modeling.
Wright, G. (1985). *Behavioral Decision Making.* New York: Plenum Press.
Wyer, R. S. and Srull, T. K. (1994). *Handbook of Social Cognition, Volume 1: Basic Processes (2nd Edition).* Hillsdale, NJ: Lawrence Erlbaum Associates.
Wyer, R. S. and Srull, T. K. (1994). *Handbook of Social Cognition, Volume 2: Applications (2nd Edition).* Hillsdale, NJ: Lawrence Erlbaum Associates.
Yapko, M. D. (1990). *Trancework: An Introduction to the Practice of Clinical Hypnosis (2nd Edition).* New York: Brunner/Mazel.
Zaidel, D. W. (1994). *Neuropsychology: Handbook of Perception and Cognition (2nd Edition).* San Diego: Academic Press.
Zeig, J. K. (1994). *Ericksonian Methods: The Essence of the Story.* New York: Brunner/Mazel.
Zeki, S. (1993). *A Vision of the Brain.* Cambridge, MA: Blackwell Scientific Publications.
Zimbardo, P. and Ebbesen, E. B. (1969). *Influencing Attitudes and Changing Behavior.* Reading, MA: Addison-Wesley.

RECOMMENDED JOURNALS AND OTHER RESOURCES

Brain and Cognition: A Journal of Clinical, Experimental, and Theoretical Research. New York: Academic Press.
Brain and Language: A Journal of Clinical, Experimental, and Theoretical Research. New York: Academic Press.
Brain, Behavior, and Immunity. New York: Academic Press.
Cerebral Cortex. Cary, NC: Oxford University Press.
Cognitive Science. NJ: Ablex Publishing.
Consciousness and Cognition. New York: Academic Press.
Cybernetics and Systems. Washington, D.C.: Taylor & Francis.
Experimental Neurology: A Journal of Neuroscience Research. New York: Academic Press.
Frontiers in Neuroendocrinology: Official Journal of the International Society of Neuroendocrinology. New York: Academic Press.
Human Brain Mapping. New York: Wiley-Liss.
Journal of Behavioral Decision Making. West Sussex, England: Wiley-Liss.
Journal of Cognitive Neuroscience. Cambridge, MA: MIT Press Journals.
Journal of Comparative Neurology. New York: Wiley-Liss.
Journal of Neurobiology. New York: John Wiley & Sons.
Journal of Neurophysiology. Bethesda, MD: American Physiological Society.
Journal of Neuroscience. Cary, NC: Oxford University Press.
Journal of Neuroscience Research. New York: Wiley-Liss.
Journal of Psychophysiology. Seattle, WA: Hogrefe & Huber Publishers.
Learning and Motivation: New York: Academic Press.
MCN: Molecular and Cellular Neuroscience: New York: Academic Press.
Memory & Cognition. Austin, TX: Psychonomic Society.
Neurobiology of Learning and Memory: An Interdisciplinary Journal. New York: Academic Press.
Neurodegeneration: A Journal for Neurodegenerative Disorders, Neuroprotection, and Neuroregeneration. New York: Academic Press.
NeuroProtocols: A Companion to Methods in Neurosciences. New York: Academic Press.

Neuropsychobiology. Basel, Switzerland: S. Karger.
Neuropsychologia. Kidlington, Oxford, U.K.: Elsevier Science Ltd.
Neuroscience Research Communications. West Sussex, England: John Wiley & Sons.
Perception & Psychophysics. Austin, TX: Psychonomic Society.
Psychobiology. Austin, TX: Psychonomic Society.
Science. Washington, D.C.: American Association for the Advancement of Science.
Seminars in the Neurosciences: Molecular and Cellular Basis of Learning. New York: Academic Press.
Social Cognition. New York: Guilford Press.

For readers who would like to continue in-depth research beyond the scope of this book, we have found the following branches of science essential for the development of the theory and applications discussed. Practitioners in the field of neurocognitive intervention will also find the following sources useful for determining the entropy effect, scope, and pattern behavior of any intervention, tool, or method.

NEUROSCIENCE

- Functional Human Brain Mapping
- Functional Neuro-Imaging
- Cyto-Architectonics
- Functional Neuroanatomy
- Cognitive Neuroscience
- Neurophysiology
- Neurobiology
- Neuroendocrinology
- Psychoneuroimmunology

STATISTICAL PHYSICS

- Dynamical Self-Organizing Systems Theory
- Quantum Mechanics/Physics
- Thermodynamics
- Synergetics

PSYCHOLOGY AND BEHAVIORAL SCIENCE

- Energy Psychology
- Neuropsychology
- Evolutionary Psychology
- Cognitive Behavior Modification
- Behavioral Engineering
- Neurolinguistic Programming

LINGUISTICS

- General Semantics
- Psycholinguistics
- Neurolinguistics
- Memetics
- Semiotics
- Pragmatics

SYSTEMS SCIENCE AND MODELING

- Dynamic Systems Modeling
- General Systems Science
- Nonlinear Systems Science
- Expert Systems Modeling
- Neural Network Modeling

Index

A

B

C

D

E

F

G

H

I

J

K

L

M

N

O

P

Q

R

S

T

V

W